Handbook of Petrochemicals and Processes

Dedicated to my parents who first
encouraged my love of chemistry.

Handbook of Petrochemicals and Processes

Second Edition

G Margaret Wells BSc, FIM

Ashgate

First edition published 1991 by Gower Publishing Company Limited

This edition published by
Ashgate Publishing Limited
Gower House
Croft Road
Aldershot
Hampshire GU11 3HR
England

Ashgate Publishing Company
Old Post Road
Brookfield
Vermont 05036
USA

British Library Cataloguing in Publication Data
Wells, G. Margaret
Handbook of petrochemicals and processes. — 2nd ed.
1. Petroleum chemicals 2. Petroleum chemicals industry
I. Title
661.8′04

ISBN 0 566 08046 X

Library of Congress Cataloging-in-Publication Data
Wells, G. Margaret, 1931–
Handbook of petrochemicals and processes / G. Margaret Wells. – 2nd ed.
p. cm.
Includes indexes.
ISBN 0-566-08046-X
1. Petroleum chemicals – Handbooks, manuals etc. I. Title.
TP692.3.W46 1999
661′.804–dc21

Typeset in 9.5/13 Palatino by Saxon Graphics Limited, Derby and printed in Great Britain at the Universtiy Press, Cambridge.

Contents

List of Figures

Preface

In the period since the first edition of this book, major changes have continued to take place in the petrochemical industry. Producers are trying to reduce costs wherever possible and this has resulted in the development of new processes, catalysts and more efficient energy-recovery systems. Environmental and legislative pressures have led to added attention being paid to reducing the volume of waste products and the better utilization of by-products by the increasing use of balanced routes.

This new edition has been extensively revised, updated and expanded to include the most recently introduced processes together with details of those due to be commercialized by the year 2000.

Application breakdown details for each product have been brought fully up to date with forecasts of future growth rates. Major plant revisions reflect the changing ownership of existing plants as well as expansions and new construction worldwide. The section on international classifications has been enlarged to help identify the changing legislation and to facilitate the selection of the data required.

Petrochemical companies are trying to find opportunities for higher growth and to lessen the swings of profitability by integrating further down stream, joining competitors by acquisition, joint ventures and mergers to share the high costs of research, to exploit proven technology and to become the dominant player for a given product in the world market. This restructuring can be expected to continue as financial pressures force companies to find alternative ways of maintaining profitability.

Because of the current economic crisis in Asia, expansion in this region will be limited in the short term. However, it is expected to recover and Asia will account for over half of the world's growth in petrochemicals before 2010.

The aim of this book is to provide a concise coverage of these changes with the stress on petrochemicals which are of major importance, together with the technology used or being commercialized for their manufacture. One of the greatest problems was the choice of products so the emphasis is on the olefins and aromatics and their main derivatives. Ammonia has been included because of its significance as a consumer of methane.

For each product there is a summary of the changes that have occurred, the processes in commerical use, physical properties, applications, major producers and major licensors worldwide. Yield figures reflect results achievable on plants in commercial operation rather than calculated theoretical values. The objective is to provide the reader with an overview of the product rather than a detailed treatise. A licensors as well as a subject index has been include for quick reference. The data refers to 1997 unless otherwise stated, with forecasts of future growth where possible. Metric units have been used throughout. A list of abbreviations used in the book has been added.

Because of the increasing movement of petrochemicals worldwide, their transportation is regulated in most countries. Since this data is frequently difficult to identify, the international classifications given should facilitate this task. A list of the major organizations responsible for legislation relating to the movement of dangerous goods by land, sea or air, and health and safety has been included.

It must be remembered that this legislation is being continually updated. While the data quoted were obtained from published sources available in 1997, the reader is advised to verify the latest position with the appropriate authorities in their respective country.

Increasing concern over workers' exposure to certain chemicals has led to even more stringent government regulations on health and safety in most countries of the world. Although the latest published exposure levels are given for the US (ACGIH) and UK (COSHH), they are constantly under revision. The reader should therefore verify the current levels in force with the appropriate organization, details of which can be found at the end of this book.

The author would like to acknowledge the help she has received from her ex-colleagues in BP Chemicals, especially Ann McRae for her co-operation with the updating of statistics on the industry, to friends in other major chemical companies and to the librarian, Imperial College, University of London, for access to new process information. Particular thanks are due to Maurice Taylor for the major contribution he has made in converting my sketches into the flow diagrams which accompany the process information, as well as his valuable advice throughout the updating of this book.

List of Abbreviations (non-chemical)

ACGIH	American Conference of Governmental Industrial Hygenists
ACTS	Advisory Committee on Toxic Substances
ADNR	Regulation concerning the transportation of dangerous goods on the Rhine and all national waterways of the countries concerned
ADR	Accord Européen relatif au transport international des merchandises dangereuses par route
BFW	Boiler Feed Water
CAA	Civil Aviation Authority
CAS	Chemical Abstracts Service
CEL	Ceiling Exposure Limit
COSHH	Control of Substances Hazardous to Health
CW	Cooling Water
DAGAS	Dangerous Goods Advisory Service
EC	European Community
EINECS	European Inventory of Existing Commercial Chemical Substances
ELINCS	European List of Notified Chemical Substances
FC	Flow Controller
HE	Heat Exchanger
HI	Hazard Identification Number
HP	High Pressure Steam
IATA–DGR	International Air Transport Association Dangerous Goods Regulations
IMDG	International Maritime Dangerous Goods Code
LP	Low Pressure Steam
MEL	Maximum Exposure Limit
MP	Medium Pressure Steam
NIOSH	National Institute for Occupational Safety and Health
OSHA	Occupational Safety and Health Administration
PEL	Permitted Exposure Limit
RID	Règlement International concernant le transport des merchandises dangereuses par chemin de fer
STEL	Short Term Exposure Limit
STM	Steam
TLV	Threshold Limit Value

TWA	Time Weighted Average
UN No.	United Nations Number
WATCH	Working Group on the Assessment of Toxic Chemicals

Acetaldehyde

CH_3CHO

SYNONYMS

ACETALDEHYDE ethanal, acetic aldehyde, ethyl aldehyde, acetyl hydride

Historically, acetaldehyde was made by the silver gauze catalytic oxidation or dehydrogenation of ethyl alcohol using copper activated with chromium oxide.

These two routes replaced the earlier acetylene-based processes, and have, in their turn, been mostly superseded by the direct liquid-phase oxidation of ethylene.

The widescale availability of ethylene, its lower price, and high selectivity of the catalyst used, which conserves raw materials, have ensured the success of this route. Additionally, lower operating temperatures and pressures have led to significant energy savings. Today, it is the principal process used in the Western world. Some acetylene-based plants are still in operation in Eastern Europe and where available, fermentation ethyl alcohol is used in a few Third World countries.

Acetaldehyde can be produced from lower hydrocarbons or synthesis gas but neither route has been developed industrially because of low yields. Although higher chemical prices have refocused interest on processes based on methyl alcohol or methyl acetate, none is yet commercial. The non-catalysed oxidation of propane or butane or a mixture of the two is only economic if there is a demand for the range of oxidation products – formaldehyde, methyl alcohol, acetone, and mixed solvents – which are formed.

Purification of the end products is complex and expensive. Process conditions, raw materials and recovery procedures can be varied so that different ratios of end products can be produced.

Most recent research has involved the reduction of acetic anhydride with hydrogen over a platinum or palladium catalyst.

Acetaldehyde is also obtained as a by-product in a number of processes. These are the production of vinyl acetate from ethylene, and acrolein and acrylic acid from propylene.

Capacities range from 10 000 to 300 000 tonnes per year.

PROCESSES

1. From ethylene by direct oxidation: The one-stage process

In the single-stage process, high-purity ethylene (99.8%) and oxygen (99%) enter into the lower part of the reactor at a pressure of 40 bar. An aqueous catalyst solution, consisting of copper chloride ($CuCl_2$) containing small quantities of palladium chloride ($PdCl_2$) is fed into the reactor where it mixes with the feed gases. The reaction takes place at 130°C and the exit gases, containing acetaldehyde, water, and unreacted ethylene, pass to a separator where lighter gases collected overhead are cooled and separated by washing with water. Any unreacted gas, which contains ethylene, is recycled to the reactor. It is necessary to purge the recycle gas to prevent the build-up of inert gases, for example carbon dioxide, formed as a by-product. (*See Figure 1*)

Crude acetaldehyde is double distilled, first extractively with water to remove lower-boiling-point products, and second by fractionation to remove higher-boiling fractions, to give pure product.

The catalyst stream from the separator is heated to 160°C in the regenerator, where by-products decompose and the cuprous chloride is reoxidized by oxygen

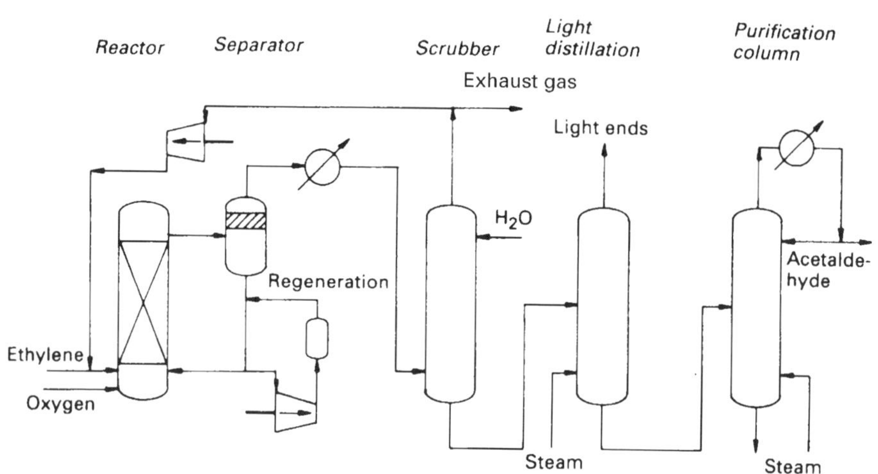

FIGURE 1 Acetaldehyde from ethylene by direct oxidation (one-stage process)

fed into the system. The addition of a buffering agent, such as copper sulphate, prevents the formation of acid caused by the reduction of palladium chloride to palladium.

2. From ethylene by direct oxidation: The two-stage process

Two-stage versions of the process are on stream, where reduction and oxidation are carried out in separate tubular reactors. Air is used instead of oxygen and the ethylene reaction takes place at 110°C and 90 bar in the presence of a solution of cupric chloride containing palladium chloride which acts as the catalyst. The gaseous reaction mixture containing acetaldehyde passes to a flash tower where an acetaldehyde–water mixture passes overhead. (*See Figure* 2)

The catalyst solution is introduced into the second reactor where the cuprous chloride is oxidized to cupric chloride with oxygen at 100 bar. Exhaust gases are separated and the regenerated catalyst solution is recycled.

The acetaldehyde–water vapour is concentrated and double distilled to remove light and heavy by-products as in the one-stage process. Because the reaction is carried out in two separate stages, ethylene obtained from naphtha cracking can be used instead of high-purity ethylene.

The choice of route depends on the source of the ethylene, availability of oxygen and the energy costs, as production costs and yields are similar.

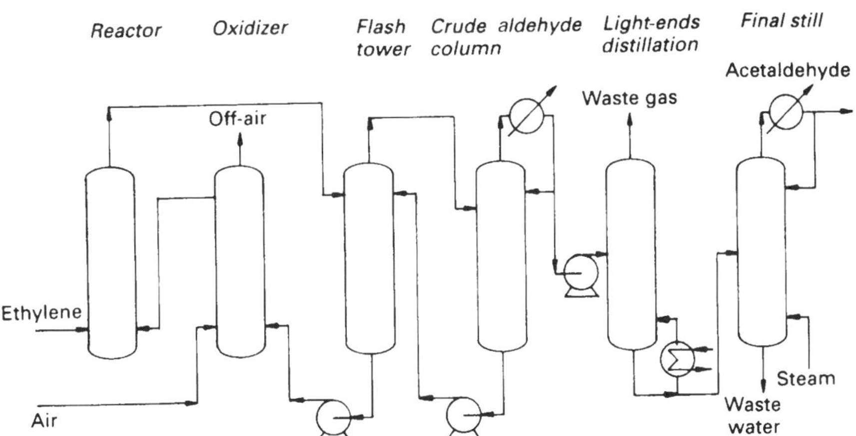

FIGURE 2 Acetaldehyde from ethylene by direct oxidation (two-stage process)

Reaction

$C_2H_4 + PdCl_2 + H_2O \rightarrow CH_3CHO + Pd + 2HCl$

$Pd + 2CuCl_2 \rightarrow PdCl_2 + 2CuCl$ catalyst

$2CuCl + \frac{1}{2}O_2 + 2HCl \rightarrow 2CuCl_2 + H_2O$ regeneration

Raw material requirements and yield

Raw materials required per tonne of acetaldehyde:

Ethylene (99% pure)	670kg
Oxygen	275Nm^3
Catalyst $PdCl_2$	1kg
$CuCl_2 + H_2O$	150kg

Yield 95%

3. From *n*-butane by oxidation

Butane or propane mixed with air and recycle gases (in the ratio 1:2:7 by volume and at a pressure of 7 bar) are passed through a tubular furnace where they are heated to 370°C, prior to entering the reactor. The exothermic reaction which takes place raises the temperature of the exit gases to 450°C. Any peroxides formed are decomposed by passing the hot gases through a ceramic-packed tube. (*See Figure 3*)

The reacted gases are cooled by passage through a quencher–scrubber through

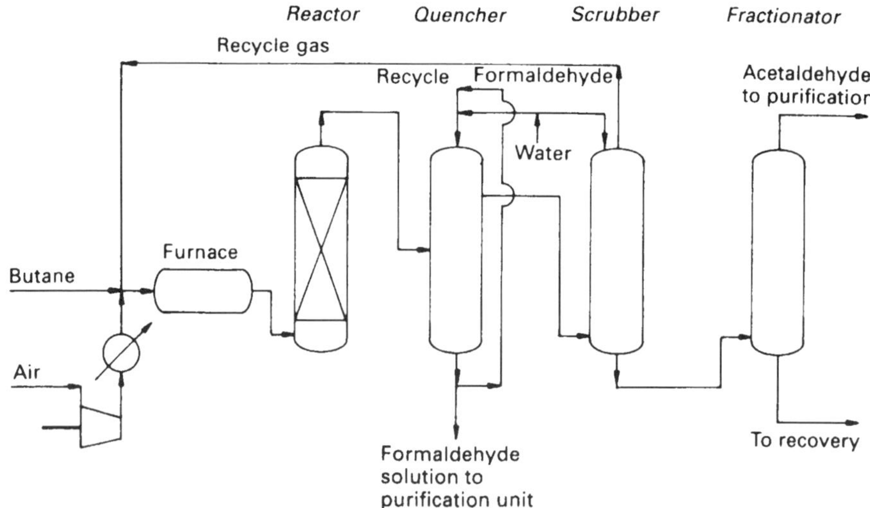

FIGURE 3 Acetaldehyde from *n*-butane by oxidation

which dilute aqueous formaldehyde circulates. The gases are absorbed by water in the scrubber and any unreacted gases are recycled.

Liquid from the bottom of the tower is fractionated to separate and recover the many compounds formed. *n*-Butane yields about 31% acetaldehyde, 33% formaldehyde, 20% methyl alcohol, 4% acetone and 12% of mixed solvents. Yields from propane are similar. Other alcohols and aldehydes can be recovered or hydrogenated to alcohols and paraffins.

By using different raw materials, reaction conditions and recovery methods, varying ratios of products can be obtained.

Reaction and yield

C_4H_{10} + O2 →	CH_3CHO +	mixed +	CO_2
	HCHO	organics	CO
			N_2

Yield 31%

4. From ethyl alcohol by dehydrogenation

Ethyl alcohol vapour is passed over a catalyst consisting of copper activated with chromium oxide. External heating is used to maintain the reaction temperature at 290°C. (*See Figure 4*)

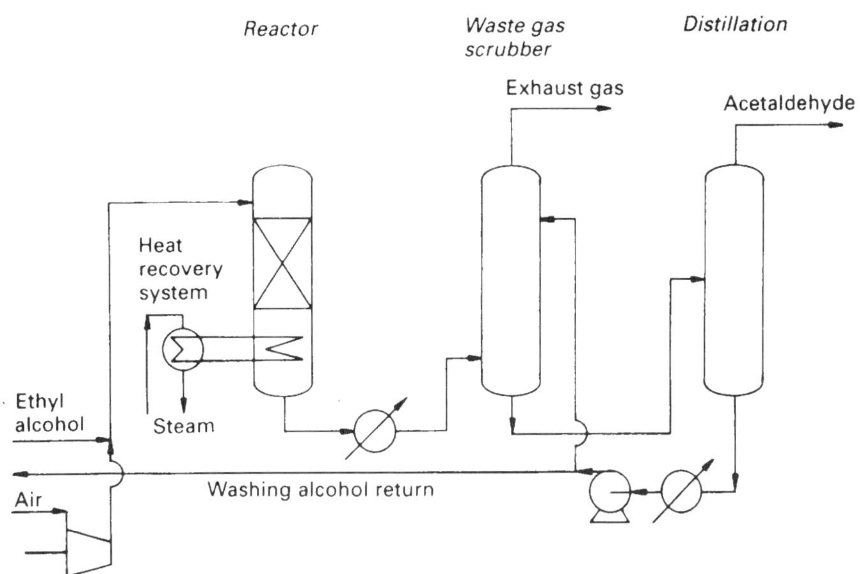

FIGURE 4 Acetaldehyde from ethyl alcohol by dehydrogenation

The exit gases from the reactor are cooled by scrubbing with cold ethyl alcohol. Acetaldehyde and any unreacted ethyl alcohol dissolve and the remaining gases, consisting mainly of hydrogen, are recovered and used as fuel.

The alcohol–acetaldehyde solution is distilled and acetaldehyde is collected overhead. The remaining ethyl alcohol can be recycled for further use. Conversion per pass is between 25% and 35%.

Reaction

$$CH_3CH_2OH \rightarrow CH_3CHO + H_2$$

Raw material requirements and yield

Raw materials required per tonne of acetaldehyde:

Ethyl alcohol (100%)	1140kg

Yield 90–95%

OTHER PROCESSES

Methyl alcohol can be hydroformylated with carbon monoxide and hydrogen in the presence of bromides or iodides of copper, nickel or iron, or iron–cobalt carbonyls. Alternative catalyst combinations are cobalt–nickel in the presence of tertiary amines, phosphines or nitriles. Reaction conditions are a temperature of 180–200°C and a pressure in the range of 300–400 bar.

Methyl acetate can be hydrocarbonylated using catalysts based on palladium, rhodium or iron salts, cobalt–nickel in the presence of tertiary amines, phosphines or nitriles, or cobalt–ruthenium catalysts promoted by methyl and sodium iodides. Most recent developments involve the reduction of acetic anhydride with hydrogen over a platinum catalyst. Although selectivities of up to 80% are claimed, none of these processes has yet been commercialized.

PROPERTIES

Mobile, colourless liquid with a strong pungent odour. Flammable. Soluble in water and most organic solvents.

Molecular Weight	44.05
Density at 20°C	0.779
Melting Point	–123.5°C
Boiling Point	20.4°C
Autoignition Temperature	140°C
Flammability Limits in air	
lower	4 vol%
upper	60 vol%

Flash Point Closed Cup	–40°C
Vapour Density (air = 1)	1.52

Exposure Limit COSHH

The Working Group on the Assessment of Toxic Chemicals (WATCH) is recommending an Occupational Exposure Standard to the Advisory Committee on Toxic Substances (ACTS).

Exposure Limit ACGIH 25 ppm TLV-STEL (Ceiling, Exposure Limit)

Classified A3, animal carcinogen.

POLYMERS OF ACETALDEHYDE

Acetaldehyde forms two polymers – paraldehyde which is a cyclic trimer, and metaldehyde, a cyclic tetramer.

In the presence of mineral acids such as sulphuric or hydrochloric, acetaldehyde polymerizes to paraldehyde. The acid is neutralized with sodium acetate or sodium bicarbonate and any acetaldehyde is recovered by distillation.

Acetaldehyde will polymerize in the gaseous phase in the presence of Al_2O_3, SiO_2 or $ZnSO_4$ which catalyse the reaction to give metaldehyde.

PROPERTIES

Paraldehyde

Colourless with a pungent odour. Soluble in most organic solvents.

Molecular Weight	132.16
Density at 20°C	0.992
Melting Point	12.5°C
Boiling Point	124.4°C
Flammability Limits in air	
lower	1.3vol%
upper	1.7vol%

Metaldehyde

White crystals. Insoluble in water but soluble in acetic acid and carbon disulphide.

Molecular Weight	176.2
Melting Point	246°C (sublimes)

GRADES

Technical >98%.

INTERNATIONAL CLASSIFICATIONS

Acetaldehyde

UN No.	1089
CAS Reg No.	75–07–0
EINECS No.	200–836–8
EC No.	605–003–00–6
Description	Flammable liquid
Packing Group	I
Emergency Action Code	2YE
HI (Kemler Code)	33

Paraldehyde

UN No.	1264
CAS Reg No.	123–63–7
EINECS No.	204–639–8
EC No.	605–003–00–1
Description	Flammable liquid
Packing Group	III
Emergency Action Code	2YE
HI (Kemler Code)	30

Metaldehyde

UN No.	1332
CAS Reg No.	108–02–3
EINECS No.	Not allocated
EC No.	605–005–00–7
Description	Flammable solid
Packing Group	III
Emergency Action Code	1Z
HI (Kemler Code)	40

APPLICATIONS

Acetaldehyde is a very reactive material and its main use is as an intermediate in chemical synthesis. Amongst the compounds that can be produced are:

- peroxyacetic acid, an intermediate in the manufacture of epoxides;
- pentaerythritol for the production of alkyd resins;
- 2-ethylhexyl alcohol, an intermediate for dioctyl phthalate used as a plasticizer for polyvinyl chloride (PVC);
- *n*-butyl alcohol which is converted to butyl acetate, a solvent for nitrocellulose lacquers;
- polyvinylacetals, used in the manufacture of laminated safety glass and washable primer paints;
- acetic acid, whose main outlet is for the manufacture of vinyl acetate;
- chloral, a raw material for the preparation of DDT.

With phenol or urea, acetaldehyde forms thermosetting resins. It will react with amines to produce compounds used in the rubber industry such as accelerators and antioxidants. Acetaldehyde is an effective disinfectant for the prevention of mould growth on leather, and is a hardener for glue, gelatin and casein. It can be used as a denaturant for ethyl alcohol.

The need for acetaldehyde as an organic intermediate has declined in recent years because of the switch to C_1-based raw materials. This fact, together with the recovery of by-product acetaldehyde, has led to the closure of a number of plants, a trend which is expected to continue.

HEALTH AND HANDLING

Acetaldehyde is irritating to the eyes, nose and throat in high concentrations, and these effects usually serve as adequate warning of its presence in the atmosphere. Adequate extraction systems must be provided. Contact with the eyes or skin should be avoided by the use of goggles and protective clothing. Any contaminated clothing must be laundered before reuse.

Containers should be constructed of stainless steel or aluminium and pressure tested because of vapour build-up caused by the low boiling point of acetaldehyde. Copper or mild steel must not be used due to the risk of corrosion from traces of acetic acid which are normally present. Acetaldehyde must be stored in a cool location under nitrogen at 1 bar pressure. Air must never be used because explosive peroxides may be formed. Storage tanks should be earthed to prevent static build-up.

Spills can be absorbed by dry sand or earth, or if small, diluted with large quantities of water to reduce the potential risk of explosion. The liquid is a dangerous fire and explosion hazard. Clean-up personnel must wear full protective clothing.

Fires should be blanketed by carbon dioxide, dry chemical or foam. Water must be avoided as it may spread the fire. Because acetaldehyde vapours can travel long distances, flashback is a hazard. All firefighters must wear protective clothing and self-contained breathing apparatus because above 400°C carbon monoxide and methane gases are given off.

MAJOR PLANTS

Plants with capacities greater than 70 000 tonnes per year:

Celanese	Lillebonne	France
Celanese	Frankfurt	Germany
	Hürth	Germany
EniChem	Priolo	Italy
IQA	Tarragona	Spain
Eastman Chemical	Longview	US
Pemex	La Grangrejera	Mexico
	Morelos	Mexico
Japan Aldehyde	Chiba	Japan
Kyowa Yuka	Yokkaichi	Japan
Showa Denko	Oita	Japan
Tokuyama Petrochemical	Shin Nanyo	Japan

MAJOR LICENSORS

Ethyl alcohol	*Veba Chemie, Huels*
Ethylene	*Hoechst/Uhde, Huels*

Acetic Acid

CH_3COOH

SYNONYMS

ACETIC ACID ethanoic acid, methane-carboxylic acid, vinegar acid

Vinegar, a dilute solution of acetic acid made by the fermentation of ethyl alcohol, has been known since earliest times and some is still produced by this route for human consumption.

Small quantities of acetic acid are recovered from pyroligneous acid liquor, obtained by the destructive distillation of hardwood. Difficulties of concentration and purification of the acid and the high capital cost involved make this route unattractive. However, a few small plants do remain.

The modern acetic acid industry began with the commercial availability of acetylene which was converted to acetaldehyde and then oxidized to acetic acid.

There are three commercial synthetic processes for the manufacture of acetic acid:

- oxidation of acetaldehyde;
- liquid-phase oxidation of *n*-butane or naphtha;
- carbonylation of methyl alcohol.

The carbonylation of methyl alcohol is the dominant technology because of its low material and energy costs and the absence of significant amounts of by-products. First developed by BASF, Monsanto's discovery of an iodide-promoted rhodium catalyst with high selectivity which enabled the reaction to take place at lower temperatures and pressures, has revolutionized this route. Synthesis gas, the raw material for the carbonylation process, can be obtained from a wide range of sources.

In 1996, BP Chemicals announced new technology, called Cativa, based on methyl alcohol and carbon monoxide feedstocks, with a more active catalyst system giving lower by-product formation. The first plant to use this process has been commercialized in Texas City. UOP and Chiyoda have also developed a new process based on carbonylation technology using methyl alcohol and carbon monoxide as feedstock.

There has been extensive research on finding alternative routes to acetic acid including the direct oxidation of ethylene, ethane or methane. Showa Denko have built an ethylene-based plant in Japan.

Of the major processes used worldwide, over 60% of acetic acid is produced from synthesis gas by the carbonylation route, around 5% from acetaldehyde, and the remainder from naphtha or *n*-butane. Of the naphtha and butane feedstocks, *n*-butane is preferred in the US, but in Europe, light naphtha is more frequently used. BP Chemicals is the world's largest producer of acetic acid for the merchant market.

Capacities range from 30 000 to 840 000 tonnes per year.

PROCESSES

1. From acetaldehyde by oxidation

Acetaldehyde solution (95%) is fed into a reactor where oxygen or air is bubbled through the liquid containing 0.1–0.5% managanese acetate which catalyses the reaction and minimizes the risk of explosion from the intermediate peracetic acid which is formed. Other catalysts, such as cobalt, and copper–cobalt mixtures can be used but yields are generally inferior. (*See Figure 5*)

The reaction takes place at 60–80°C and a pressure of 3–10 bar. Heat from the exothermic reaction is removed in a heat exchanger. The overhead gases are cooled, scrubbed with recycle acetaldehyde and then with water before being flared.

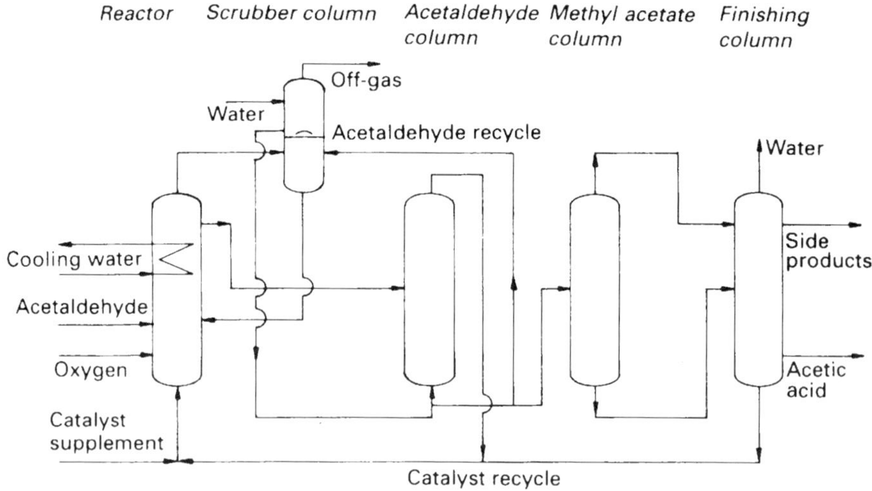

FIGURE 5 Acetic acid from acetaldehyde by oxidation

The reactor mixture is fed into a distillation column where acetaldehyde is recovered and recycled. Methyl acetate is removed in the second distillation column and the residual acetic acid is distilled in the finishing column. High-purity acetic acid vapour is obtained as a side stream, and water is removed as an azeotrope overhead. Final distillation takes place in the presence of potassium permanganate or a similar oxidant.

The liquid from the bottom of the third distillation column, containing the catalyst, is returned to the reactor.

Reaction

$$CH_3CHO + \tfrac{1}{2}O_2 \rightarrow CH_3COOH$$

Raw material requirements and yield

Raw materials required per tonne of acetic acid:

Acetaldehyde (100%)	770kg
Oxygen (100%)	205Nm^3
Steam (14 bar)	700kg
Nitrogen	4Nm^3

Yield 95%

2. From *n*-butanes or naphtha by oxidation

A butane stream, containing 95% *n*-butane with small amounts of isobutane, is fed into a reactor containing a solution of acetic acid and cobalt, manganese, nickel or chromium acetate. A cobalt acetate catalyst yields 6wt% each of formic and propionic acids, but with manganese acetate more formic acid (up to 13wt%) and less acetic acid (around 61wt%) are obtained. (*See Figure 6*)

Air or oxygen is bubbled through the solution; the liquid-phase oxidation takes place at 150–200°C with a pressure in the region of 56 bar.

The reaction gases, containing a mixture of organic acids, lower alcohols, acetone and methyl ethyl ketone, are cooled and pass to a gas–liquid separator. Here, the upper phase containing mainly hydrocarbons is recycled to the reactor, while the lower aqueous phase, which contains acetic acid together with a range of by-products, is successively distilled. Water is removed by azeotropic distillation and formic, acetic, and propionic acids are recovered overhead.

Bottoms from the light-ends column pass to a solvent column where ketones and esters are recovered either individually or as mixtures. The vent gases are expanded, and the energy formed is used to generate electricity or to compress air feed into the reactor.

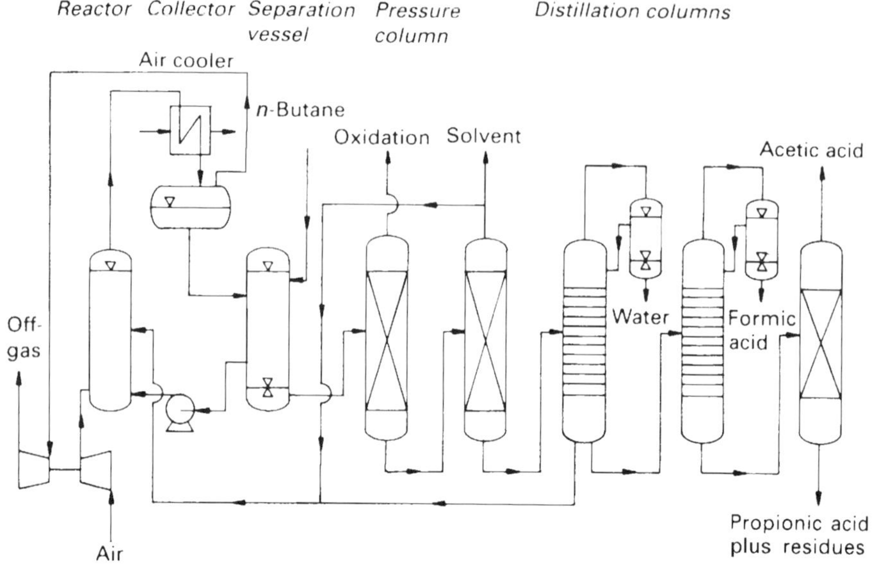

FIGURE 6 Acetic acid from *n*-butane by oxidation

Light naphtha, comprising hydrocarbons in the C_3–C_7 range can be used as feedstock instead of butane.

The oxidation reaction, which is highly exothermic, is carried out in the liquid phase at a temperature of 120–180°C and a pressure of 50 bar. The reaction mixture containing organic acids, intermediate products, and water is successively distilled. Formic, acetic, and propionic acids are recovered overhead. Succinic acid can be obtained from the residual heavy ends, and acetone is optionally recovered from the intermediate stream or returned to the reactor. Other by-products obtained include acetaldehyde, methyl ethyl ketone and ethyl acetate.

The chief advantage of naphtha is its cheaper cost and the less severe reaction conditions required, but the number of by-products is greater, and their separation more complex. Lower yields of acetic acid are obtained.

Reaction

$$C_2H_4 + O_2 \rightarrow CH_3COOH \begin{array}{l} + \text{ alcohols} \\ + \text{ ketone} \\ + \text{ acids} \end{array} \left.\right\} \begin{array}{l} + \text{ methane} \\ + \text{ carbon dioxide} \\ + \text{ carbon monoxide} \\ + \text{ nitrogen} \end{array} \left.\right\}$$

Raw material requirements and yield

Raw materials required per tonne of acetic acid:

n-butane	965kg
Air	3 750kg

Yield 75–80%

3. From methyl alcohol by carbonylation

Two commercial processes for the carbonylation of methyl alcohol have been developed. In the initial high-pressure route by BASF, conversion took place at 250°C and 700 bar in the presence of a cobalt iodide catalyst. Subsequent improvements to the process have enabled the pressure to be reduced to 5–10 bar. (*See Figure 7*)

In 1968, Monsanto announced an alternative low-pressure process, which utilized an iodide-promoted rhodium catalyst. All new plants tend to employ this process, because of the high selectivity of the catalyst.

Methyl alcohol and carbon monoxide are fed into a stainless steel reactor containing the catalyst. The reaction takes place in the liquid phase at a temperature of 150–200°C and a pressure of 15 bar. Spectroscopic studies have shown that the reaction takes place in a five-step mechanism, the main by-products being carbon dioxide and hydrogen.

Crude acetic acid is sent to a light-ends column, where light ends are recovered overhead. These are combined with the overhead gases from the reactor in a scrubber before being recycled. Off-gases from the scrubber are flared.

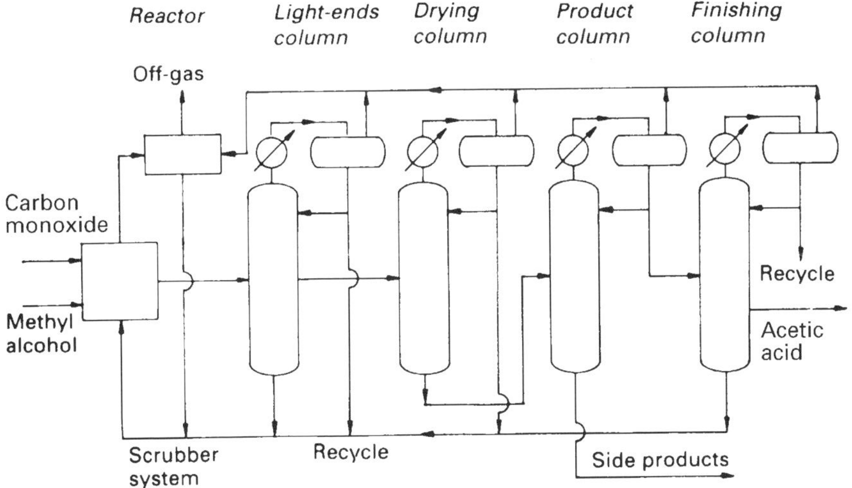

FIGURE 7 Acetic acid from methyl alcohol by carbonylation

The acetic acid stream passes to a drying column where water, removed as an azeotrope, is returned to the reactor. Dry acetic acid is sent to the product column, and propionic acid is removed as a base stream. Overheads pass to the finishing column, and high-purity acetic acid is obtained as a vapour side stream. Overheads from the distillation columns are recycled to the reactor. Because of the milder operating conditions, the amounts of by-products are small, being mainly carbon dioxide and hydrogen.

Reaction

$C_2H_4 + O_2 \rightarrow CH_3COOH$
$CO + H_2O \rightarrow CO_2 + H_2$

Raw material requirements and yield

Raw materials required per tonne of acetic acid:

Methyl alcohol	533kg
Carbon monoxide	467kg

Yield

based on methyl alcohol	99%
based on carbon monoxide	90%

BP Chemicals have developed new technology for the manufacture of acetic acid. Based on methyl alcohol and carbon monoxide as feedstock, the process utilizes a promoted iridium-based catalyst, which gives higher reaction rates and less by-product formation, especially propionic acid, resulting in lower separation costs. Savings of up to 30% are claimed over conventional methyl alcohol carbonylation routes, and the process has been installed for the first time at Sterling Chemicals', Texas City, US plant. The UOP and Chiyoda process uses rhodium complexed to polyvinyl pyridine promoted with methyl iodide as the catalyst. The reaction is carried out in a bubble reactor and yields of 93% have been achieved.

OTHER PROCESSES

Two processes have been developed utilizing *n*-butene fractions as raw materials.

In the Bayer process, *n*-butene and acetic acid are converted into *sec*-butyl acetate in a cascade reactor. The aqueous-phase reaction is carried out at 110–200°C and 19 bar, using finely ground acid ion exchange resin as catalyst.

After centrifuging to remove the catalyst, the reaction mixture is fed into an oxidizing reactor and treated with air at 200°C and 60 bar pressure. No catalyst is required.

Butyl acetate breaks down to give three molecules of acetic acid, one of which is recycled to the first reactor. The crude acid is distilled, water being removed as an azeotrope, and formic acid and other by-products are recovered by fractionation. Yield 58%.

Vapour-phase oxidation of *n*-butene with air has been developed by Huels. The reaction takes place at 180–240°C and a pressure of 2–30 bar over a titanium–vanadium oxide catalyst. Steam is added to improve selectivity and minimize explosive risks. The resultant aqueous acetic acid is concentrated and purified. Formic acid is the major by-product and can be recovered. Yield 46%.

Although both processes have been developed to pilot plant stage, neither has been commercialized to date.

PROPERTIES

Clear, colourless, mobile liquid with pungent odour and acidic taste. Corrosive and flammable. Soluble in water and most organic solvents. Not soluble in mineral oils and carbon disulphide. Solidifies into colourless, lamellar crystals on freezing.

Molecular Weight	60.05
Density at 20°C	1.049
Melting Point	16.7°C
Boiling Point	118.2°C
Autoignition Temperature	450°C
Explosive Limits in air	
lower	4.0vol%
upper	17.0vol%
Flash Point Closed Cup	42°C
Vapour Density (air = 1)	2.07
Exposure Limit COSHH	15ppm 15 minutes 10ppm 8 hour TWA
Exposure Limit ACGIH	15ppm TLV-STEL 10ppm TLV-TWA

GRADES

Technical 80% and 99% minimum, pure (pharmacopeia standard) 80% and 99+%, glacial 99.5%.

INTERNATIONAL CLASSIFICATIONS

UN No.	2789 (>80% by weight)
CAS Reg No.	64–19–7
EINECS No.	200–580–7
EC No.	607–002–00–3 (glacial) 607–002–00–6 conc. >90%
Description	Corrosive substance, flammable liquid
Packing Group	II
Emergency Action Code	2P
HI (Kemler Code)	83
UN Number	2790 (>10% but <80% by weight)

CAS Reg No.	64–19–7
EINECS No.	200–580–7
EC No.	607–002–00–3 25% >conc. <90%
Description	Corrosive substance
Packing Group	II
Emergency Action Code	2R
HI (Kemler Code)	80

APPLICATIONS

Acetic acid is one of the most important organic chemicals produced today with many industrial applications.

Globally, its largest outlet, accounting for 44% of total consumption, is in the manufacture of vinyl acetate leading to polyvinyl acetate and polyvinyl alcohols used for paints, adhesives and plastics.

Around 12% of acetic acid production is converted to acetic anhydride which is used to manufacture cellulose acetate, paper sizing agents, a bleach activator (TAED), and aspirin. Cellulose acetate is used for fibres, films, plastics and cigarette filters.

Another 13% of acetic acid is consumed in the production of a range of acetates which have many uses. The esters are used extensively as solvents for coatings, inks, resins, gums, flavourings and perfumes. Sodium acetate is used in blood dialysis, and the dyeing and photographic industry. Chromium and copper acetates are employed in leather tanning and as paint dryers, while the zinc salt is found in wood preservatives and porcelain printing. Ammonium acetate is used to manufacture acetamide, as a diuretic, and in textile dying and metal treatment.

Acetic acid is an excellent solvent for chemical reactions and 12% is used in the manufacture of terephthalic acid (TPA), used for polyethylene terephthalate (PET) bottles and polyester film and fibre.

Other minor outlets are in textile processing, the manufacture of chloroacetic acid for the manufacture of herbicides, carboxymethyl cellulose, and dyestuffs. It is also used in the food industry as a flavouring agent and preservative.

World demand for acetic acid is increasing at 4% per year with the Far East as the fastest growing region.

HEALTH AND HANDLING

Acetic acid is very corrosive. Splashes of the acid will cause moderate–severe burns. Protective clothing, gloves, rubber boots and goggles must be worn during handling operations. Because of the irritant vapour, good ventilation is essential and breathing apparatus should be used.

Acetic acid of 80% or greater concentration can be stored in stainless steel, high-density polyethylene or glass containers. In the presence of water, acetic acid is more corrosive.

Spills should be absorbed with sand or earth but small amounts can be diluted with large volumes of water. Acetic acid will react violently with strong alkalis, so dilution is essential before neutralization with lime. Even so, care should be exercised to avoid a dangerous reaction.

Fires should be extinguished with carbon dioxide, dry chemical, water spray or polymer foam. Personnel should wear protective clothing and self-contained breathing apparatus when dealing with fires or spills.

MAJOR PLANTS

Plants with capacities greater than 200 000 tonnes per year:

Acetex	Pardies	France
BP Chemicals	Hull	UK
Complex	Severodonetsk	Ukraine
Eastman Chemical	Kingsport	US
Hoechst Celanese	Clear Lake	US
	Pampa	US
Millenium Petrochemicals	Deer Park	US
Sterling Chemicals	Texas City	US
Koyo Chemical	Aboshi	Japan
Showa Denko	Oita	Japan
Samsung/BP Chemicals	Ulsan	South Korea

MAJOR LICENSORS

Naphtha	*BP Chemicals*
Acetaldehyde	*Daicel*
	Hoechst
	Huels
n-Butane	*Celanese*
	Huels
Methyl alcohol	*BASF*
	BP Chemicals
	Monsanto
	UOP Chiyoda
n-Butene	*Bayer*
	Huels
Ethylene	*Showa Denko*

Acetic Anhydride $(CH_3CO)_2O$

SYNONYMS

ACETIC ANHYDRIDE ethanoic anhydride, acetic acid anhydride, acetyl oxide

The oldest process for the manufacture of acetic anhydride was based on the conversion of sodium acetate in the presence of an inorganic chloride, such as thionyl chloride, to give acetyl chloride which is reacted with further sodium acetate to give acetic anhydride. This route was modified by using acetic acid and phosgene in the presence of aluminium chloride.

Subsequently, two other processes were introduced:

- the cleavage of ethylidene diacetate in the presence of acid catalysts such as zinc chloride to give acetaldehyde and acetic anhydride;
- the reaction between vinyl acetate and acetic acid using a palladium catalyst to yield acetaldehyde and acetic anhydride.

None of these processes is now of any industrial importance.

Rapid increases in ethylene and energy costs in the 1970s led Halcon to the development of a new route to acetic anhydride based on methyl alcohol and carbon monoxide via methyl acetate. Synthesis gas produced from coal can be used as the raw materials source. In 1983, the first plant to use this carbonylation of methyl acetate process started production. Within five years, methyl acetate processes had captured 25% of capacity in the US.

Today, acetic anhydride is produced primarily from ketene and as the economics of ketene-based plants are better than those employing acetaldehyde, the latter are in decline. The growth of methyl acetate based processes is expected to continue with the greater interest in C_1 feedstocks.

Capacities range from 30 000 to 840 000 tonnes per year.

PROCESSES

1. From acetic acid via ketene

Glacial acetic acid vapour is passed into a multi-coil reactor containing small amounts of triethyl phosphate catalyst. At a temperature of 700–750°C, acetic acid is cleaved to ketene and water. Reduced pressure is used in order to isolate the ketene before it reacts with either acetic acid or water formed by the reaction. (*See Figure 8*)

Ketene from the separator is absorbed by acetic acid in a scrubber containing Raschig rings, or in a liquid ring pump containing acetic acid at 45–55°C and 0.1–0.2 bar. The resultant mixture, containing 85–90% acetic anhydride, is cooled and the off-gases are separated. Pure acetic anhydride is recovered by fractionation, either continuously or discontinuously, the former method being the preferred option in newer plants.

Most cellulose acetate producers use the liquid pump (Wacker) process to recover acetic anhydride from concentrated waste acids, and this route is gaining in popularity because of its energy efficiency. The off-gases are normally burnt to preheat the gases entering the reactor.

Ketene can also be prepared by the cleavage of acetone, but this process is no longer of any economic importance.

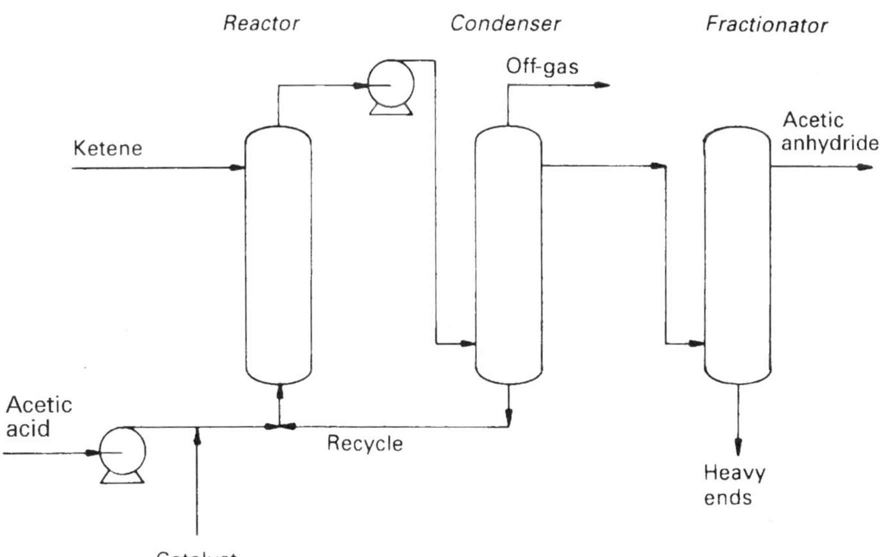

FIGURE 8 Acetic anhydride from acetic acid

Reaction

$$CH_3COOH \rightarrow CH_2CO + H_2O$$

$$CH_3COOH + CH_2CO \rightarrow (CH_3CO)_2O$$

Raw material requirements and yield

Raw materials required per tonne of acetic anhydride:

Acetic acid	1250kg
Catalyst	small

Yield 96%

2. From acetaldehyde by oxidation

Acetaldehyde is reacted with air or oxygen in the liquid phase in a reactor containing a catalyst consisting of a mixture of metallic salts. These can be manganese and copper acetates, cobalt and nickel acetates, or cobalt and copper salts of higher fatty acids. Manganese acetate is particularly important as it inhibits the formation of explosive amounts of peracetic acid that may be formed during the reaction. (*See Figure 9*)

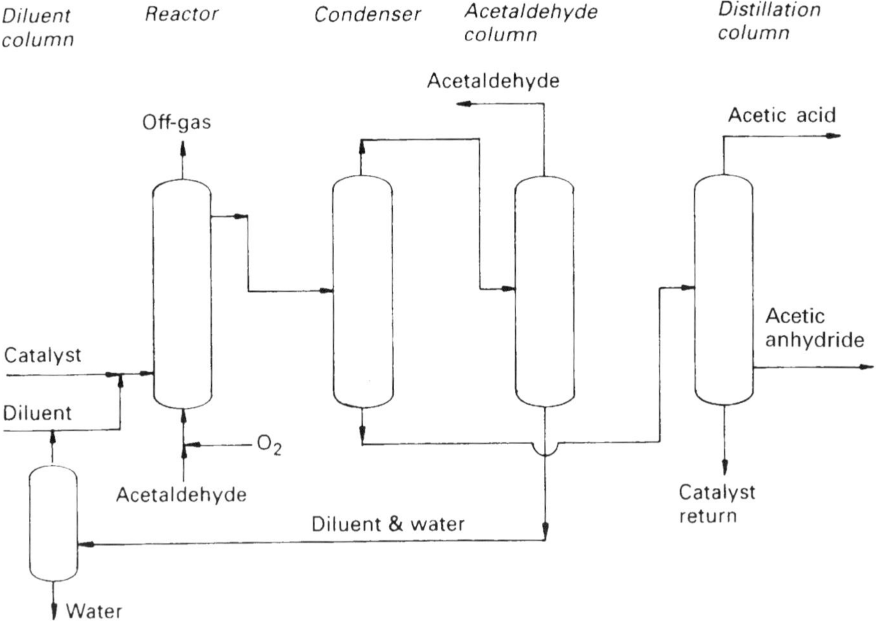

FIGURE 9 Acetic anhydride from acetaldehyde by oxidation

The process operates at 2–3 bar and 40–60°C, because at temperatures above 60°C acetic anhydride decomposes rapidly. The strongly exothermic reaction is controlled by the addition of a diluent, ethyl acetate, which is mixed with acetaldehyde in the ratio of 2:1. The lower acetates, such as ethyl acetate, form azeotropic mixtures with the water formed in the reaction and provide a rapid method for its separation and removal. Other diluents that can be used are methylene chloride, diisopropyl ether, cyclohexane, alkyl benzoates, and alkyl phthalates.

The reactor products pass through a water-cooled condenser where non-condensable gases proceed to a packed column fed from the top with fresh acetaldehyde. The dehydrated mixture from the base of the reactor is distilled to recover acetic anhydride. Acetaldehyde is reclaimed from the side stream and recycled to the reactor. The off-gas containing acetates is flared.

Yields depend on the catalyst and diluent used, acetaldehyde ratios, and operating temperature and pressure. With optimal conditions, conversion rates of 70% have been achieved.

Reaction

$$CH_3CHO + O_2 \rightarrow CH_3COOOH$$

$$CH_3COOOH + CH_3CHO \rightarrow (CH_3CO)_2O + H_2O$$

Raw material requirements and yield

Raw materials required per tonne of acetic anhydride:

Acetaldehyde	1200kg
Catalyst	1kg
Diluent (less if the diluent is recovered)	2400kg

Yield 70%

3. From methyl acetate by carbonylation

Methyl acetate, produced by the esterification of acetic acid with methyl alcohol or as a by-product from the methyl alcohol-carbon monoxide process for the manufacture of acetic acid, is dried and sent to a reactor. Carbon monoxide is compressed to 200–500 bar before being fed in. The reaction takes place in the liquid phase at a temperature of 160–190°C in the presence of a rhodium chloride catalyst activated by methyl iodide, hydrogen iodide, lithium iodide, iodine, or other iodides. Although nickel can be used as a catalyst, rhodium is ten times more active and is the preferred choice. (*See Figure 10*)

Heat from the exothermic reaction is removed by a heat exchanger and used to preheat the methyl acetate feed and to produce low-pressure steam.

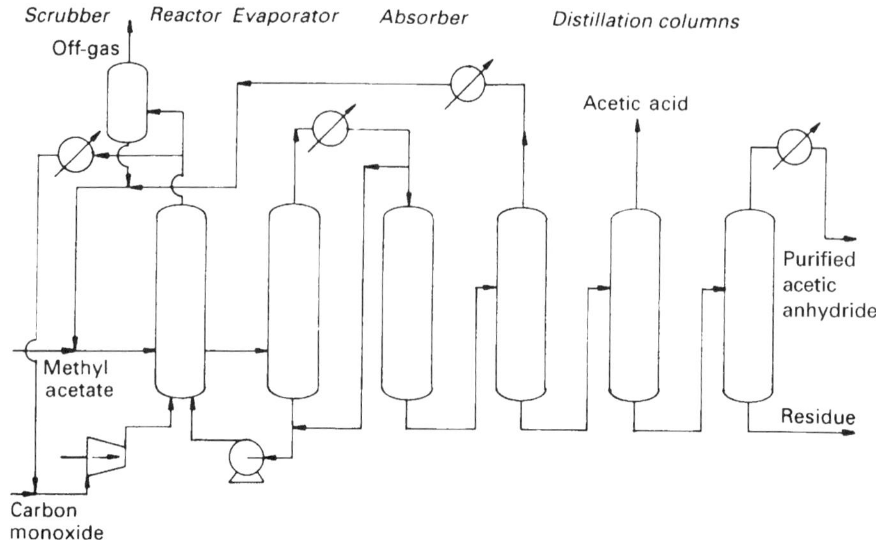

FIGURE 10 Acetic anhydride from methyl acetate by carbonylation

Unreacted carbon monoxide gas from the top of the reactor is cooled to remove any iodides, esters and acetic anhydride before being recycled. Part of the recycle stream, bled off to prevent the build-up of inert gases, is washed with acetic anhydride to remove any iodides before being vented as off-gas. The acetic anhydride–iodide mixture is returned to the reactor where additional catalyst is added as required.

The liquid reaction products are flash distilled in a carbon monoxide–hydrogen atmosphere at 50 bar to prevent catalyst decomposition. Bottoms are recirculated to the carbonylation reactor. Gases from the top of the evaporator are condensed, and pass through an absorber where traces of rhodium and iodide compounds are removed.

Acetic anhydride is purified by triple distillation. In the first column, methyl iodide and methyl acetate are removed overhead and returned to the carbonyl reactor. The bottoms pass to the second column where acetic acid is distilled overhead. Pure acetic anhydride is obtained from the third distillation column and high boiling compounds are removed from the base. A solution of potassium acetate in acetic anhydride is added to the top of the purification column to remove any remaining iodides from the pure anhydride. Heavy residues are burnt.

The selectivity and lifetime of the catalyst can be increased by the addition of 2–7% of hydrogen to the carbon monoxide reactor feed.

Reaction

With recycle acetic acid:

$$CH_3OH + CH_3COOH \rightarrow CH_3COOCH_3 + H_2O$$

$$CH_3COOCH_3 + CO \rightarrow (CH_3CO)_2O$$

Without recycle acetic acid:

$$CH_3COOCH_3 + CO \rightarrow (CH_3CO)_2O$$

$$(CH_3CO)_2O + 2CH_3OH \rightarrow CH_3COOCH_3 + H_2O$$

Raw material requirements and yield

Raw materials required per tonne of acetic anhydride:

Methyl alcohol	660kg
Carbon monoxide	620kg

Yield 95–99%

PROPERTIES

Clear, colourless liquid with pungent odour. Corrosive. Soluble in a wide range of organic liquids. Slightly soluble in water but reacts rapidly as the temperature increases or in the presence of traces of mineral acids.

Molecular Weight	102.09
Density at 20°C	1.081
Melting Point	–73.1°C
Boiling Point	140°C
Flash Point Closed Cup	48.9°C
Autoignition Temperature	392°C
Explosive Limits in air	
lower	2.0vol%
upper	10.3vol%
Vapour Density (air = 1)	3.52

Exposure Limit COSHH 5ppm 15 minutes
Substance on the current Advisory Committee on Toxic Substances (ACTS) and the Working Group on the Assessment of Toxic Chemicals (WATCH) work programme.
Exposure Limit ACGIH 5 ppm 8 hour TLV-TWA
(Ceiling Exposure Limit)

GRADES

Commercial 98%, pure over 99% (pharmaceutical grade).

INTERNATIONAL CLASSIFICATIONS

UN No.	1715
CAS Reg No.	108–24–7
EINECS No.	203–564–8
EC No.	607–008–00–9
Description	Flammable liquid
Packing Group	II
Emergency Action Code	2W
HI (Kemler Code)	83

APPLICATIONS

The major uses of acetic anhydride in the chemical industry are as acetylating and dehydrating agents.

The acetylation of cellulose to cellulose acetate, used for the manufacture of films, fibres, plastics, cigarette filters and coatings, is the largest single outlet. Another important use is as a binding agent for ammonia released in the production of polymethyl acrylimide hard foam.

Acetic anhydride is employed in the manufacture of a range of pharmaceutical intermediates, particularly aspirin and paracetamol.

It has a number of industrial uses:

- in the food industry, to improve the solubility and clarification of oils and fats by acetylation;
- in the detergents industry, to produce ethylenediaminetetraacetate, a cold bleaching activator in washing powder formulations;
- in the paper industry, to produce acetylated starch for sizing agents;
- in the plastics industry, for the manufacture of plasticizers such a glycerol triacetate, acetyl tributyl citrate and acetyl ricinolate.

Because of its reactivity, acetic anhydride is used in the synthesis of a wide range of chemical intermediates including acetates, amides, carboxylic anhydride and acetyl peroxide. Small quantities are used in the manufacture of the explosive, cyclonite.

The market for acetic anhydride is mature and its future will depend on the demand for cellulose acetate fibres and plastics because such a considerable volume of production is captive to the large fibre manufacturers.

HEALTH AND HANDLING

Rubber gloves, clothing and boots should be worn when handling acetic anhydride because of its irritant effect on the skin, and breathing apparatus is required in confined areas. The vapour will cause eye irritation and lacrimation, and contact lenses must be avoided because they concentrate the effects. Excessive levels will cause permanent damage.

Acetic anhydride should be stored in closed containers made of stainless steel or aluminium, in a cool, dry place, away from heat. It is incompatible with acids, bases, oxidizing and reducing agents, and finely divided metals.

If acetic anhydride is spilled, it should be contained or if small amounts, absorbed with activated carbon, and the residues placed in closed containers and disposed of by incineration according to local regulations. Contaminated surfaces should be neutralized with sodium bicarbonate or soda ash, sprayed with water and the resultant solution kept in a holding area. Acetic anhydride must not be allowed to contaminate waterways or sewers.

Fires can be extinguished with carbon dioxide, dry chemical or polymer foam. Water must be used with caution because acetic anhydride reacts, generating much heat. Firefighters must wear protective clothing and self-contained breathing apparatus.

MAJOR PLANTS

Plants with capacities over 120 000 tonnes per year:

BP Chemicals	Hull	UK
Novaceta	Spondon	UK
Eastman Chemical	Kingsport	US
Hoechst Celanese	Narrows	US
	Pampa	US
	Rock Hill	US
Daicel Chemical	Hyogo	Japan

MAJOR LICENSORS

Methyl acetate	*Air Products & Chemicals*
	Halcon-Scientific Design
	Hoechst
	Mitsubishi Gas Chemical
	Rhone-Poulenc
Methyl alcohol	*Monsanto*
	Wacker Chemie
Ketene	*Wacker Chemie*
Acetaldehyde	*Rhone-Poulenc*

Acetone

CH_3COCH_3

SYNONYMS

ACETONE 2-propanone, dimethyl ketone, ketone propane, dimethyl-formaldehyde, pyroacetic acid, pyroacetic ether

Early industrial manufacture of acetone was based on the thermal decomposition of calcium acetate, produced by treating pyroligneous acid, the condensate from the destructive distillation of wood, with lime.

As consumption of acetone increased, carbohydrate fermentation of corn starch or molasses by the bacteria *Clostridium acetobutylicum* was developed. This method co-produces acetone and *n*-butyl alcohol in a 1:2 ratio. Although important in the past, this route is only used to a limited extent in a few countries, such as South Africa, Brazil, Egypt, and India.

The oxidation of natural gas was employed in the US in the 1930s but the wide range of products produced required careful fractionation to separate them.

The ready availability of propylene in the 1960s led to the development of acetone processes using either isopropyl alcohol or cumene as the intermediate. Because of its lower cost, the cumene route is preferred, now accounting for the greater share of acetone manufacture. Small quantities of acetone are produced as a by-product of the oxidation of naphtha or *n*-butane to acetic acid. The oxidation of *p*-diisopropylbenzene and of *p*-cymene can also yield small quantities of acetone as a by-product.

The latest process to be developed is the direct oxidation of propylene and this route has attained importance in Japan. In the rest of the world, cumene plants built to meet phenol demand have provided by-product acetone, without the corrosion problems and high capital and operating costs associated with the direct propylene oxidation route. Isopropyl alcohol based acetone has declined because cumene-based material is cheaper.

Currently, around two-thirds of acetone is produced from cumene, with most of the remainder coming from isopropyl alcohol and toluene. Small quantities are obtained as by-product from various oxidation processes used in the manufacture of acetic acid, hydroquinone and hydrogen peroxide.

There are great differences in these percentages between countries. In the US, the isopropyl alcohol route is more important than in Europe, while in Japan the catalytic oxidation of propylene is the second most important process. In the future, hydrocarbons are expected to remain the major source of acetone, but recent advances in biochemistry could change this situation.

Capacities range from 10 000 to 315 000 tonnes per year.

PROCESSES

1. From cumene by oxidation

Cumene is oxidized in the liquid phase to cumene hydroperoxide which in turn is cleaved, in the presence of sulphuric acid, to yield phenol and acetone (see Phenol). (*See Figure 11*)

The cleavage mixture is cooled rapidly to prevent further reaction by the acetone formed, and then the residual acid is either neutralized with alkali or removed by passage through an ion exchange column. Crude acetone, together with some α-methylstyrene and cumene, is recovered overhead. Acetone is purified by distillation with steam in a second column.

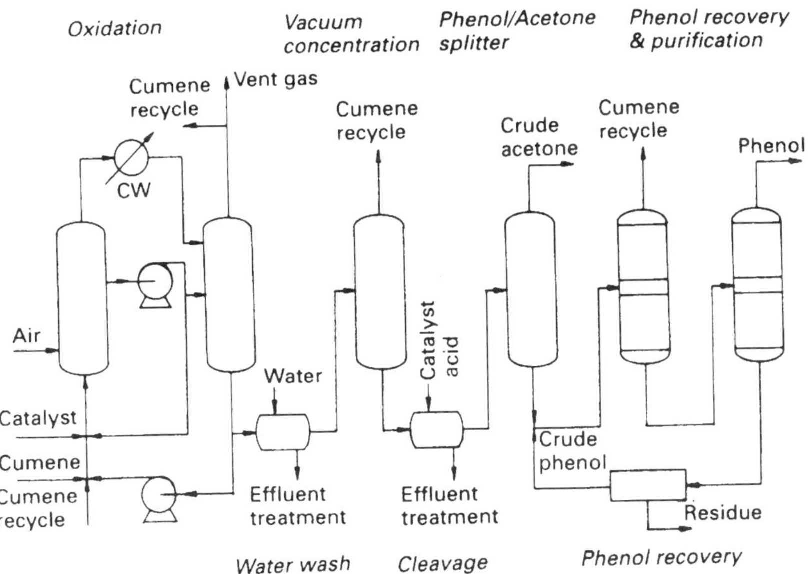

FIGURE 11 Acetone from cumene by oxidation

Reaction

$$C_6H_5CH(CH_3)_2 + O_2 \rightarrow C_6H_5C(CH_3)_2OOH$$

$$C_6H_5C(CH_3)_2OOH \rightarrow C_6H_5OH + CH_3COCH_3$$

Raw material requirements and yield

Raw materials required per tonne of acetone:

Cumene	2300kg

Yield 90%

2. From isopropyl alcohol by dehydrogenation

Isopropyl alcohol can be dehydrogenated to acetone over a catalyst which can be a metal, metal oxide or salt. (*See Figure 12*)

Isopropyl alcohol vapour, preheated by passing through a heat exchanger, is fed into a multi-tubular reactor containing a catalyst on an inert support such as pumice. A large number of catalyst systems have been developed, but copper or zinc or their oxides are most commonly employed.

If an azeotropic mixture of aqueous isopropyl alcohol is used as the feedstock, then platinum and ruthenium, or 0.25% platinum on sodium-activated alumina, have been found to be most effective. The presence of steam acts as a diluent and reduces the formation of hot spots and local coking. Hydrogen can also be mixed with the feed to reduce the build-up of carbon deposits.

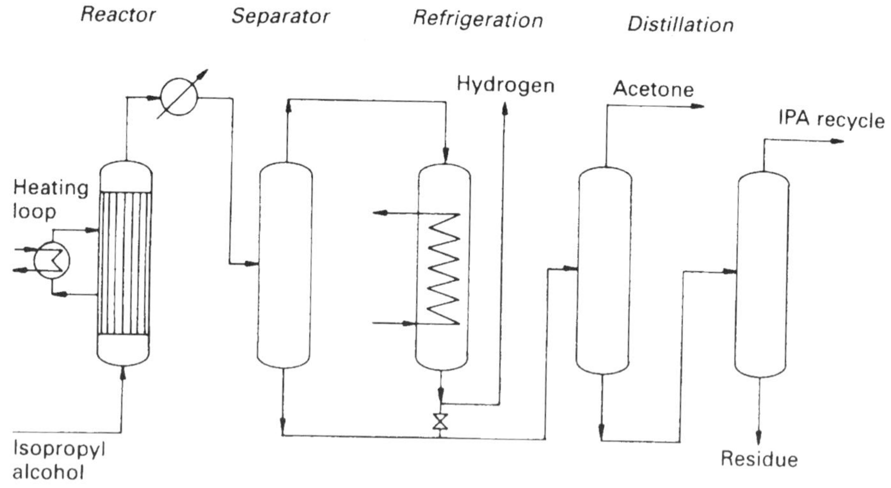

FIGURE 12 Acetone from isopropyl alcohol by dehydrogenation

The reactor tubes are heated by oil or high-pressure steam and the reaction takes place at 300–450°C and a pressure of 2–3 bar. The exit gases containing isopropyl alcohol, acetone and hydrogen are cooled and condensed. Hydrogen passes overhead and is scrubbed to remove any residual isopropyl alcohol or acetone before being used as fuel. The aqueous solution is added to the condensate which is then fractionated. Acetone is collected from the top of the column. The bottoms are redistilled under reduced pressure, and unconverted isopropyl alcohol is separated from water and other high boiling by-products and recycled.

Catalyst regeneration is carried out periodically in situ using a mixture of steam and air at 500°C to burn off the coke. Catalyst life varies from 6 to 9 months.

The reaction can be carried out at a lower temperature (around 150°C) and atmospheric pressure if a high boiling point inert solvent is used as a diluent. Anhydrous isopropyl alcohol is required, while the catalyst is kept in solution by stirring.

Reaction

$(CH_3)_2CHOH \rightarrow CH_3COCH_3 + H_2$

Raw material requirements and yield

Raw materials required per tonne of acetone:

Isopropyl alcohol	1100kg
Yield 90–95%	

3. From propylene by direct oxidation

The process can be operated in one stage or two stages depending on the feedstock quality. Generally, the two-stage process is employed as it allows low-quality feedstock to be used and permits optimum operating conditions in each stage. (*See Figure* 13)

In the first stage, propylene (minimum 90% concentration) and the catalyst, an aqueous solution of cupric chloride with palladium chloride and acetic acid, are fed into the base of the reactor. The aqueous catalyst solution within the reactor also contains suspended palladium and cuprous chloride.

Propylene is rapidly absorbed. During the reaction palladium chloride is reduced to hydrogen chloride and palladium, which is then reoxidized by the cupric chloride. The reaction conditions are 115°C and a pressure of 14–20 bar. A cooling jacket or coil coolers remove heat generated by the highly exothermic reaction.

The reaction products are removed from the top of the reactor and the pressure reduced to atmospheric. Steam stripping removes the acetone and by-product

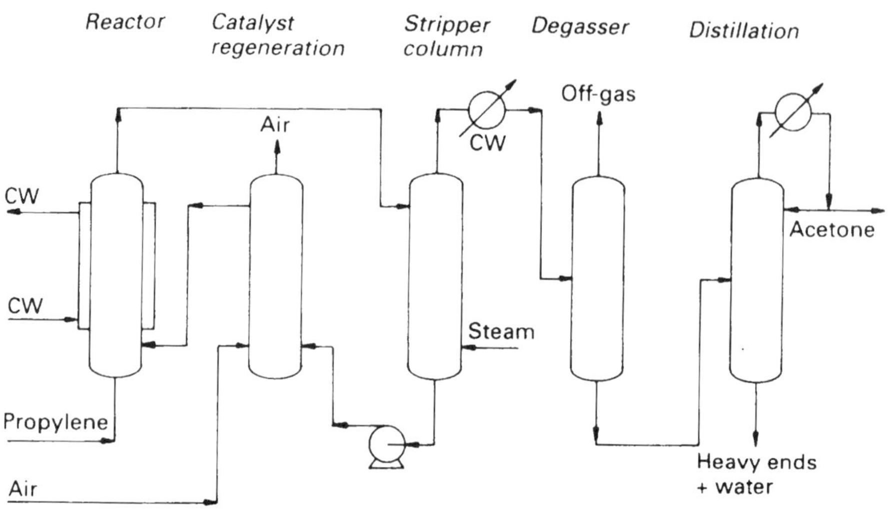

FIGURE 13 Acetone from propylene by direct oxidation

propionaldehyde and the overhead product is condensed. Any gases are vented. Light ends are removed from the crude condensate and acetone is recovered by fractionation in a second column.

Acetone is recovered overhead and water and heavy ends from the base. Make-up hydrochloric acid is added to the catalyst solution before it is fed into the catalyst regenerator.

Air sparged into the base oxidizes the cuprous chloride to cupric chloride, and most of the hydrochloric acid is absorbed. Catalyst concentration ranges from 300 to 400mg/l for palladium and 50 to 100 mg/l for cupric chloride. The conversion rate is controlled by the palladium concentration, and cupric chloride governs the conversion.

Corrosion is the major problem of the process. It is aggravated by the presence of oxygen in the regenerator, and the use of high-resistance metals in the construction of the plant adds to the capital costs.

Reaction

$$2C_3H_6 + O_2 \rightarrow CH_3COCH_3 + CH_3CH_2CHO$$

$$Pd + 2CuCl_2 \rightarrow PdCl_2 + 2CuCl$$

$$4CuCl_2 + 4HCl + O_2 \rightarrow 4CuCl_2 + 2H_2O$$

Raw material requirements and yield

Raw material required per tonne of acetone:

Propylene (100%)	781kg
Air	1100Nm^3
HCl (100%) (used as 30% solution)	15kg
Catalyst $PdCl_2$	0.9kg
$CuCl_2 + H_2O$	150kg

Yield 93%

PROPERTIES

Colourless, volatile liquid, slightly hygroscopic, with a characteristic sweetish odour. Highly flammable. Soluble in water, ethyl alcohol and ether. Will react explosively with strong oxidizing agents.

Molecular Weight	58.08
Density at 20°C	0.79
Melting Point	–94.7°C
Boiling Point	56.2°C
Autoignition Temperature	465°C
Explosive Limits in air	
lower	2.9 vol%
upper	12.8 vol%
Flash Point Closed Cup	–17°C
Vapour Density (air = 1)	2.0
Exposure Limit COSHH	1500ppm 15 minutes 750ppm 8 hour TWA
Exposure Limit ACGIH	750ppm TLV-TWA

GRADES

Technical 99.5%.

INTERNATIONAL CLASSIFICATIONS

UN No.	1090
CAS Reg No.	67–64–1
EINECS No.	200–662–2
EC No.	606–001–00–8
Description	Flammable liquid
Packing Group	II
Emergency Action Code	2YE
HI (Kemler Code)	33

APPLICATIONS

Acetone, the simplest of the aliphatic ketones, is an important chemical intermediate and solvent.

Its largest outlet, accounting for just over a quarter of total consumption, is in the manufacture of methyl methacrylate and methyl acrylate. The production of methyl methacrylate from butyl alcohol, which commenced in Japan in 1982, has greatly reduced the volume of acetone required for this outlet. Around 23% of acetone is used as a solvent for surface coatings and as a spinning agent for cellulose acetate fibre.

The third most important outlet is as an intermediate for the production of C_6 solvents, such as methyl isobutyl ketone, methyl isobutyl carbinol, diacetone alcohol, hexylene glycol, isophorone and mesityl oxide. Together these products account for 10% of acetone demand. A growing outlet, consuming 8% of total demand, is for the production of bisphenol A.

Acetone is used in a wide range of miscellaneous outlets including pharmaceuticals, intermediates for vitamins, toiletries and cosmetics, and rubber antioxidants.

Future demand for acetone is expected to follow the state of the economy, and is forecast to be around 2% in the US and Europe. Bisphenol A is the fastest growing outlet. Because phenol consumption is growing at a faster rate than acetone, there will be a ready availability of by-product acetone.

In Japan, changes in feedstocks used for the manufacture of methyl methacrylate from acetone to butyl alcohol, which began in 1982, have caused demand in that country to remain relatively constant.

HEALTH AND HANDLING

Acetone vapour has a slight narcotic effect and excessive exposure can lead to dizziness, nausea and drowsiness. At high concentrations the vapour is irritating to the eyes, nose and throat leading to coughs and headaches. There appears to be no long-term effects and acetone is not a carcinogen.

Because acetone is a good solvent, it has a defatting effect on skin which can lead to irritation and dermatitis. When handling, protective goggles, butyl or natural rubber gloves, boots and aprons should be worn. Contact lenses should be avoided as they can absorb acetone vapour leading to eye injury. Any contaminated clothing must be removed and laundered before being reused.

Acetone should be stored in closed containers made of mild steel, copper or aluminium, in a cool, well-ventilated, explosion-proof area. Copper is to be avoided for recovered acetone which may have developed some acidity. In the presence of sunlight, acetone can decompose releasing carbon monoxide.

Because of its low flash point, containers must be earthed to prevent static build-up, and kept away from oxidizing agents. All containers must carry a flammable liquid label.

Acetone is a high fire risk. Fires should be blanketed by carbon dioxide, dry chemical or alcohol foam. As acetone vapours are heavier than air, they can roll back considerable distances making flashback a hazard. Spills should be contained and if small absorbed with sand or earth, then disposed of in accordance with local regulations.

MAJOR PLANTS

Plants with capacities greater than 100 000 tonnes per year:

Phenolchemie	Gladbeck	Germany
Shell Nederland Chemie	Pernis	Netherlands
EniChem	Mantua	Italy
AlliedSignal	Frankford	US
Aristech Chemical	Haverhill	US
Dow Chemical	Freeport	US
Eastman Chemical	Kingsport	US
GE Plastics	Mount Vernon	US
Georgia Gulf	Plaquemine	US
Shell Chemical	Deer Park	US
Sasol	Sasol	South Africa
Mitsubishi Chemical	Kashima	Japan
Mitsui Petrochemical Industries	Chiba	Japan
Mitsui Toatsu Chemical	Senboku	Japan
Formosa Chemical & Fibre	Maillo	Taiwan

MAJOR LICENSORS

Cumene	*Allied Signal*
	BP Chemicals/Hercules
	Dow Chemical/Kellogg
	Hoechst
	Mitsui Petrochemical
	Monsanto
	Rhone-Poulenc
	UOP
IPA	*Edeleanu*
	IFP
	Shell Development
	Toyo Engineering
Toluene	*Stamicarbon*
Propylene	*BP Chemicals*
	Hoechst/Uhde

Acetylene

$HC \equiv CH$

SYNONYMS

ACETYLENE ethyne, ethine

Acetylene was the main feedstock for the chemical industry prior to the advent of oil. The traditional route to acetylene is from calcium carbide, produced by the interaction of coke and lime at high temperatures. This process has become increasingly uneconomic because of the high costs of the energy required. For this reason, carbide production tends to be concentrated in locations where coal and cheap electricity are readily available.

Subsequently, electric arc processes, followed by partial oxidation of natural gas or the thermal decomposition of methane and other hydrocarbons, were introduced. With the growth of the petrochemical industry, demand for acetylene declined rapidly because of the ready availability of naphtha-based ethylene and other olefins at low cost.

Ethylene-rich process streams containing 0.1–0.8wt% of acetylene, depending on the source of the feedstock and severity of the cracking, also provide a source of the material. In the US, the consumption of ethane as a feedstock for ethylene has resulted in low acetylene production.

Although many variants exist, there are only three major commercial routes for the manufacture of acetylene. These routes are:

- calcium carbide;
- arc process;
- partial oxidation of natural gas.

Older processes involving thermal cracking by heat carriers have become uneconomic.

Recent advances in technology resulting in plasma arc processes, crude oil cracking, and the advanced cracking reactor, have only reached the pilot-plant stage due to lack of demand for acetylene. They have failed to halt the trend towards ethylene as the prime petrochemical feedstock.

Acetylene will continue to be used as a raw material for chemical manufacture owing to its high conversion rates and absence of by-products using proven technology. An additional factor is that most acetylene is produced in old plants which have a low capital cost.

In the Western world, the dominant feedstock is natural gas, except in Germany where some naphtha, and Japan where calcium carbide, are used.

Capacities range from 3 000 to 130 000 tonnes per year.

PROCESSES

1. From calcium carbide

Powdered calcium carbide is introduced into a generator containing a large number of circular trays by means of a feed screw. Water is added to the carbide in the ratio of 1:1 by weight. Paddles push the mixture to the outside of the top tray. It falls onto the tray below where paddles push it towards the centre and the process is repeated. (*See Figure 14*)

Acetylene is released and the heat of the reaction vaporizes any excess water. The temperature of the reaction is controlled below 150°C to avoid polarization reactions and the risk of explosion.

Crude acetylene gas is scrubbed with water to remove dust. Impurities, mainly sulphur and phosphorus compounds, are removed by scrubbing with 98% sulphuric acid followed by caustic soda to absorb sulphur dioxides formed during the oxidation.

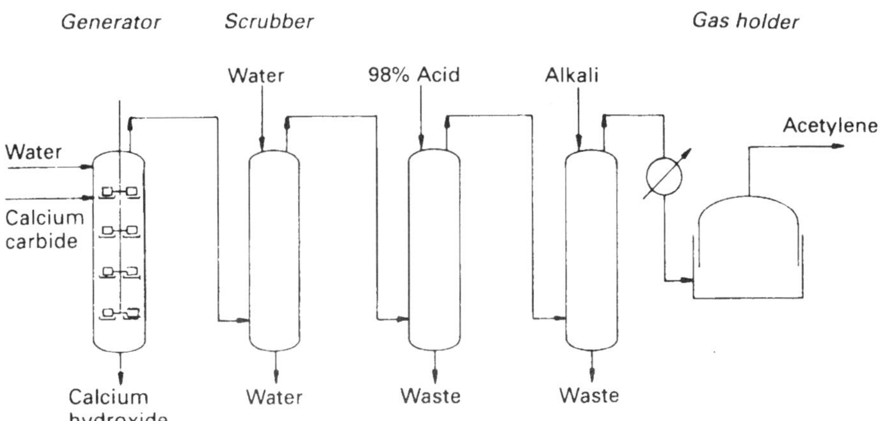

FIGURE 14 Acetylene from calcium carbide

After cooling to below 40°C, the water vapour present condenses, and the dry gas is fed into storage containers or used directly in other chemical processes. The advantage of limiting the amount of water used is that the volume of waste calcium hydroxide is greatly reduced, and being in a dry state, it can be used as a source of lime.

Reaction

$$CaC_2 + 2H_2O \rightarrow HC \equiv CH + Ca(OH)_2$$

Raw material requirements and yield

Raw materials required per tonne of acetylene:

Calcium chloride	3230kg
Yield 76%	

2. From hydrocarbons by electric arc process

Natural gas, liquid hydrocarbons or coal can be used as feedstock, the exact design of the arc furnace being dependent on the feedstock used. The furnace consists of a cathode, vortex chamber and anode. Both anode and cathode have water-jacketed tubes made of carbon steel. (*See Figure 15*)

Gas is forced tangentially into the vortex chamber so that the arc between anode and cathode is stabilized. The arc burns in the dead zone and, in order to increase

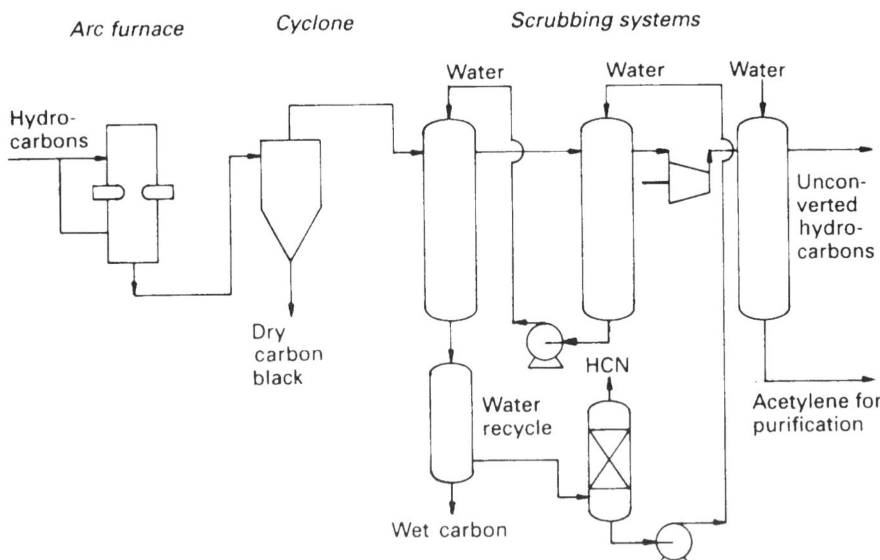

FIGURE 15 Acetylene from hydrocarbons by electric arc process

the life of the electrodes, the striking point is moved around rapidly. A temperature gradient of 600°C to 20 000°C can be achieved. The gas is cracked to yield acetylene with some ethylene, hydrogen and soot. Residence time is a few milliseconds.

The exit gases are quenched with liquid hydrocarbons to lower the temperature to around 1200°C, and then with water to 200°C to prevent decomposition of acetylene to soot and hydrogen. The gases pass through a cyclone which removes around 70% of the soot formed.

If liquid hydrocarbons are used, a two-step process is usually employed. Hydrogen is heated in the arc furnace, prior to the introduction of the feedstock into the hydrogen plasma. Acetylene, ethylene, hydrogen, soot and other by-products (depending on the feedstock type) are formed in the cracking reaction. The ratio of acetylene to other products can be varied by altering the residence time.

The exit products are quenched with oil to 300°C and the heat recovered is used to generate steam. Soot is removed by the quench oil and the scrubbers take out any unconverted feedstock which is recycled. Acetylene is purified further by absorption (see following section).

Reaction

$2CH_4 \rightarrow HC{\equiv}CH + 3H_2$

Raw material requirements and yield

Raw materials required per tonne of acetylene:

From Naphtha		*From Crude Oil*	
Naphtha	1920kg	Crude oil	3670kg
Ethylene	500kg		480kg
Hydrogen	130kg		10kg
Cracking composition			
Acetylene (vol%)	13.7		14.5
Ethylene (vol%)	6.4		6.5
Yield (cracking gas)	78%		56%

3. From natural gas by controlled oxidation (Sachsse process)

Natural gas (rich in methane) and oxygen are preheated separately to 600°C before being mixed rapidly in a pressure burner in the molar ratio of 1:0.6. The deficiency of oxygen prevents the reaction from going to completion and the controlled oxidation provides the energy necessary for the main reaction. (*See Figure 16*)

The gaseous mixture is fed through a special burner, where the heat of combustion raises the reaction temperature to about 1500°C and methane is cracked to acetyl-

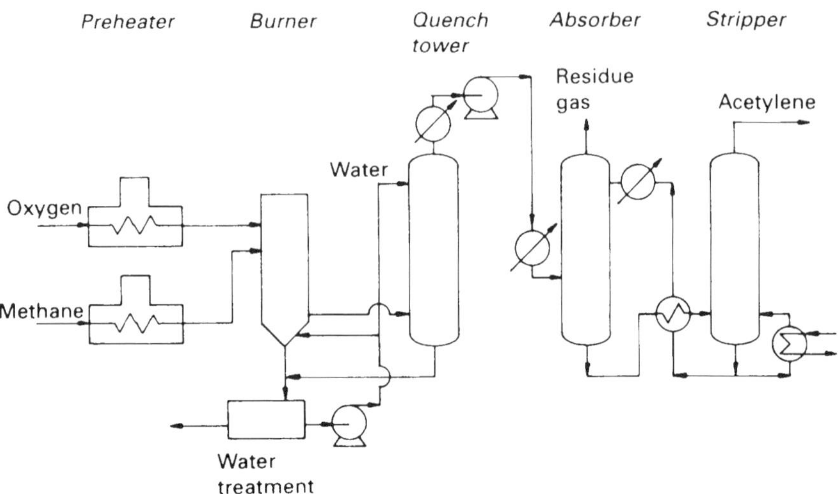

FIGURE 16 Acetylene from natural gas by controlled oxidation (Sachsse process)

ene. Back mixing of the gases between the mixing and the reaction zones is prevented by the use of a diffuser.

The reaction gases are immediately quenched with sprays of water or oil from a series of nozzles to prevent decomposition of the acetylene formed. If oil is used the temperature drops to 200–250°C and with water it drops to 80°C. Water is most commonly employed as the coolant.

Soot formed by the reaction is partially removed by the water quench and the remainder is trapped by passing the gases through an electrofilter. If oil is used the soot settles in the quench chamber and is withdrawn from the base. Heat recovered by the quench water is used in the process.

The cooled gases, containing around 8% of acetylene, pass through an absorption unit where a solvent such as N-methyl pyrrolidone extracts the acetylene. Diacetylene and other higher homologues, being more soluble than acetylene, are removed first by scrubbing with a small volume of solvent. Acetylene is then absorbed under a pressure below 1 bar and at low temperatures to prevent polymerization.

The remaining off-gases, principally carbon monoxide and hydrogen, may be sent to a synthesis plant or used as fuel gas. Low-solubility products are stripped from the absorbent solvent by stage depressurization prior to acetylene removal in a second stripping column. The extraction solvent is purified before being recycled. The residual gas is normally burnt as fuel.

Reaction

$CH_4 + 1\frac{1}{2}O_2 \rightarrow CO + 2H_2O$

$2CH_4 \rightarrow HC \equiv CH + 3H_2$

Raw material requirements and yield

Raw materials required per tonne of acetylene:

Natural gas	5950m^3
Oxygen (95%)	5300kg

Yield 30%

4. From hydrocarbon feedstock by absorption

Ethylene-rich process streams, derived from naphtha cracking and containing less than 42wt% of acetylene (to avoid decomposition), are brought into contact with a solution of dimethylformamide (DMF) which absorbs acetylene and some other gases. Acetone or N-methyl pyrrolidine can be used as alternative extractants.

The DMF solution passes to a flash drum at reduced pressure, and co-absorbed gases are collected and recycled. Any remaining gases are removed in an ethylene stripper. The DMF extractant solution is fed into a stripper column where acetylene is recovered.

OTHER PROCESSES

Three companies, Kureha Chemical Industry, Chiyoda Chemical and Union Carbide have developed an Advanced Cracking Reactor (ACR) process.

The multi-port burner is fired by fuel and oxygen to create a heat carrier gas at a temperature of 2000°C in the presence of preheated steam. Oil is sprayed into the carrier gas stream before being fed into the Advanced Cracking Reactor. Cracking takes place at 900–1600°C and 5 bar pressure, with a residence time of 0.01–0.02 seconds.

The reaction gases are cooled immediately in an oil quencher and the heat is used to generate steam. Yields of 11% hydrogen and methane, 4% acetylene and 32% ethylene are obtained, but operating conditions can be varied to give a range of yields.

Dow Chemical has developed a partial combustion process aimed at reducing the formation of soot. The feedstock is preheated and cracked by the heat generated by burning fuel in the presence of oxygen.

Several other companies have reported various processes for crude oil or heavy oil cracking involving fluidized technology and thermal cracking. Although these processes accept a wide range of feedstocks and have product range flexibility, their economics in full-scale operation have yet to be established.

PROPERTIES

Colourless, flammable, non-toxic, odourless gas when pure, but usually with a garlic odour due to the presence of impurities. Soluble in acetone, ethyl alcohol and water. Burns with a hot and sooty flame.

Molecular Weight	26.04
Density at 20°C	0.618
Melting Point	–80.85°C
Boiling Point	–84°C
Explosive Limits in air	
lower	2.5vol%
upper	81.5vol%
Flash Point Closed Cup	Gaseous at room temperature
Vapour Density (air = 1)	0.91
Exposure Limit COSHH	Asphyxiant
Exposure Limit ACGIH	Asphyxiant

(2500ppm 10 hour TWA has been set by OSHA due to traces of toxic impurities.)

GRADES

Technical 97–99%, pure 99.5–99.9%.

INTERNATIONAL CLASSIFICATIONS

UN No.	1001
CAS Reg No.	74–86–2
EINECS No.	200–816–9
EC No.	601–015–00–0
Description	Flammable gas
Emergency Action Code	2SE
HI (Kemler Code)	239

APPLICATIONS

Considerable quantities of acetylene are consumed for non-chemical uses such as cutting and welding of metals and the production of carbon for batteries.

Although acetylene is a highly reactive chemical because of its triple bond, its use has declined as new routes using ethylene, propylene and other petroleum hydrocarbons have been developed. Acetylene still finds a use in the manufacture of vinyl chloride, vinyl acetate, vinyl esters, acrylic acid, 1,4-butynediol and acetylinic alcohols.

Except where special conditions apply, such as the ready availability of coal and cheap power, future demand for acetylene is expected to continue to decline.

HEALTH AND HANDLING

Acetylene is a simple asphyxiant and anaesthetic. It produces narcosis with a weak and irregular pulse. The effects of repeated exposure do not appear to be cumulative, but as acetylene frequently contains impurities, these can be toxic.

Because of the danger of explosion and fire, care must be taken in handling. Storage cylinders should be kept upright, away from oxidizing agents, in a cool, well-ventilated area containing explosion-proof equipment. Cylinders should be protected from physical damage and personnel must wear leather gloves, safety shoes and goggles.

Liquid acetylene can decompose explosively on heating, impact or in the presence of catalysts. For this reason, acetylene is dissolved in a solvent, usually acetone or dimethylformamide, for storage. Pressures in excess of 1 bar must not be used and approved valves and manifolds are needed.

Acetylene reacts with copper, silver and mercury to form explosive acetylides. Distribution systems require flame traps, leak detectors, non-return valves and protection from fire. Leaks in transportation lines can be stopped with heavy-duty cloth or putty.

Leaking cylinders should be moved outside and their contents allowed to discharge slowly into the atmosphere. All personnel should be evacuated and sources of heat and ignition eliminated. Water can be used to cool cylinders and to protect clean-up staff.

High flammability, low flash point and a wide explosive range in air makes acetylene a dangerous fire and explosion hazard. In the event of fire, attempts should be made to shut off the source if possible while using water to keep containers and surroundings cool. Unless the source can be located and sealed, it is safer to let the gas burn. Firefighting staff must wear protective clothing and self-contained breathing apparatus.

MAJOR PLANTS

Plants with capacities greater than 40 000 tonnes per year:

Paradies Acetiques	Pardies	France
EniChem	Porto Marghera	Italy
Complex	Berente	Hungary
Polsin Karbid	Katowice	Poland
Complex	Novaky	Slovakia
Complex	Ufa	Russia
Complex	Severodonetsk	Ukraine
Complex	Navoi	Uzbekistan
Borden	Geismar	US
Shawinigan Chemical	Shawinigan	Canada
Denki Kagaku	Omi	Japan
State	Chang-Shou	China

MAJOR LICENSORS

ABB Lummus Global
BASF
Chiyoda
Dow Chemical
EniChem
Hoechst
Huels
Kureha Chemical Industry
Soc. Belge de L'Azote
Stone & Webster
Union Carbide (Wulff)

Acrolein

$CH_2{=}CHCHO$

SYNONYMS

ACROLEIN propenal, acrylaldehyde, allyl aldehyde, ethylene aldehyde, acrylic aldehyde, 2-propen-1-one

Acrolein was initially made on a commercial scale by the vapour-phase condensation of acetaldehyde and formaldehyde in the presence of a sodium silicate catalyst.

The discovery by Shell that propylene could be oxidized to acrolein in the presence of a cuprous oxide catalyst led the move towards current commercial processes.

Because of low conversion rates, Sohio developed catalysts based on bismuth molybdate–bismuth phosphomolybdate. Since then, several companies have modified the multi-component catalyst systems based on bismuth, molybdenum and iron oxides with cobalt, nickel and other metals. With propylene availability at reasonable prices, all acrolein commercially produced uses this technology.

Plant capacities range from 2000 to 30 000 tonnes per year.

PROCESSES

From propylene by vapour-phase oxidation

A propylene, air and steam mixture in the molar ratio of 1:8:2–6 is fed into a multi-tubular fixed-bed reactor which operates at 300–400°C and 1.5–2.5 bar pressure. The catalyst consists of multi-component metal oxides based on molybdenum, iron and bismuth on a silica support. Residence time is 0.8 seconds. Heat from the exothermic reaction is removed by a circulating molten salt stream and is recovered as steam. (*See Figure 17*)

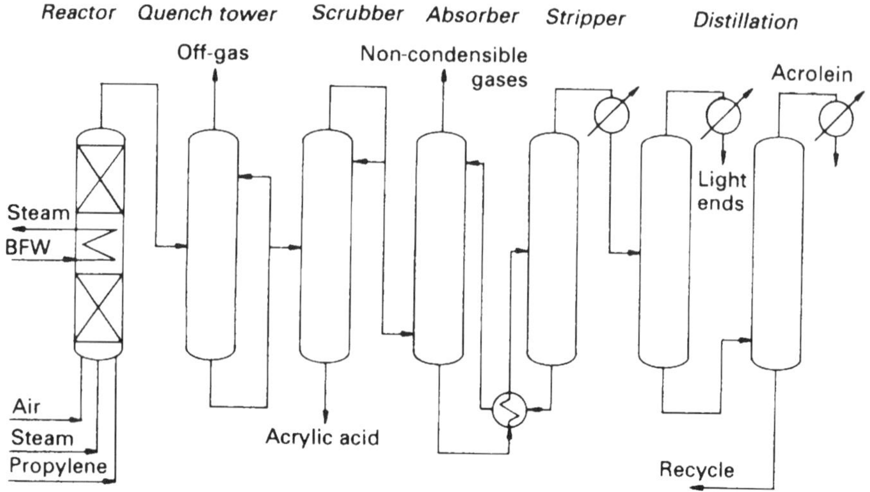

FIGURE 17 Acrolein from propylene by vapour-phase oxidation

The reactor gases pass to a quench tower to prevent subsequent reactions and then to a scrubber where acrylic acid and other high boiling point by-products are removed with water at a temperature above the boiling point of acrolein.

An aqueous solution of acrolein is obtained by passing the exit gases through cold water. Aqueous acrolein is pumped to the stripper column, where it is distilled and most of the acrolein is recovered overhead as its water azeotrope. Further purification is carried out by fractionation to remove low-boiling acetaldehyde and heavy ends. By-product acetaldehyde can be recovered.

The resultant acrolein, 95–96% pure, is stabilized with hydroquinone to minimize polymerization. Acrylic acid, formed in 5–10mol% based on propylene, can be recovered as a useful by-product from the scrubber effluent.

Acrolein yields of up to 85% are claimed based on propylene with 5–10% of acrylic acid. Catalyst life tends to be between 2 and 4 years after which yields tend to drop. A stabilizer has to used in the acrolein process to minimize the risk of polymerization.

Reaction

$$CH_2 = CH - CH_3 + O_2 \rightarrow CH_2 = CHCHO + H_2O$$

Raw material requirements and yield

Material requirements per tonne of acrolein:

Propylene	1160kg
Yield 85%	

PROPERTIES

Colourless, flammable, lacrimatory liquid with distinct odour. Volatile and toxic. Partially soluble in water. Soluble in many organic solvents such as alcohols, ethers and aliphatic or aromatic hydrocarbons.

Molecular Weight	56.06
Density at 20°C	0.843
Melting Point	–86.9°C
Boiling Point	52.5°C
Autoignition Temperature	220°C
Flammability Limits in air	
lower	2.8vol%
upper	31vol%
Flash Point Closed Cup	–26°C
Vapour Density (air = 1)	1.94
Exposure Limits COSHH	0.3ppm 15 minutes 0.1ppm 8 hour TWA
Exposure Limits ACGIH	0.3ppm TLV-STEL 0.1ppm 8 hour TLV-TWA

GRADES

Commercial 95%, inhibited with 0.1–0.2% hydroquinone to prevent radical initiated polymerization.

INTERNATIONAL CLASSIFICATIONS

Acrolein inhibited	
UN No.	1092
CAS Reg No.	107–02–8
EINECS No.	203–453–4
EC No.	605–008–00–3
Description	Toxic substance, flammable liquid
Packing Group	I
Emergency Action Code	2WE
HI (Kemler Code)	663
Acrolein dimer, stabilized	
UN No.	2607
CAS	100–73–2
EINECS No.	Not allocated
EC No.	Not allocated
Description	Flammable liquid
Packing Group	III
Emergency Action Code	2Y
HI (Kemler Code)	39

APPLICATIONS

The major outlet for acrolein is in the manufacture of DL methionine, an essential amino acid additive to animal feeds. It is also used as an intermediate in the production of acrylic acid and glycerol, where it is not isolated from the vapour-phase reactor products.

Acrolein in very low concentrations, around 10ppm, is a very effective biocide. When condensed with formaldehyde, the water-soluble polymers formed can be used in canals to control aquatic weeds, to control algae in recirculating water systems and oilfield brines, and to deodorize sulphur compounds in oilfield waters.

Copolymers of acrolein and acrylic acid are useful sequestering agents. Acrolein polymers can be used in textile treatment, paper reinforcement, and photography.

HEALTH AND HANDLING

Liquid acrolein vaporizes easily; its vapours are twice as heavy as air and highly toxic. Skin contact must be avoided. Personnel coming into contact with acrolein should receive regular medical checks with particular attention to the heart and lungs.

Acrolein must be stored and transported in the dark under a blanket of nitrogen at temperatures below 20°C. Equipment used for the handling of acrolein must be clean to reduce the risk of polymerization by contaminants. On account of its flammability and toxicity, transportation is closely controlled in most countries and labelling requirements and local regulations must be consulted.

Because of its tendency to polymerize easily – the dimer is present in all except freshly prepared or stabilized monomers – it should be stored for no longer than three months. The inhibitor should be checked regularly and additions made if necessary. Acrolein will react explosively in the presence of alkalis, concentrated mineral acids, amines, sulphur dioxide, metal salts and oxidants.

In the event of spills, sources of ignition must be extinguished, the liquid absorbed with vermiculite or sand and placed in a container. Care must be taken to avoid run-off into sewers because of the danger of explosion.

Acrolein is a dangerous fire hazard because of the risk of flashback and explosion caused by polymerization. Carbon dioxide, non-alkaline powder or foam can be used to blanket fires. Water should be avoided as it tends to scatter the flames. Protective clothing and self-contained breathing apparatus must be worn because toxic gases are given off on burning.

MAJOR PLANTS

Plants with capacities greater than 9000 tonnes per year:

Elf Atochem	Pierre-Benite	France
Degussa	Wesseling	Germany
Complex	Volgograd	Russia
Union Carbide	Taft	US
Daicel Chemical	Otake	Japan

MAJOR LICENSORS

Degussa
Elf Atochem
Nippon Shokubai
Sohio

Acrylic Acid

$CH_2=CHCOOH$

SYNONYMS

ACRYLIC ACID 2-propenoic acid, ethylene carboxylic acid, vinylformic acid, acroleic acid

Although acrylic acid has been known for over a hundred years, it was only when routes were discovered from acetylene that it became of any commercial significance.

Acetylene will react with water and carbon monoxide in the presence of nickel bromide-cuprous bromide as catalyst to give acrylic acid, or with an alcohol and carbon monoxide to give acrylic esters. Both routes have been largely abandoned because of safety and pollution control problems.

Another process with major disadvantages is the reaction of ketene, obtained by the pyrolysis of acetone or acetic acid, with formaldehyde in the presence of an aluminium chloride catalyst. The 2-propiolacetone formed is reacted with alcohol to give acrylic esters. This process is obsolete.

An alternative route via ethylene cyanohydrin has fallen into disuse because of environmental concerns over the handling of hydrogen cyanide and ammonium sulphate by-product disposal. Production via the hydrolysis of acrylonitrile has similarly been largely abandoned because of low yields and disposal problems caused by by-product ammonium sulphate.

Currently, the manufacture of acrylic acid by the gas-phase catalytic oxidation of propylene, either in a single- or two-stage process, has supplemented all other routes. Earlier attempts were made to develop liquid-phase propylene-based processes but none have been used commercially. The greater part of acrylic acid production is converted to esters; because of the similarity of the esterification processes, dual-product plants are frequently constructed.

Acrylic acid is produced as a by-product in the manufacture of acrolein from propylene (see Acrolein).

Capacities range from 15 000 to 400 000 tonnes per year. Several of the largest producers consume acrylic acid captively. Several major projects are scheduled for start-up by 2000 in the Asia–Pacific region.

PROCESSES

From propylene by catalytic oxidation

The reaction can be carried out in a one-step or two-step reaction, the latter being favoured because of the higher yields obtained. Different catalysts and reaction conditions are used in order to optimize conversion and selectivity at each step. (*See Figure 18*)

Technical-grade propylene of 90–95% purity, steam and air are premixed before entering the first-stage reactor. Tail gases are added to prevent the formation of a flammable mixture. The fixed-bed reactor can be of a shell or tubular type, surrounded by a transfer bath to remove heat generated by the exothermic reaction, which is recovered as steam.

The catalyst for the first stage, consisting of multi-component metal oxides (based on molybdenum, bismuth and iron, with cobalt, nickel and tungsten, on silica support), is highly acrolein selective. Reaction conditions are 300–350°C and pressures of 1.5–2.5 bar.

The effluent gases pass to the second-stage multi-tube reactor containing catalysts based on cobalt–molybdenum–vanadium oxides with mixtures of copper, chromium, strontium, aluminium or tungsten. The temperature is maintained at

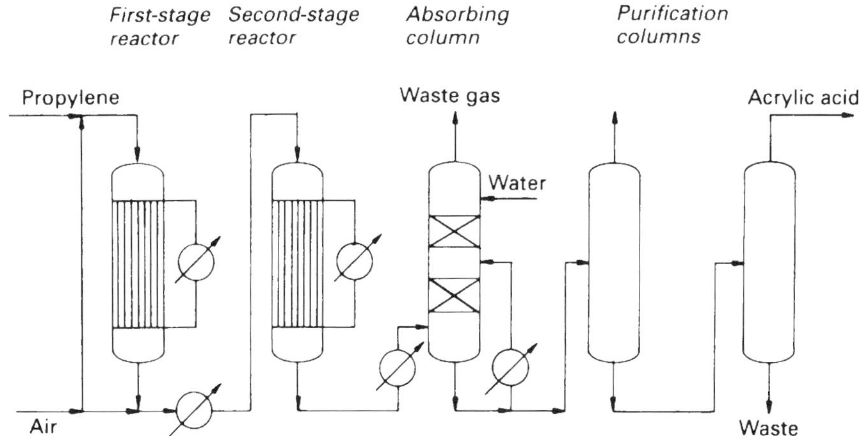

FIGURE 18 Acrylic acid from propylene by catalytic oxidation

250–300°C and near atmospheric pressure, with contact times of 1–3 seconds. Almost 100% conversion of acrolein to acrylic acid is obtained.

The reactor exit gases are cooled to 200°C and scrubbed with cold water. Acrylic acid in the aqueous solution is extracted with a countercurrent of an organic solvent in an absorbing column. Organic solvents used either have a lower boiling point than acrylic acid (such as butyl or ethyl acetates, 2-butanone, or ethyl acrylate) or have boiling points higher than the acid, in which case *tert*-butyl phosphate, isophorone or aromatic hydrocarbons are commonly used.

The solvent extract is distilled and water and solvent are obtained overhead and separated. The solvent is recycled to the absorbing column. The bottom stream passes to the light-ends column where acetic acid is distilled overhead and recovered. Acrylic acid is obtained from the bottoms by further distillation in the product column and collected overhead.

Any dimer in the residues is decomposed to monomer and the remainder is incinerated. As acrylic acid polymerizes readily, all distillation is carried out under reduced pressure in the presence of an inhibitor.

In the one-step process, the catalyst used consists of polyvalent oxides and molybdenum oxide promoted by tellurium oxide. Yields are lower and catalyst life is short due to the loss of tellurium oxide by sublimation.

Reaction

One-step process:

$$CH_2{=}CHCH_3 + 1\tfrac{1}{2}O_2 \rightarrow CH_2{=}CHCOOH + H_2O$$

Two-step process:

$$CH_2{=}CHCH_3 + O_2 \rightarrow CH_2{=}CHCHO + H_2O \quad \text{first stage}$$

$$2CH_2{=}CHCHO + O_2 \rightarrow 2CH_2{=}CHCOOH \quad \text{second stage}$$

Raw material requirements and yield

Raw materials required per tonne of acrylic acid:

Propylene (100%)	880kg (one-step process)
	680kg (two-step process)

Yields

one-step process	50–60%
two-step process	90%

PROPERTIES

Flammable, volatile, mildly toxic, colourless liquid which is inhibited to prevent polymerization. Forms crystalline needles in the solid state. Miscible with water, alcohols, esters and many organic solvents.

Molecular Weight	72.06
Density at 20°C	1.051
Melting Point	13.5°C
Boiling Point (1 atm)	141.6°C
Autoignition Temperature	412°C
Flammability Limits in air	
lower	2vol%
upper	8vol%
Flash Point Open Cup	50°C
Vapour Density (air = 1)	2.5
Exposure Limits COSHH	20ppm 15 minutes 10ppm 8 hour TWA
Exposure Limit ACGIH	5ppm TLV-STEL 2ppm TLV-TWA

GRADES

Technical 99%, aqueous 80%, (inhibited with 50–500 ppm phenothiazine, hydroquinone or hydroquinone monomethyl ether).

INTERNATIONAL CLASSIFICATIONS

UN No.	2218
CAS Reg No.	79–10–7
EINECS No	201–177–9
EC No.	607–061–00–8
Description	Corrosive substance, flammable liquid
Packing Group	II
Emergency Action Code	2WE
HI (Kemler Code)	839

APPLICATIONS

Acrylic acid is used as an intermediate for the production of acrylates. Acrylate esters, of which methyl, ethyl, butyl and 2-ethyl hexyl are the most important, account for 55% of global demand for acrylic acid.

Butyl acrylate, the most widely used ester, which makes up half of this total, is used in paints, varnishes and inks. Due to regulatory and market pressures to reduce the volatile solvent content of conventional surface coatings and adhesives, the water-based paint market is growing at twice the rate of older types. The other important primary acrylate esters include methyl acrylate used for fibres and water treatment chemicals, and ethyl and 2-ethyl hexyl acrylates, which find outlets in textile coatings and adhesives.

Polyacrylic acid outlets are growing quickly especially in copolymers used as co-builders for detergents and for super-absorbent resins. The latter, used for baby diapers and adult incontinence pads, together amount to over 26% of world demand for acrylic acid.

Adhesives and binders are a major consumer of acrylic ester copolymer latices especially in Europe and Japan. They are widely used for bonding fibres, polyvinyl chloride (PVC) films and metal foil to substrates such as wood, metal, leather and canvas.

Other miscellaneous outlets for acrylates include oil-resistant acrylic rubbers, flocculants, dispersants and thickeners in the form of their aqueous solutions of sodium or ammonium salts, and as additives for rigid PVC to improve its surface finish and impact strength.

Global demand for acrylic acid is forecast to increase at 5% per year up to 2000.

HEALTH AND HANDLING

Acrylic acid vapour is irritating to the eyes and the respiratory tract. The liquid is very corrosive and will cause burns in contact with skin. Protective clothing and vapour-proof goggles and breathing apparatus must be used when handling the product.

Acrylic acid is normally stabilized with inhibitors to prevent polymerization. It should be stored in polypropylene or stainless steel containers lined with glass, polyethylene or polypropylene, under a blanket of oxygen to activate the inhibitor. Containers should be kept at a temperature of 15–30°C away from direct sunlight. Freezing must be avoided as this can lead to localization of the inhibitor. Heat must never be used to thaw acrylic acid as polymerization could occur. Acrylic acid is often used as a 80% aqueous solution.

Spills can be absorbed with sand or earth using non-sparking tools. The waste material should be disposed of rapidly as it can polymerize. Acrylic acid must not be allowed to pollute sewers or waterways.

Carbon dioxide, dry chemical or foam can be used to fight fires. As acrylic acid is flammable and its vapour is heavier than air, flashback is a hazard. Firefighting staff should wear protective clothing and breathing apparatus.

MAJOR PLANTS

Plants with capacities greater than 90 000 tonnes per year:

BASF	Antwerp	Belgium
Elf Atochem	St Avold	France
BASF	Ludwigshafen	Germany
BASF	Freeport	US
Hoechst Celanese	Clear Lake	US
Rohm & Haas	Deer Park	US
Union Carbide	Taft	US
Mitsubishi Chemical	Yokkaichi	Japan
Nippon Shokubai	Himeji	Japan

MAJOR LICENSORS

BASF
Mitsubishi Chemical
Mitsui Petrochemical
Nippon Shokubai
Sohio/BP Chemicals
Sumitomo Chemical
Toyo Engineering

Acrylonitrile

$CH_2{=}CHCN$

SYNONYMS

ACRYLONITRILE   2-propenenitrile, propenoic acid nitrile, vinyl cyanide

Interest in acrylonitrile first developed when Buna rubber was introduced into Germany in the late 1930s. During World War II, acrylonitrile containing polymers was developed in the US and West Germany due to its resistance to oils and lack of access to natural rubber. In recent years, the main growth has come from acrylic fibres which were introduced by Du Pont in the 1950s.

The ammoxidation process for the manufacture of acrylonitrile from propylene, ammonia and air was developed in the 1960s. Prior to that date, the major commercial process employed was the addition of acetylene to hydrogen cyanide to give ethylene cyanohydrin which was dehydrated over a catalyst. This earlier process has been replaced by the propylene-based route, because of its advantages of a higher conversion rate, no recycling of unreacted raw materials and resultant lower production cost.

Four major company groups, Sohio, Nitto, EniChem/UOP and BP Chemicals/ Ugine have developed ammoxidation processes of commercial importance. Sohio, now BP Chemicals, is the world's largest licensor with, in the late 1990s, 95% of total installed capacity.

BP Chemicals has announced new technology for the production of acrylonitrile based on the one-step ammoxidation of propane in a fluidized-bed reactor. The economic advantages of the process are claimed to be the price advantage of propane over propylene, increased production of valuable acetonitrile and hydrogen cyanide by-products and lower effluent treatment cost. It is hoped to introduce the process at the beginning of the 21st century. Mitsubishi Chemical and BOC have also developed a propane-based ammoxidation process which has higher selectivity than typical propylene-based systems.

Plant capacities range from 30 000 to 420 000 tonnes per year. The US and Europe each account for one-third of total world installed capacity, but many new plants are scheduled for the Far East to meet a steady increase in demand in those countries.

PROCESSES

From propylene

The reaction mixture, consisting of chemical-grade propylene (95% pure) and ammonia, both preheated to 150°C, is fed into a fluidized-bed catalytic reactor. The original catalyst used consisted of bismuth phosphomolybdate on silica, but oxides of bismuth, molybdenum, iron, nickel and cobalt are quoted as alternatives. (*See Figure 19*)

Filtered air is introduced into the base of the reactor. The preferred reaction conditions are a temperature of 420–480°C and pressure of 0.34–2 bar. With a contact time of up to 20 seconds, yields of around 70% of theoretical are obtained.

The gases from the highly exothermic reaction are cooled by means of internal water coils. Unreacted ammonia is removed by water acidified with sulphuric acid as aqueous ammonium sulphate, which can be recovered by crystallization.

Scrubbing in an absorber column separates off-gases which are passed overhead, and an aqueous solution of acrylonitrile passes into the recovery column where it is steam stripped. Hydrogen cyanide is removed by distillation in the lights column before acrylonitrile is purified by azeotropic and conventional distillation. By-product acetonitrile and hydrogen cyanide can be recovered if desired.

Other commercial ammoxidation processes differ from that of Sohio mainly in reactor design, catalyst composition and yield. Nitto uses Sohio technology but employs its own iron–antimony catalyst which produces less of the by-product hydrogen cyanide. Moderate pressures are employed in the EniChem process and efficient heat recovery minimizes purification energy requirements.

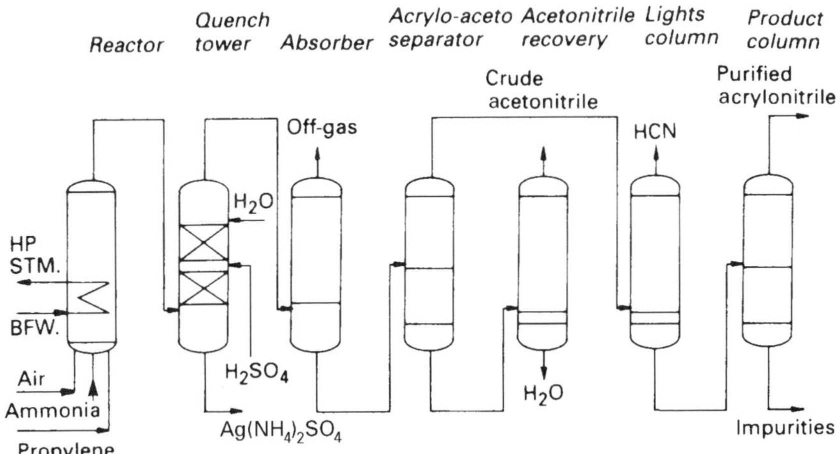

FIGURE 19 Acrylonitrile from propylene

BP Chemicals/Ugine technology differs from all the others in using a fixed-bed reactor. Catalyst systems, based on antimony–tin or more recently on combinations of iron–molybdenum–cobalt, give an enhanced yield. The process is claimed to be very suitable for smaller capacity plants because of its flexibility and lack of significant effluent problems.

Reaction

$$CH_3{=}CHCH_2 + NH_3 + 1\tfrac{1}{2}O_2 \rightarrow CH_2{=}CHCN + 3H_2O$$

Raw material requirements and yield

Raw materials required per tonne of acrylonitrile 100%:

	Sohio	*Nitto*	*BP/Ugine*	*EniChem/UOP*
Propylene	1180kg	1080kg	1100kg	1180kg
Ammonia	490kg	460kg	520kg	480kg
HCN	150kg	110kg	120kg	60kg
Acetonitrile	100kg	10kg	50kg	30kg
Yield	70%	73%	72%	70%

PROPERTIES

Clear, colourless and volatile liquid, with slightly pungent odour. Very toxic and highly flammable. Partially soluble in water. Soluble in most organic solvents including ethyl alcohol, acetone, benzene, carbon tetrachloride and ethyl acetate. Unstable and will polymerize violently in the presence of alkalis or peroxides.

Molecular Weight	53.1
Density at 20°C	0.806
Melting Point	–83.6°C
Boiling Point	77.3°C
Autoignition Temperature	481°C
Explosive Limits in air	
lower	3.5vol%
upper	17vol%
Flash Point Closed Cup	–1°C
Vapour Density (air = 1)	1.83

Exposure Limit COSHH (Schedule 1) 2ppm 8 hour TWA (Maximum Exposure Limit)
Defined as a carcinogen under COSHH regulations.
Exposure Limit ACGIH 1ppm TLV-STEL
2ppm TLV-TWA
Classified A2, suspected human carcinogen.

GRADES

Commercial 99% (usually contains a polymerization inhibitor).

INTERNATIONAL CLASSIFICATIONS

UN No.	1093
CAS Reg No.	107–13–1
EINECS No.	203–466–5
EC No.	608–003–00–4
Description	Flammable liquid, toxic substance
Packing Group	I
Emergency Action Code	3WE
HI (Kemler Code)	33

APPLICATIONS

Acrylonitrile is a highly reactive molecule and polymerizes readily. The major outlet, accounting for approximately 53% of total world consumption, is synthetic fibres formed by polymerization or copolymerization to give acrylic or modacrylic fibres.

It will form copolymers with styrene, or butadiene and styrene, to yield styrene–acrylonitrile, or acrylonitrile-butadiene-styrene (ABS) thermoplastics, known for their strength and resistance to light, heat, and a wide range of solvents. These copolymers utilize around 30% of acrylonitrile production. Copolymerization with butadiene to nitrile rubbers consumes a further 4%.

Other monomers, for example vinyl chloride, vinylidene chloride, vinyl acetate and acrylates, will copolymerize with acrylonitrile to form resins used in paints, surface coatings and packaging. Because of its reactivity, acrylonitrile can be used as a chemical intermediate. Examples are acrylic acid and acrylamide by hydrolysis, adiponitrile, a nylon intermediate, by electrolytic coupling and amines by cyanoethylation.

Recent growth in fibres and ABS resins have resulted in a strong market for acrylonitrile. Fibres will play the dominant role in determining future increases in acrylonitrile production, as most fibre manufacturers are captive users.

World consumption is growing at 4% per year. Additional plants will be required in the Asia–Pacific region up to 2000 to meet the derivatives capacity growth in this area.

HEALTH AND HANDLING

Acrylonitrile vapour is irritating to the eyes, nose and lungs leading to laboured breathing, dizziness and nausea. Contact with the toxic, volatile liquid leads to irritation and blistering of the skin. Acrylonitrile is defined as a carcinogen. In high concentrations, inhalation of vapour and skin absorption of the liquid can lead to death. Protective clothing, a respirator, goggles and gloves must be worn. Contact lenses must be avoided as they concentrate the vapour. Any contaminated clothing must be removed immediately and disposed of.

Store in closed containers in a well-ventilated, cool, explosion-proof area away from oxidizing agents, alkalis and naked lights. Equipment must be earthed to prevent static build-up. Inhibitor levels should be checked weekly. Acrylonitrile will polymerize on exposure to light.

In the event of spillage, evacuate personnel and issue protective clothing to clean-up staff before containment. The waste material should be disposed of in accordance with local regulations. Acrylonitrile must be kept away from waterways and drains.

Acrylonitrile is a severe fire and explosion hazard, and the heavy vapour can lead to flashback. Alcohol foam or dry chemical can be used to fight fires. Protective clothing and self-contained respirators must be worn because of the toxic gases formed on burning.

MAJOR PLANTS

Plants with capacities greater than 100 000 tonnes per year:

Erdoelchemie	Cologne	Germany
DSM	Geleen	Netherlands
EniChem	Gela	Italy
Repsol Quimica	Tarragona	Spain
BASF	Middlesbrough	UK
Complex	Saratov	Russia
BP Chemicals	Green Lake	US
	Lima	US
Cytec Industries	Fortier	US
Du Pont	Chocolate Bayou	US
Monsanto Chemicals	Chocolate Bayou	US
Sterling Chemicals	Texas City	US
Asahi Chemical	Kurashiki	Japan
Mitsubishi Chemical	Mizushima	Japan
Chinese Petroleum	Kaohsiung	Taiwan
Shanghai Petrochemical	Jenshan	China

MAJOR LICENSORS

Badger
BP Chemicals/Ugine
EniChem/UOP
Nitto Chemical Industry
Sohio/BP Chemicals

Acrylonitrile-Butadiene-Styrene (ABS) Resins

SYNONYMS

ABS Styrene will polymerize readily with acrylonitrile and butadiene to form acrylonitrile-butadiene-styrene (ABS) resins. These polymers have a better impact resistance and higher tensile strength than polystyrene which has led to their use in applications where these properties are desirable. The properties of the wide range of grades available are determined by molecular and morphological parameters.

Three types of polymerization process can be employed for the production of ABS resins:

- emulsion;
- suspension;
- bulk.

Historically, emulsion and suspension processes were the most widely used but now the bulk polymerization route is more important. This is because the reaction does not take place in the aqueous phase and as a result there are no large aqueous streams for disposal and less energy is required as no drying of the polymer formed is involved. Major disadvantages of the bulk process are higher equipment costs, lower conversion rates and less product flexibility. Batch emulsion processes are used for the production of high-impact resins while lower-impact materials are formed by bulk and emulsion routes.

ABS polymers are expected to come under threat from metallocene thermoplastics particularly polypropylene, with their enhanced engineering properties.

Capacities range from 20 000 to 1 000 000 tonnes per year.

PROCESSES

From styrene by bulk polymerization

A solution of lightly linked polybutadiene dissolved in styrene and acrylonitrile and a diluent are fed into a reactor. An organic peroxide and chain transfer agent such as terinolene are added. The mixture is heated to around 100°C to begin the polymerization, with continuous agitation to prevent cross-linking. When 30% conversion to polymer has taken place, the syrup is transferred to a bulk polymerizator where the reaction is continued at a temperature of 150°C until 50% condensation has been achieved. (*See Figure 20*).

Heat from the exothermic polymerization is removed by evaporation of the monomers which are condensed and recycled. The polymer passes to a devolatorizer where any remaining monomers and diluent are removed by heating to 150°C under vacuum. The ABS is extruded and pelletized before being stored.

Raw material requirements

Raw materials required per tonne of ABS:

Polybutadiene latex	140–210kg
Styrene	570–620kg
Acrylonitrile	220–260kg

Quantities are variable depending on grades of resins formed.

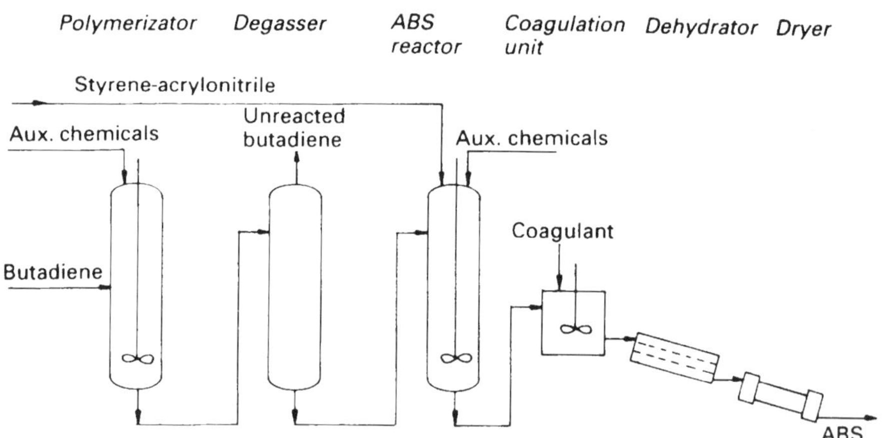

FIGURE 20 ABS resins from styrene by bulk polymerization

OTHER PROCESSES

Emulsion polymerization

The emulsion process can be separated into three stages:

- polybutadiene latex preparation;
- styrene/acrylonitrile graft onto the latex;
- styrene-acrylonitrile copolymer formation.

The last two stages are frequently combined.

In the first stage, a redox system consisting of three separate solutions of cumene hydroperoxide initiator, sodium pyrophosphate, dextrose, and ferrous sulphate activators and a sodium oleate emulsifier are fed into a polymerization reactor which has been purged of oxygen. Butadiene and mineralized water are introduced with continuous agitation and the mixture is heated to initialize the reaction. Heat from the reaction is removed by water circulation through an external jacket. The temperature is maintained below 70°C and when 70% conversion has taken place the reaction is stopped. Any excess monomer is stripped off.

Next, the polybutadiene latex formed, together with a 2% aqueous solution of potassium persulphate which acts as the initiator, are pumped into a second reactor. The mixture is heated to 55–70°C before acrylonitrile and styrene are fed in slowly with continuous stirring. The emulsion ABS formed is removed from the latex by coagulation at 100°C with acids or a dilute salt solution. Degassing and treatment of the ABS resin are carried out as for the bulk polymerization process.

Suspension polymerization

An organic peroxide such as *tert*-butyl peroxide is added to a solution of a lightly cross-linked polybutadiene latex dissolved in styrene and acrylonitrile. In a polymerizator, the mixture is heated to 100°C with continuous stirring to initialize the reaction. When 30% conversion has taken place, the resultant viscous solution is dispersed in water with the aid of a dispersing agent such as carboxyl methyl cellulose.

Polymerization is completed by heating the dispersion to 160°C under a pressure of 3.5 bar. The resultant slurry is separated by centrifuging. After washing, the resin beads are dried and conveyed to storage silos.

PROPERTIES

White powder. ABS burns slowly but is not self-extinguishing.

Density at 20°C	1.04
Vicat Softening Point	103°C
Tensile Yield MPa	41.4
Modulus MPa	2070
Elongation at rupture %	20
Impact Strength (notched Izod)	267

GRADES

High-impact, low-impact.

INTERNATIONAL CLASSIFICATIONS

UN No.	Not listed
CAS Reg No.	9003–56–9

APPLICATIONS

The most important outlet for ABS resins is in transportation which accounts for 25% of total world consumption. Another 20% goes into electronics. Included in this category are parts for business machines, electronics, computers, radios, TVs and telephone handsets. Domestic appliances both large and small, for example refrigerators and sewing machines, consume around 15% of total ABS production. Pipes, fittings and a wide range of products used in the construction industry take 7% of ABS demand. A further 7% goes into recreational goods such as boats and mobile homes. To the year 2000, annual growth is forecast to be around 3–4% in the US, 2–2.5% in Europe and 10% in the Far East.

HEALTH AND HANDLING

ABS is harmful if inhaled or absorbed through the skin as it is a suspected carcinogen. The dust can lead to eye and skin irritation, and contact lenses must not be worn. Care must be taken to prevent contact by wearing protective clothing, goggles, gloves, boots and a mask to prevent inhalation of dust. Long or repeated exposure should be avoided.

Store in a cool, dry place with good ventilation and away from strong oxidizing agents and strong bases. Powder spills should be collected and placed in a closed container for disposal by mixing with a solvent and incineration.

ABS burns slowly giving off carbon monoxide, carbon dioxide and nitrogen oxides as it decomposes. Carbon dioxide and water should be used to extinguish fires. Clean-up and firefighting staff should wear protective clothing, goggles and self-contained respirators when carrying out their duties. Contaminated clothing should be laundered before reuse and boots cleaned.

MAJOR PLANTS

Plants with capacities greater than 80 000 tonnes per year:

Bayer	Antwerp	Belgium
BASF	Ludwigshafen	Germany
Bayer	Dormagen	Germany
Dow Chemical	Terneuzen	Netherlands
GE Plastics	Grangemouth	UK
Bayer	Addyston	US
Dow Chemical	Midland	US
GE Plastics	Ottawa	US
	Port Bienville	US
	Washington	US
Asahi Chemical	Yokkaichi	Japan
Japan Synthetic Rubber	Yokkaichi	Japan
Mitsubishi Chemical	Yokkaichi	Japan
Ube Cycon	Ube	Japan
Cheil Plastics	Yeochon	South Korea
LG Chemical	Yeochon	South Korea
Miwon Chemical	Ulsan	South Korea
Chi Mei Industrial	Hsian	Taiwan
Grand Pacific	Kaohsiung	Taiwan
Toray Plastics	Penan	Malaysia

Most plants are capable of producing styrene-acrylonitrile resins which can be sold separately or converted to ABS.

MAJOR LICENSORS

Bayer
Dow Chemical
GE Plastics
International Synthetic Rubber
Japan Synthetic Rubber
Mitsui Petrochemical
Monsanto
Toray Industries

Adipic Acid $HOOC(CH_2)_4COOH$

SYNONYMS

ADIPIC ACID hexanedioic acid, 1,4-butanedicarboxylic acid, adipinic acid

Adipic acid is the most important aliphatic carboxylic acid because of its use as an intermediate in the production of nylon 66.

In early processes, adipic acid was prepared by the air oxidation of cyclohexane. Later, a boric acid assisted process was developed. Although it made the first stage more effective, it suffered from two major disadvantages; the recovery of boric acid, and as cyclohexanol was the principal product formed, an excessive consumption of nitric acid.

Today, most production of adipic acid is from cyclohexanone via the nitric acid oxidation of a cyclohexanol–cyclohexanone mixture or ketone–alcohol (KA) oil, although cyclohexanol or cyclohexanone can be used separately.

The advantages of the nitric acid process over the air oxidation processes are:

- good yield;
- high conversion rate;
- short reaction time;
- high product purity, suitable for nylon 66.

It does, however, utilize large quantities of nitric acid, which because of its corrosive nature, requires high-cost stainless steel equipment.

Considerable effort has been spent developing a one-step oxidation of cyclohexane using nitric acid, air or nitrogen dioxide. Asahi Chemical, Gulf Oil Chemicals and others have investigated single-stage liquid-phase air oxidation processes using cobaltous acetate and cobaltic acetylacetonate catalysts in an acetic acid solvent, but there is considerable by-product formation. Benzene is partially dehydrogenated to cyclohexanone which is then hydrolysed to cyclohexanol and adipic acid is formed by nitric acid oxidation.

Hydrocarboxylation processes for the synthesis of adipic acid have been patented by BASF and Texaco. The stepwise carbonylation of butadiene in the liquid phase to esters, using a rhodium chloride catalyst promoted with methyl iodide, has been proposed with yields of 60%.

More recently, Monsanto has reported the dicarbonylation of 1,4-dimethoxy-2-butene at 100°C in the presence of palladium halides. The unsaturated dimethyl ester formed is hydrogenated and then hydrolysed to adipic acid. These processes have had little impact on the commercial production of adipic acid to date.

Over 90% of adipic acid is manufactured from cyclohexane, with the remainder divided between cyclohexanol or phenol as the starting material.

Capacities range from 40 000 to 400 000 tonnes per year.

PROCESSES

From cyclohexane by air and nitric acid oxidation

Cyclohexane of 98% purity is oxidized in the presence of air using a cobalt salt, either naphthenate or oleate, as a catalyst (see Cyclohexanol & Cyclohexanone).

The cyclohexanol–cyclohexanone (KA) mixture is fed into a reactor together with a 50–60% nitric acid stream containing 0.2% ammonium metavanadate–cupric nitrate catalyst. Temperature is controlled at 60–80°C and pressure at 1–4 bar. Heat produced by the highly exothermic reaction is removed by heat exchangers. The reaction is controlled by maintaining an excess of nitric acid to KA feed in a ratio between 3:1 and 300:1 After a residence time of a few minutes, the mixture passes to a second reactor kept at a higher temperature of around 115°C. (*See Figure 21*)

The reaction mixture is passed through a column where excess nitrogen oxides are removed with air overhead. These gases are absorbed by water and the nitric acid produced is recycled. Any remaining nitrogen oxides are removed from the off-gases by scrubbing with cyclohexanol–cyclohexanone feed prior to entering the reactor.

Water contained in the reaction mixture is flashed off under vacuum distillation and adipic acid is recovered from the concentrated product stream by crystallization at 5°C, followed by centrifuging. Part of the filtrate is recycled, the remainder being processed to recover the copper–vanadium catalyst.

Succinic and glutaric acid by-products in the filtrate can be removed by ion exchange resins, or converted to their esters by the addition of methyl alcohol. Individual esters can be separated by subsequent distillation. Crude adipic acid is purified by recrystallization from water.

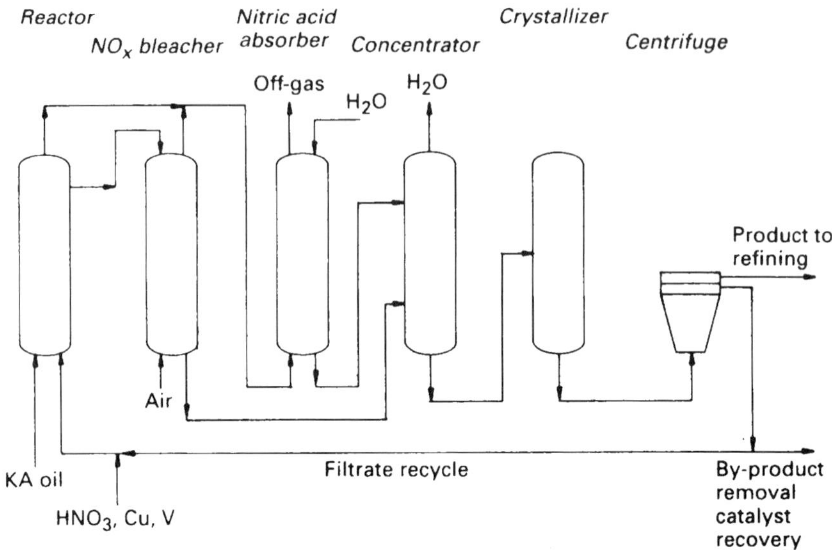

FIGURE 21 Adipic acid from cyclohexane by air and nitric acid oxidation

Reaction

$2C_6H_{12} + 1\frac{1}{2}O_2 \rightarrow C_5H_{10}CO + C_6H_{11}OH + H_2O$ — 1st stage

$C_6H_{11}OH + C_5H_{10}CO + 3\frac{1}{2}O_2 \rightarrow 2HOOC(CH_2)_4COOH + H_2O$ — 2nd stage

Raw material requirements and yield

Raw materials required per tonne of adipic acid:

Cyclohexanol–cyclohexanone	730–750kg
Nitric acid	1000kg
Ammonium metavanadate	0.25kg
Copper	0.2kg

Yield 92–94%

PROPERTIES

Colourless, odourless crystals with an acidic taste. Soluble in methyl alcohol, ethyl alcohol, water and acetone. Forms water-soluble salts with alkalis and ammonia.

Molecular Weight	146.1
Density at 20°C	1.36
Melting Point	152°C
Boiling Point (decomposes at 10 bar pressure)	337°C
Autoignition Temperature	420°C
Explosion Limits in air	
lower	0.035kg/m^3 (dust)
Flash Point Closed Cup	196°C
Vapour Density (air = 1)	5.04

Exposure Limit COSHH	No limits set
Exposure Limit ACGIH	5mg/m^3 for dust

GRADES

Fibre-grade 99.6%.

INTERNATIONAL CLASSIFICATIONS

UN No.	Not listed
CAS Reg No.	124–04–9
EINECS No.	204–673–3
EC No.	607–144–00–9

APPLICATIONS

The major outlet for adipic acid, accounting for over 80% of total world demand, is in the manufacture of nylon 66 fibres, and nylon 66 plastics. A small amount is consumed captively for the production of adiponitrile.

Adipic acid esters are used as plasticizers for polyvinyl chloride (PVC) and in polyurethane resins, which find outlets in speciality foams, adhesives, lacquers, and surface coatings. Small amounts of adipic acid are used in the food industry to acidify jams and to act as a buffering agent in other foods. Minor outlets are for the manufacture of insecticides, dyes, tanning products and chemicals used in the textile industry. Adipic acid is being employed for flue gas desulphurization.

Most production of adipic acid is used captively by the major fibre manufacturers. Future demand is dependent on the nylon fibre market which is declining due to competition from other synthetic fibres, such as polyesters and polypropylene, and from natural fibres.

HEALTH AND HANDLING

Care must be taken when handling adipic acid to avoid exposure to dust which can cause irritation to the skin, nose and throat. There is also a danger of dust explosions.

For these reasons, most adipic acid is transferred mechanically to storage or transport containers. Stainless steel is used for storage because adipic acid is corrosive to steel. Adipic acid can cake during storage, so good stock control must be practised.

Spills should be cleared with non-sparking tools. Scoop or vacuum into containers for disposal. Care must be taken not to generate dust because a concentration in excess of 10–15mg of dust per litre of air is an explosion hazard. The contaminated area should be sprinkled with soda ash, moistened to neutralize any remaining acid, and then flushed with large volumes of water.

Water, carbon dioxide, dry chemical or alcohol foam can be used to fight fires. All clean-up and firefighting staff must wear self-contained breathing apparatus.

MAJOR PLANTS

Plants with capacities greater than 120 000 tonnes/year:

Rhone-Poulenc	Chalampe	France
BASF	Ludwigshafen	Germany
Bayer	Leverkusen	Germany
Radici Chimica	Novara	Italy
Du Pont	Wilton	UK
Monsanto Chemicals	Pensacola	US
Du Pont	Orange	US
	Victoria	US
Du Pont	Palau Sakra	Singapore

MAJOR LICENSORS

ABB Lummus Global
Asahi Chemical
BASF
Du Pont
Halcon-Scientific Design
ICI
Monsanto
Rhone-Poulenc
Texaco Development

Ammonia

NH_3

Ammonia is one of the most important chemicals used in the petrochemical industry. Ammonium salts have been known since earliest times and free ammonia was first produced in the late 18th century.

The synthesis of ammonia from its elements commenced commercially prior to World War I. Since then there have been major improvements and a reduction in production costs with the development of the completely integrated large-scale processes used today. Current ammonia plants employ a high-capacity single-train operation, centrifugal compressors, higher-activity catalysts and efficient waste-heat recovery to generate high-pressure steam. Recent efforts have concentrated on reducing energy requirements per tonne of ammonia by integrating energy needs and heat recovery at all stages of the reaction process.

As the principal raw material source for ammonia plants is natural gas for producing hydrogen by steam reforming, its cost is crucial to the economics of ammonia production. Naphtha or residual oil are alternatives, but a partial oxidation process is normally used in this case. Larger feed requirements and greater generation of by-products lead to higher operating costs. Hydrogen streams from catalytic reformers are another valuable source of hydrogen, but the volumes available are too small to meet the requirements of an average-sized ammonia plant. Nitrogen is obtained by the liquefaction of air.

The new low-cost ammonia (LCA) process developed by ICI has revolutionized ammonia production. By separating the elements of the process and introducing sophisticated system control and energy recovery, small plants can now be as competitive as large ones.

Over 90% of ammonia synthesis from nitrogen and hydrogen is based on the Haber–Bosch process. Although ammonia is formed as a by-product of the destructive distillation of coal, this route is of little commercial importance. Recent growth in capacity has taken place in Asia and the Middle East where sources of cheap natural gas are available.

Capacities range from 30 000 to 2 100 000 tonnes per year.

PROCESSES

From nitrogen and hydrogen by catalytic synthesis

Nitrogen obtained by the liquefaction of air or from producer gas is mixed with hydrogen in the molar ratio of 3:1. Hydrogen is normally produced from natural gas, by steam reforming over a nickel catalyst, or by non-catalytic partial oxidation of naphtha or residual oil using air or oxygen. Because of the increase in the price of oil in the 1980s, natural gas is the preferred feedstock. Whatever the source of the hydrogen, it must be free from the oxides of carbon (which reduce catalyst activity) and from phosphorus, sulphur and arsenic compounds (which are catalyst poisons in the subsequent reaction). (*See Figure* 22)

The purified nitrogen–hydrogen gases are compressed to 150–350 bar, mixed with recycle feed before being fed into a tubular or multiple-bed reactor. The reaction takes place at 450–600°C over an iron catalyst promoted with oxides of aluminium and potassium. Conversions of 20–25% per pass are achieved. A wide range of newer, more reactive catalyst systems to replace the older iron-based ones have been proposed, which have the added advantage of operating at lower synthesis pressures. Iron–titanium metals, ruthenium–alkali metals or ruthenium promoted by potassium and barium on activated carbon, as well as phthalocyamine complexes, have shown considerable efficacy.

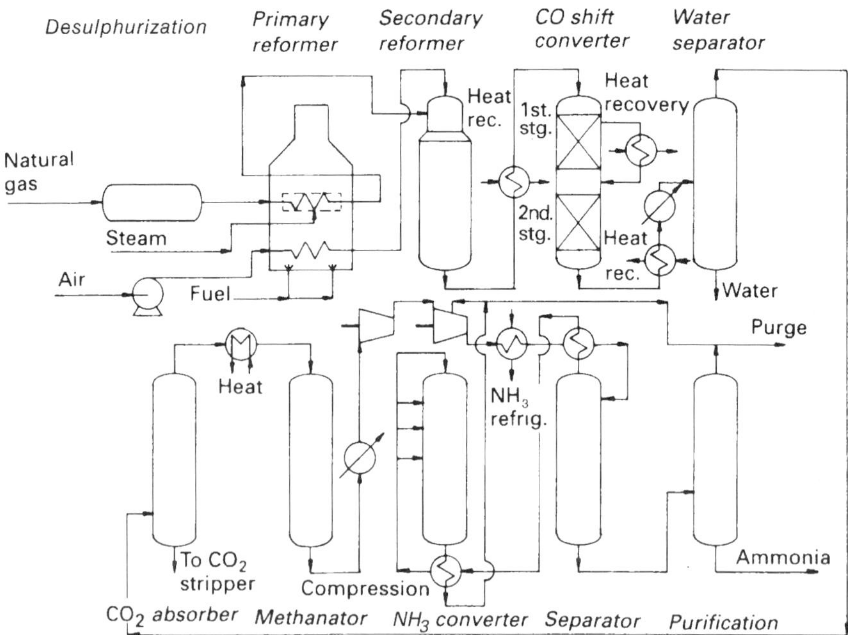

FIGURE 22 Ammonia from nitrogen and hydrogen by catalytic synthesis

The reacted gases are cooled to –20°C and most of the ammonia is liquefied. The unreacted gases are purged to prevent the build-up of diluents before being compressed and recycled to the reactor.

Reaction

$N_2 + 3H_2 \rightarrow NH_3$

Raw material requirements and yield

Raw materials required per tonne of ammonia:

Natural gas	$810m^3$
Yield 85%	

PROPERTIES

Colourless gas with a very pungent odour. Soluble in water, ethyl alcohol and ether. It is only flammable in extreme conditions in confined areas.

Molecular Weight	17.03
Density at 20°C (gas)	0.771
Melting Point	–77.7°C
Boiling Point	–33.4°C
Autoignition Temperature	615°C
Explosive Limits in air	
lower	16 NH_3 vol%
upper	27 NH_3 vol%
Vapour Density (air = 1)	0.597
Exposure Limit COSHH	35ppm 15 minutes 25ppm 8 hour TWA
Exposure Limit ACGIH	35ppm TLV-STEL 25ppm TLV-TWA

GRADES

Anhydrous, pure 99.9%, technical 99%, refrigeration 99.98%

INTERNATIONAL CLASSIFICATIONS

Anhydrous

UN No.	1005
CAS Reg No.	7664–41–7
EINECS No.	231–635–3
EC No.	007–001–00–5
Description	Toxic gas, corrosive
Packing Group	Not allocated
Emergency Action Code	2RE
HI (Kemler Code)	268

Solution (specific gravity <0.88 at 15°C in water, with >50% ammonia)

UN No.	3318
CAS Reg No.	1336–21–6
EINECS No.	215–647–6
EC No.	007–001–02–2
Description	Non-flammable, non-toxic gas
Packing Group	Not allocated
Emergency Action Code	2RE
HI (Kemler Code)	268

Solution (specific gravity <0.88 at 15°C in water, with >35% but <50% ammonia)

UN No.	2073
CAS Reg No.	7664–41–7
EINECS No.	231–635–3
EC No.	007–001–01–2
Description	Non-flammable, non-toxic gas
Packing Group	Not allocated
Emergency Action Code	2RE
HI (Kemler Code)	20

Solution (specific gravity between 0.88 and 0.957 at 15°C in water, with <35% but >10% ammonia)

UN No.	2672
CAS Reg No.	1336–21–6
EINECS No.	215–647–6
EC No.	007–001–02–2
Description	Corrosive substance
Packing Group	III
Emergency Action Code	2R
HI (Kemler Code)	80

APPLICATIONS

The major outlet for ammonia is in agriculture, either by conversion to fertilizers, of which urea and ammonium nitrate are the most important, or in the form of liquid ammonia for direct soil injection. As ammonia is difficult to store, fertilizers are a useful way of mitigating the cyclical nature of demand and pressure on prices.

Ammonia is used as a refrigerant in large industrial and commercial plants and in air-conditioning equipment. In metallurgy, ammonia provides an inert atmosphere for metal-working processes. Other applications include:

- removal of sulphur dioxide and the oxides of nitrogen from steam boiler flue gases;
- pulping of wood;
- stabilizer for natural and synthetic rubber latices;
- curing agent for animal hides;
- ingredient in household cleaners and drugs;
- preparation of ammonium–copper hydroxide for the solubilization of cotton linters in rayon production.

Ammonia has a wide range of chemical uses in the manufacture of caprolactam, hexamethylenediamine, acrylonitrile, methylamines, alkylamines and nitric acid.

It is a useful form in which to transport hydrogen as ammonia can be broken down to its elements by thermal cracking.

World growth rate is expected to be around 1.5% in the late 1990s, with the largest increase taking place in Asia.

HEALTH AND HANDLING

Ammonia vapour is very irritating to the eyes, nose and throat, and at high concentrations can cause coughing, chest pain and pulmonary oedema due to its alkalinity. In contact with the skin, the liquid will lead to burns and blisters, while damage to the eyes may be permanent. No contact lenses should be worn. Ammonia is not a carcinogen.

Protective rubber clothing, boots, aprons, gloves, safety goggles and face shields must be used. For short-period vapour concentrations below 300ppm, respirators with chemical cartridges can be utilized, but above this level self-contained breathing apparatus is required.

Anhydrous ammonia is stable at room temperature but decomposes above 450°C. It should be stored in steel containers fitted with pressure relief valves. The vessels must be welded not brazed, as contact with copper, bronze, brass or galvanized steel must be avoided. Store in a well-ventilated, cool, fire-resistant area away from sunlight, oxidizing agents, and combustible materials. Separate outside storage is desirable. Rigid pipe must be used for the transfer of ammonia, and all electrical services must be explosion-proof. Workers should be given training on correct handling methods.

If a leak occurs, personnel should be evacuated and all forms of ignition eliminated. Water can be used to absorb the ammonia; the aqueous ammonia solution should be collected and not discharged into rivers or sewers. Very small spills can be diluted with large volumes of water and if concentrations do not exceed regional limits, they may be allowed to enter drains with the agreement of the local authorities. Spills must not be neutralized with hydrochloric acid as copious fumes will result.

Ammonia is a moderate fire and explosion hazard when exposed to heat or flames. Water spray or fog should be used to extinguish fires. If possible the flow of gas should be stopped. Firefighting and clean-up staff must wear protective clothing, full face shields and self-contained breathing apparatus when carrying out their duties. Regular training on emergency procedures is advised.

MAJOR PLANTS

Plants with capacities greater than 850 000 tonnes per year:

BASF	Ludwigshafen	Germany
DSM	Geleen	Netherlands
Hydro Agri	Sluiskil	Netherlands
Zaktady Azotowe	Putawy	Poland
Berezniki Nitrogen	Berezniki	Russia
Complex	Tolyatti	Russia
Kemerovo Nitrogen	Kemerovo	Russia
Nevinnomyssk Chemical	Nevinnomyssk	Russia
Novogrod Chemical	Novogrod	Russia
Novomoskovsk Chemical	Novomoskovsk	Russia
Complex	Dneprodzerzhinsk	Ukraine
Complex	Gorlovka	Ukraine
Complex	Grigoryevka	Ukraine
Complex	Severodonetsk	Ukraine
Agrico Chemicals	Verdigris	US
Arcadian	Geismar	US
CF Industries	Donaldsonville	US
Farmland Industries	Enid	US
MCCI	Sterlington	US
PCS	Augusta	US
Unocal	Kenai	US
Canadian Fertilizers	Medicine Hat	Canada
Sherritt	Fort Saskatchewan	Canada
Pemex	Casoleacaque	Mexico
Arcadian	Point Lilas	Trinidad
Pequiven/Koch	El Tablazo	Venezuela
SABIC	Al Jubail	Saudi Arabia
Pupuk Kalimantan	Bontang	Indonesia
Pupuk Sriwidjaja	Palembang	Indonesia

MAJOR LICENSORS

Braun
Brown & Root
Cascale
Haldor Topsoe
ICI Katalco
Kellogg
Krupp-Koppers
Linde
Lurgi
Uhde

Aniline

SYNONYMS

ANILINE aminobenzene, phenylamine, benzenamine, aminophen, arylamine, aniline oil

Aniline was identified as a constituent of coal tar over a hundred years ago, and it became one of the major chemicals used in the manufacture of synthetic dyestuffs.

There are two routes to aniline in commercial production today. These are:

- catalytic hydrogenation of nitrobenzene;
- ammonolysis of phenol.

The hydrogenation process can take place in the liquid or vapour phase.

Comparison of the two routes show that the nitrobenzene process has a higher conversion rate and lower energy requirement, while the phenol process has a longer catalyst life, minimal waste disposal, and lower capital cost. Because large quantities of aniline are used captively, many producers have integrated their process back to the oxidation of cumene for the production of phenol, or have the facility to nitrate benzene to nitrobenzene.

Most world production of aniline is based on nitrobenzene, but Mitsui Petrochemical Industries (Japan) utilizes the phenol process. The ammonolysis of chlorobenzene process is no longer used commercially.

Capacities range from 20 000 to 400 000 tonnes per year. Over half the aniline produced is consumed captively.

PROCESSES

From phenol by ammonolysis

Phenol and ammonia are vaporized separately and preheated before being fed into a fixed-bed reactor containing a catalyst consisting of silica-alumina. The vapour-phase reaction takes place at a temperature of 370°C and a pressure of 17 bar. Ammonia is present in excess – in a molar ratio of 20:1 – to inhibit the formation of diphenylamine, triphenylamine and carbazole by-products. (*See Figure 23*)

After cooling, the effluent gases are condensed and any unreacted ammonia gas is recovered and recycled. The liquid fraction is separated into aqueous and organic layers. The organic layer has any remaining water removed in a drying tower before high-purity aniline is recovered by distillation. A small amount of unconverted phenol in the residue is recovered as a phenol–aniline azeotrope and recycled.

Regeneration of the catalyst is accomplished by passing an inert gas through to flush the system and then air to burn off any organic deposits. Catalyst life is claimed to be in excess of 7 years.

Reaction

$$C_6H_5OH + NH_3 \rightarrow C_6H_5NH_2 + H_2O$$

Raw material requirements and yield

Raw materials required per tonne of aniline:

Phenol	1030kg
Ammonia	1000kg

Yield

based on phenol	96%
based on ammonia	80%

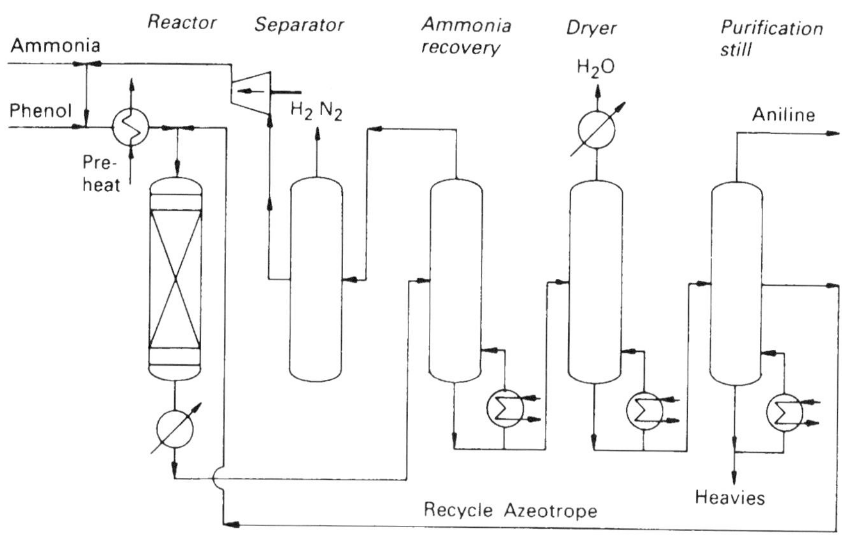

FIGURE 23 Aniline from phenol by ammonolysis

2. From nitrobenzene by hydrogenation

In the vapour-phase process, nitrobenzene is atomized in a hot hydrogen stream and circulated through a reactor containing a fluid bed of copper on pumice catalyst. Hydrogen is present in excess in a molar ratio of up to 6:1 to nitrobenzene. (*See Figure 24*)

The highly exothermic reaction takes place at 150–300°C, and a pressure of 2–15 bar. Excess heat is absorbed (either by recycling hydrogen or circulating a heat-absorbing liquid through tubes in the reactor) and used for steam generation.

The gases from the reactor are filtered to remove any catalyst, condensed and cooled. Crude aniline collects in the separator and any unreacted hydrogen is recycled. The aniline mixture is purified using vacuum distillation to remove high-boiling compounds. The catalyst is regenerated by firstly flushing the system with an inert gas, followed by air at 300°C to burn off any deposits.

Several different catalyst systems have been patented. BASF utilizes a 15wt% of copper on silica promoted with nickel, zinc, and barium, while Bayer uses ferrous chloride and iron filings to reduce the nitrobenzene to aniline. The exact operating conditions vary according to the type of catalyst used.

Nitrobenzene can be hydrated in the liquid phase, a process carried out at 80–250°C under pressure. Aniline is used as the solvent and the catalyst consists of finely divided nickel on kieselguhr. By recirculating some aniline to the reactor, the process can be operated on a continuous basis.

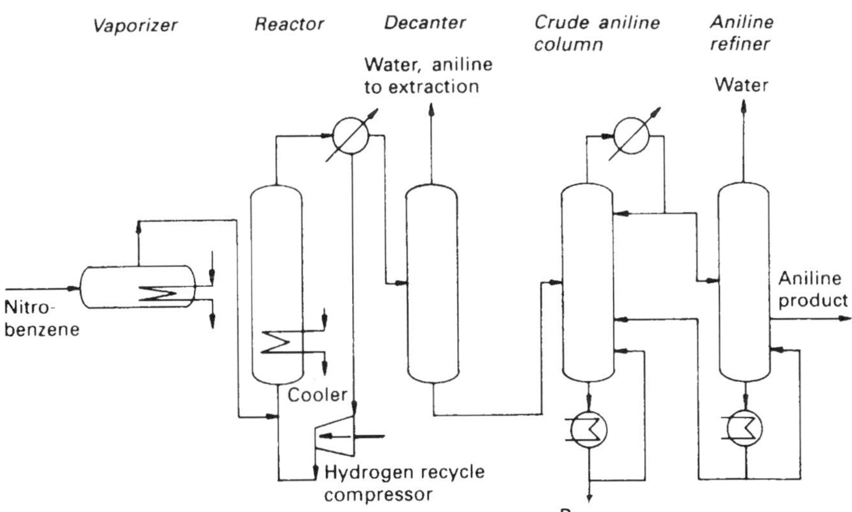

FIGURE 24 Aniline from nitrobenzene by hydrogenation

The advantage of the liquid-phase process is that it has a greater production capacity than the vapour-phase process for a given reactor size.

Reaction

$C_6H_5NO_2 + 3H_2 \rightarrow C_6H_5NH_2 + 2H_2O$

Raw material requirements and yield

Raw materials required per tonne of aniline (vapour-phase process):

Nitrobenzene	1350kg
Hydrogen	80m^3
Copper catalyst	0.7kg

Yield 98–99%

PROPERTIES

Colourless, oily liquid which turns brown on exposure to light and air. Soluble in ether, ethyl alcohol, and most organic solvents. It is a weak base and forms water-soluble salts with strong acids.

Molecular Weight	93.13
Density at 20°C	1.022
Melting Point	–6.3°C
Boiling Point	184.4°C
Autoignition Temperature	615°C
Explosive Limits in air	
lower	1.3vol%
upper	11 0vol%
Flash Point Closed Cup	70°C
Vapour Density (air = 1)	3.30

Exposure Limit COSHH
Substance on the current Advisory Committee on Toxic Substances (ACTS) and the Working Group on the Assessment of Toxic Chemicals (WATCH) work programme.

Exposure Limit ACGIH	5ppm TLV-STEL
	2ppm 8 hour TLV-TWA (skin)

GRADES

Technical 99%

INTERNATIONAL CLASSIFICATIONS

UN No.	1547
CAS Reg No.	62–53–3
EINECS No.	200–539–3
EC No.	612–008–00–7

Description	Toxic substance Class 6 poison (UN)
Packing Group	II
Emergency Action Code	3X
HI (Kemler Code)	60

APPLICATIONS

Around 80% of aniline produced globally is consumed in the manufacture of diphenylmethane diisocyanate (MDI), formed by condensation with formaldehyde, which is used to produce rigid polyurethane foam.

Rubber chemicals, once the largest outlet for aniline, is now in second place. The range of chemicals consists of accelerators which aid vulcanization, antioxidants, antiozonants, inhibitors, and stabilizers.

Aniline is used in the manufacture of:

- dye intermediates and synthetic dyes;
- agricultural chemicals (herbicides, fungicides, insecticides, defoliants);
- textile chemicals which impart water repellency and flame retardancy.

Other minor uses include the production of hydroquinone used as a developer in the photographic industry, sulpha drugs, analgesics and antipyretics for the pharmaceutical industry, and in amino resins to improve thermosetting properties.

Although there has been growth in the polyurethane market, the demand for synthetic dyes and rubber chemicals has declined.

HEALTH AND HANDLING

Aniline is toxic, readily absorbed by the skin, and a suspected carcinogen. Vapour can affect the eyes and lungs and, by reducing the ability of blood to carry oxygen, exposure can lead to headache, drowsiness or unconsciousness. In such cases, prompt treatment is required.

Protective clothing, goggles and gloves must be worn and any splashes washed away immediately with soap and water. When handling aniline good ventilation is essential. Although it is a combustible liquid, aniline can normally be handled with little risk of fire.

Aniline is stored in carbon steel or cast iron containers except where colour is important when stainless steel must be used. Brass, copper and its alloys are attacked, and must be avoided. Because aniline is classified as a poisonous liquid, all labels should be marked poison.

Spills should be absorbed with sand or vermiculite and the waste disposed of by controlled incineration. Water spray, dry chemical, carbon dioxide or foam can be used to fight fires. Staff must wear protective clothing and self-contained breathing apparatus.

MAJOR PLANTS

Plants with capacities greater than 90 000 tonnes per year:

BASF	Antwerp	Belgium
Bayer	Antwerp	Belgium
Bayer	Krefeld	Germany
ICI	Wilton	UK
Aristech Chemicals	Haverhill	US
BASF	Geismar	US
Bayer	Baytown	US
Du Pont	Beaumont	US
First Chemical	Pascagoula	US
Rubicon	Geismar	US
Mitsui Petrochemical Industries	Chiba	Japan

MAJOR LICENSORS

Nitrobenzene	*BASF*
	Bayer
	ICI
	Sumitomo Chemical
Phenol	*Scientific Design*

Benzene

SYNONYMS

BENZENE benzole, phenebenxole cyclohexatriene, phenyl hydride, coal, naphtha

Historically, coal tar liquor, a by-product of the coking industry, was the dominant source of commercial benzene. With the development of the petroleum industry and the growth in demand for benzene, petroleum became the prime source. Catalytic cracking processes, developed mainly to provide higher octane gasolines for motor vehicles, are also an excellent source of petroleum benzene.

Crude oil contains small amounts of benzene, toluene and xylene, but recovery is uneconomic. Today, pyrolysis gasoline and catalytic reforming of naphtha are the major sources of benzene, toluene and higher aromatics.

In catalytic reforming, aromatics are formed by:

- dehydrogenation of naphthenes;
- isomerization of alkylnaphthenes followed by dehydrogenation;
- dehydrocyclization of paraffins followed by dehydrogenation.

Pyrolysis gasoline is obtained as a by-product from the cracking of naphtha or gas oil to produce ethylene and propylene (see Ethylene). The type of feedstock and the operating conditions used affect the quantity of pyrolysis gasoline produced and its aromatic content, which can vary from 40 to 70%. Where aromatics are required, heavy naphtha feedstocks and high severity cracking are employed. Because of its instability, pyrolysis gasoline must be hydrogenated prior to aromatics extraction.

Unfortunately, the percentages of benzene, toluene and xylene formed, on average 11:51:38 respectively, are not in the same proportion as demand. Benzene output can be increased, especially from toluene which is produced in the greatest quantity in the above routes by hydrodealkylation, disproportionation or combination processes. Hydrodealkylation of toluene is a more expensive route than catalytic reforming, and the volumes produced by this route are dependent on the price differential between the two products. Hydrodealkylation units are used to fill the demand gap for benzene but their use is declining.

Whichever route is chosen to produce the crude aromatics stream, benzene is recovered from the concentrates by a range of different techniques. These include solvent extraction using sulpholane, diethylene and propylene glycols, N-methylpyrrolidone or morpholine derivatives, extractive or azeotropic distillation, solid absorption employing molecular sieves or crystallization. Further distillation can be carried out to remove impurities.

A new process to convert butanes and propanes, present in liquefied petroleum gas (LPG), to aromatics in a single step has been commercialized. The products are predominately benzene, toluene and xylenes.

Other potential sources of benzene are oil shale, tar sands and liquid fuels from coal. Plants using liquid fuels from coal are in operation in South Africa and New Zealand. However, as long as petroleum is available at current prices, their commercial use, except in certain special conditions, is not expected to grow in the short term.

Although in the US the major sources of benzene are catalytic reforming, followed by pyrolysis gasoline, more is increasingly being produced by co-product routes which are driven by the demand for the other product. Some 13% is made by toluene dealkylation and 14% by toluene disproportionation. Small quantities are made from coal and light oil. Worldwide, pyrolysis gasoline is growing at the expense of catalytic reforming.

Plants, which are normally sited as part of or adjacent to a refinery complex, range from 25 000 to 900 000 tonnes per year, with growth of new installations being particularly strong in the Middle and Far East.

PROCESSES

1. From petroleum by catalytic reforming

Naphtha, a petroleum distillate boiling between 70 and 190°C, is mixed with hydrogen and preheated before passing through a reactor containing a catalyst consisting of 0.2–0.5% platinum on alumina. Operating conditions can vary but are generally 425–530°C and 7–35 bar with large amounts of hydrogen recycle. The catalyst is sensitive to sulphur compounds and these must be removed prior to reforming. (*See Figure 25*)

The reactor products are fed into a separator where hydrogen gas is removed, scrubbed to extract any sulphur compounds, and recycled. Light hydrocarbons

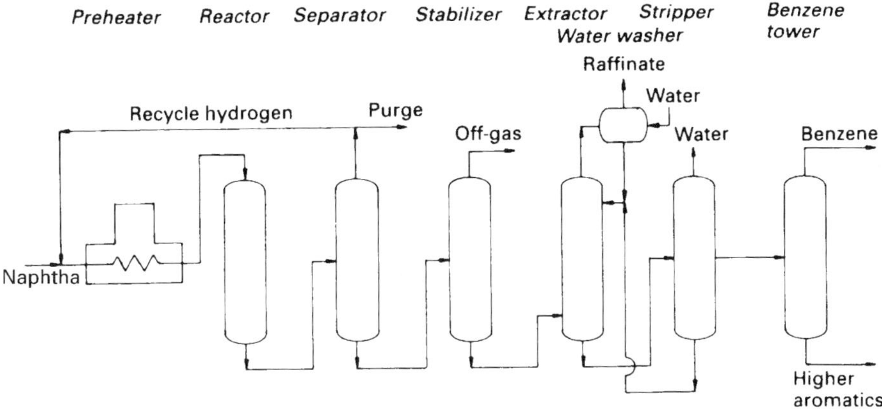

FIGURE 25 Benzene from petroleum by catalytic reforming

gases are removed from the liquid phase in a fractionator, before the reformate is sent to a tower where the aromatic-rich fraction is obtained overhead.

Numerous processes exist to extract aromatics from aromatic-rich fractions obtained from catalytic reforming or pyrolysis gasoline. In the Udex process developed by UOP/Dow Chemical, aqueous diethylene glycol is fed into the top of a multi-stage countercurrent extractor column and the aromatic-rich fraction is introduced about halfway up. The aromatic-rich solvent is drawn from the bottom of the column into a stripper where benzene, toluene and xylene are collected overhead and the solvent is returned to the extractor.

After water washing, the benzene–toluene–xylene mixture is treated with clay to remove unsaturates before fractionation to separate benzene, toluene and xylene. Aromatics recovery has been improved and capacity increased by replacing diethylene glycol with tetraethylene glycol.

Other solvents used for aromatics extraction are tetramethylene sulfone (sulfolane) by Shell, aqueous 1-methyl-2-pyrrolidone by Lurgi, dimethyl sulphoxide by IFP, methyl formamide from Leunawerke, and N-formylmorpholine by Snamprogetti and Krupp-Koppers. Operating conditions vary from process to process.

Reaction and yield

$C_6H_{12} \rightarrow C_6H_6 + 3H_2$

$C_6H_{14} \rightarrow C_6H_{12} + H_2$

Yield 95%

2. From toluene by hydrodealkylation

Hydrodealkylation of alkyl aromatics can be carried out in the catalytic or vapour phase. Toluene is normally the feedstock because it is present in excess in most large aromatic complexes, but mixtures of toluene and higher aromatics can be used. (*See Figure 26*)

Hydrogen and toluene are mixed in the molar ratio of 5–12:1 prior to being fed into a reactor. In the catalytic process, the reaction takes place at a temperature of 540–650°C and a pressure of 35–83 bar in the presence of a fixed-bed alkylation catalyst such as chromia-alumina. The thermal process is similar to the catalytic route but takes place at a temperature of 600–800°C in a hydrogen atmosphere and at the upper range of pressure. The temperature of the exothermic reaction is controlled by cold hydrogen circulation.

The reactor effluent is cooled by heat exchangers and benzene and unreacted toluene are condensed, before it is passed to a flash drum where hydrogen and lower-boiling compounds separate. Hydrogen is recycled and the remaining gas is used for fuel.

Bottoms containing benzene and toluene are passed through a fixed-bed clay treater before being distilled in a fractionator. Unconverted toluene is fed to the incoming feed to the reactor.

Several combination hydrogenation–dealkylation processes have been developed which operate at lower temperatures thus avoiding the need for hydrogenated product recycling.

Reaction

$C_6H_5CH_3 + H_2 \rightarrow C_6H_6 + CH_4$

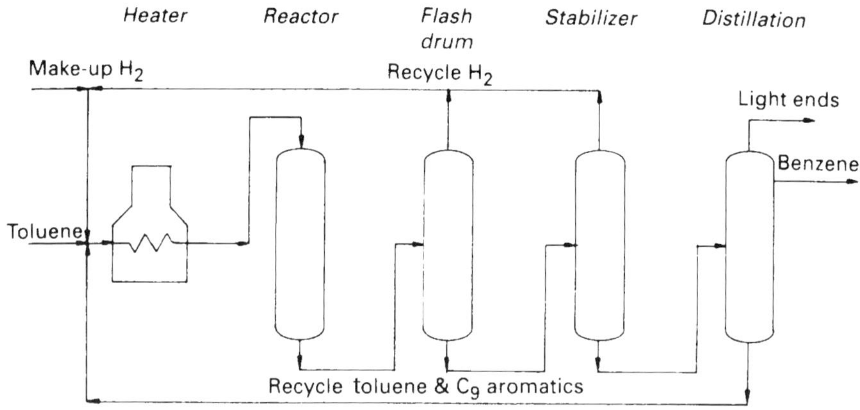

FIGURE 26 Benzene from toluene by hydrodealkylation

Raw material requirements and yield

Raw materials required per tonne of benzene:

Toluene	1220kg
Hydrogen	28kg

Yield 99%

3. From toluene by disproportionation

Recently, processes have been developed to disproportionate toluene or a toluene–C_9 mixture with low saturates content, to benzene and xylene. The reaction takes place at 350–530°C and 10–50 bar in the presence of a noble metal or rare earth fixed-bed catalyst. A zeolite-based catalyst has been developed which enables the reaction to be carried out at 260°C and 45 bar pressure.

The effluent from the reactor is triple distilled. Non aromatics are removed from the first, benzene is recovered from the second and toluene from the third. Higher aromatics remain as the bottoms stream. The ratio of benzene to xylene can be varied depending on the feedstock composition. Low hydrogen consumption and high conversion per pass are obtained (see Xylene).

Reaction

$$2C_6H_5CH_3 \rightarrow C_6H_6 + C_6H_4(CH_3)_2$$

$$C_6H_5CH_3 + C_6H_3(CH_3)_2 \rightarrow 2C_6H_4(CH_3)_2$$

Raw material requirements and yield

Raw materials required per tonne of benzene:

Toluene	2404kg
Hydrogen	5kg

Yield 97%

1340 tonnes of xylene are co-produced.

OTHER PROCESSES

From coal

Light oil, formed as a by-product of the high-temperature carbonization of coal, contains about 60vol% of benzene. Aromatics are removed by countercurrent absorption with a high-boiling petroleum fraction, in the boiling range 300–400°C. The mixture is stripped by continuous steam distillation. Light hydrocarbons are removed by distillation and the crude product is washed with concentrated sulphuric acid to remove sulphur compounds.

The hydrocarbons pass to a tank where any remaining acid is neutralized and then water washed. Hydrogenation eliminates any remaining sulphur and nitrogen compounds and any unsaturates. Benzene is obtained from the aromatics fraction by extractive distillation with a solvent such as sulpholane.

From petroleum gas

Butane and propane, which are present in LPG recovered from gas fields and refinery operations, can be converted to aromatics by the Cyclar process. The paraffins are catalytically dehydrogenated and the unsaturated products formed oligomerize and cyclicize to yield aromatics. The products are predominately benzene, toluene, ethylbenzenes and xylenes together with considerable quantities of hydrogen. The first Cyclar unit was built in Scotland by BP.

PROPERTIES

Colourless, refractive liquid with a characteristic odour. In cold weather, benzene solidifies to a white crystalline mass. Highly flammable and burns with a sooty flame. The vapour forms explosive mixtures with air over a wide range. Soluble in ethyl alcohol but only slightly soluble in water.

Molecular Weight	78.11
Density at 20°C	0.879
Melting Point	5.53°C
Boiling Point	80.1°C
Autoignition Temperature	498°C
Explosive Limits in air	
lower	1.4vol%
upper	7.1vol%
Flash Point Closed Cup	–11.1°C
Vapour Density (air = 1)	2.77

Exposure Limit COSHH (Schedule 1) 5ppm 8 hour TWA (Maximum Exposure Limit)
Defined as a carcinogen under COSHH regulations
Exposure Limit ACGIH 0.3ppm TLV-TWA
Classified A1, confirmed human carcinogen.

GRADES

Motor benzole; standard 90%, pure 99% and thiophene free.

INTERNATIONAL CLASSIFICATIONS

UN No.	1114
CAS Reg No.	71–43–2
EINECS No.	200–753–7
EC No.	601–020–00–8
Description Flammable liquid (class A2 carcinogenic potential)	
Packing Group	II
Emergency Action Code	3WE
HI (Kemler Code)	33

APPLICATIONS

In 1997, 46% of total US benzene production was alkylated with ethylene to ethylbenzene which was dehydrogenated to styrene. The second most important outlet for benzene, accounting for 21% of total demand, is cumene, the starting material for the manufacture of phenol and acetone.

Cyclohexane is the third major derivative, with 13% of benzene going into this outlet. It is used for the production of adipic acid, caprolactam and hexamethylenediamine.

Another product accounting for about 5% of total consumption is nitrobenzene which is the intermediate for aniline. Other products made from benzene are chlorobenzenes, maleic anhydride, resorcinol and detergent alkylates.

Global demand for benzene is forecast to grow at 4.5% between 1997 and 2000. The trend towards new plants being built in countries having sources of oil and natural gas is expected to continue, with capacity growing more rapidly in Asia and the Far East.

HEALTH AND HANDLING

Benzene vapour is very toxic and can be absorbed through the lungs or skin. Exposure leads to drowsiness followed by headaches and nausea. Benzene is a cumulative poison which builds up in the blood and body tissues and may lead to anaemia. It is listed as a class A2 industrial substance with suspected carcinogen potential. Tighter restrictions concerning its use, handling and exposure can be expected.

Great care must be exercised when handling benzene, and efficient ventilation provided at all times. PVC-coated gloves provide good protection, but any clothing or boots in contact with benzene must be changed and allowed to dry out in a safe area away from personnel.

Benzene has practically no corrosive effects on metals, and iron and steel containers are usually employed for storage. Any sulphur compounds present as impurities may affect copper and aluminium which precludes their use.

Wherever possible, handling of benzene should be carried out in closed containers, vented to atmosphere away from personnel to minimize vapour exposure. Equipment must be earthed to prevent static build-up. Because of its toxicity and flammability, labelling and transportation are closely controlled.

Spills should be contained and absorbed by earth or dry sand, placed in closed containers using non-sparking equipment, and disposed of in accordance with local regulations.

Fires should be blanketed with carbon dioxide, dry chemical or foam. Water is ineffective as it tends to scatter the flames. Benzene easily forms explosive, flammable mixtures with air; because of its heavy vapours, flashback is an additional hazard. Firefighting personnel must wear protective clothing, goggles and self-contained breathing apparatus.

MAJOR PLANTS

Plants with capacities greater than 500 000 tonnes per year:

Deutsche Shell	Godorf	Germany
Dow Benelux	Terneuzen	Netherlands
Agip Petroli	Priolo	Italy
ICI	Wilton	UK
Amoco	Texas City	US
BP Chemicals	Lima	US
Chevron Chemicals	Pascagoula	US
	Port Arthur	US
Dow Chemical	Plaquemine	US
Exxon Chemical	Baton Rouge	US
	Baytown	US
Koch Refining	Corpus Christi	US
Mobil Chemical	Beaumont	US
Shell Chemical	Deer Park	US
Phillips Puerto Rico	Guayama	Puerto Rico
Hess Oil	St Croix	Virgin Islands
Pequiven	Jose	Venezula
SADAF	Al Jubail	Saudi Arabia
	Toledo	Saudi Arabia
Mitsubishi Chemical	Kashima	Japan

MAJOR LICENSORS

Extraction/separation	*Glitsch*	*Leunawerke*
	HRI	*Lurgi*
	IFP	*Shell Development/UOP*
	Krupp-Koppers	*Snamprogetti*

Disproportionation/transalkylation	*Arco Technology*	*Mobil Oil*
	BASF	*Toray Industries/UOP (Tatoray)*
	Chevron	*UOP*
Dealkylation	*Arco Technology*	*Houdry/ABB Lummus Global*
	BASF	*HRI/IFP*
	Chiyoda/Mitsubishi Chemical	*UOP*
	Gulf Oil Chemicals	
Petrochemical Gas Conversion	*BP/UOP*	

Benzoic Acid

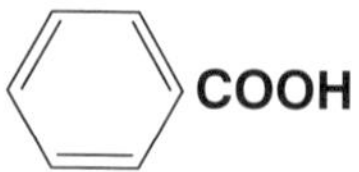

SYNONYMS

BENZOIC ACID benzenecarboxylic acid, phenyl formic acid, dracylic acid, phenol carboxylic acid

Benzoic acid is found in nature in certain gums, and the bark, foliage and fruits of cherries and plums.

Several processes for its manufacture have been developed:

- vapour-phase oxidation of toluene;
- sodium dichromate oxidation of toluene;
- vapour-phase decarboxylation of phthalic acid;

These routes now being obsolete, benzoic acid is produced by the continuous liquid-phase air oxidation of toluene in the presence of a cobalt catalyst.

Considerable effort has been spent on the development of improved catalyst systems. Higher reaction rates and better conversions have been obtained by the addition of bromine compounds in the atomic ratio to cobalt of 1:1. Slightly higher temperatures and pressures are required for the reaction. The addition of manganese salts also increases the reaction rate. The main disadvantage of bromine is that expensive metals must be used for the fabrication of the reactor in order to avoid corrosion problems.

Capacities range from 3 000 to 120 000 tonnes per year.

PROCESSES

From toluene by liquid-phase oxidation

Toluene, air and a catalyst consisting of 0.1% of cobalt naphthenate are fed into a stirred tank reactor. The toluene must be pure because sulphur and nitrogen compounds, phenol and olefins inhibit the reaction. Reaction conditions are 160–170°C and a pressure of 8–10 bar. The reaction is highly exothermic, heat being removed by external circulation of the reaction liquids through a heat exchanger. (*See Figure 27*)

After 40–50% conversion has been achieved, the reaction liquids pass to a column where the pressure is reduced to atmospheric and unreacted toluene and light gases are stripped off. The crude oxidation product is rectified to yield pure benzoic acid overhead and heavy ends, containing the metal catalyst, are sold for metal recovery. Further purification of benzoic acid can be carried out by sublimation in an inert atmosphere, or by recrystallization.

Care must be taken to prevent a runaway reaction because of the quantity of heat generated by the process. Several by-products are produced, but only benzaldehyde can be recovered by distillation.

Reaction

$$C_6H_5CH_3 + 1\tfrac{1}{2}O_2 \rightarrow C_6H_5COOH + H_2O$$

Raw material requirements and yield

Raw materials required per tonne of benzoic acid:

Toluene	840kg
Air	1600m^3

Yield 90%

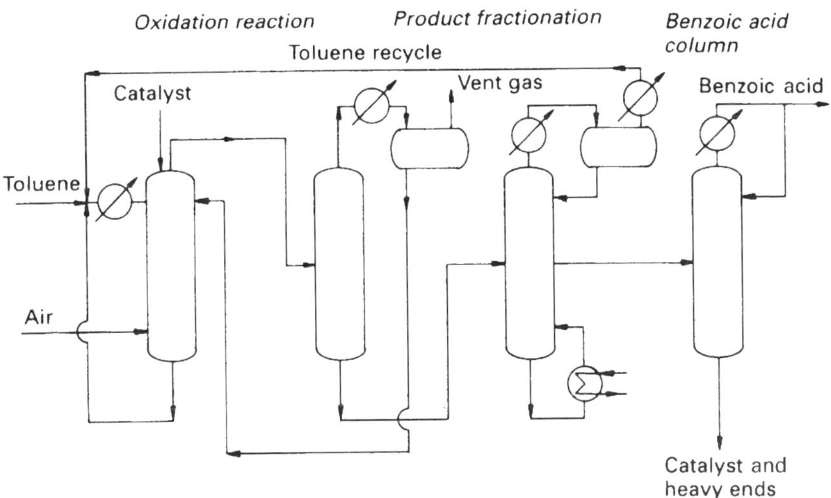

FIGURE 27 Benzoic acid from toluene by liquid-phase oxidation

PROPERTIES

White crystals which sublime at 100°C. Slightly soluble in water. Soluble in ethyl alcohol, acetone and ether. Stable to permanganate, chromic acid and hypochlorite oxidizing agents.

Molecular Weight	122.12
Density at 20°C	1.321
Melting Point	122.4°C
Boiling Point	249.2
Autoignition Temperature	573°C
Flash Point Closed Cup	121°C
Vapour Density (air = 1)	4.21
Exposure Limit COSHH	None established
Exposure Limit ACGIH	None established

GRADES

Technical 97.5–99%, pharmaceutical 99.5%.

INTERNATIONAL CLASSIFICATIONS

UN No.	Not allocated
CAS Reg No.	65–85–0
EINECS No.	200–818–2
EC No.	Not allocated

APPLICATIONS

More than 50% of benzoic acid production is used captively for the manufacture of phenol and caprolactam. Of the remainder, plasticizers are an important outlet. Glyco esters of benzoic acid are used in plasticized vinyl resins. Benzoic acid is added to alkyd resins where it regulates the degree of polymerization and improves surface properties.

It is used as a mordant for dyes and as an intermediate in the pharmaceutical, textile, herbicide and perfume industries. Benzoic acid is converted to sodium benzoate, benzoyl chloride and benzoyl benzoate. Sodium benzoate is employed as a food preservative.

Future demand will depend on the outlook for nylon for which phenol and caprolactam are important intermediates.

HEALTH AND HANDLING

Benzoic acid is stored as flakes in drums or bags, or in the molten state in stainless steel containers. It can be kept without deterioration for long periods but may solidify. It should be stored in tightly closed containers in a cool, dry area.

When handling benzoic acid, care should be taken to avoid contact or inhalation of fine particles of the product. Safety goggles and protective outerwear are recommended.

Spills should be swept up and placed into containers for disposal taking care to avoid creating dust. Benzoic acid is only a slight fire hazard but its hot vapours may form explosive mixtures with air. Fires should be extinguished with dry chemical, carbon dioxide, foam or water spray. Staff must wear protective clothing and self-contained breathing apparatus when carrying out either of these tasks.

MAJOR PLANTS

Plants with capacities greater than 30 000 tonnes per year:

DSM	Botlek	Netherlands
Soc. Chimica del Friuli	Torviscosa	Italy
BF Goodrich	Kalama	US
Velsicol Chemical	Chattanooga	US
Dow Chemical	Ladner	Canada
Liquid Quimica	Cuilatao	Mexico
Mitsubishi Chemical	Kitakyushu	Japan

MAJOR LICENSORS

Mid Century Corp.
Snia Engineering

Benzyl Chloride

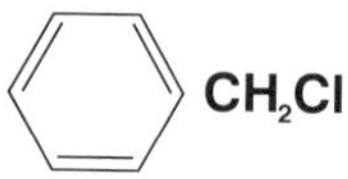

SYNONYMS

BENZYL CHLORIDE chloromethyl benzene, α-chlorotoluene

Benzyl chloride was initially obtained by the reaction of hydrochloric acid with benzyl alcohol, but today it is manufactured by the chlorination of toluene in the liquid phase. It can also be produced by the oxychlorination of toluene in the vapour phase.

A mixture of mono-, di-, and tri-chlorinated products is obtained, but the ratio can be varied by adjusting the chlorine–toluene ratio.

Capacities range from 3 000 to 35 000 tonnes per year.

PROCESSES

From toluene by chlorination

When toluene is chlorinated a mixture of three chlorides is obtained. By adjusting the toluene–chlorine ratio, the quantity of benzyl chloride can be maximized. (*See Figure 28*)

Benzyl chloride is produced by passing dry chlorine gas into boiling toluene at 100–110°C (the toluene containing ferric chloride as the catalyst) until the density increases to 1.283, when the volume of by-products (benzal chloride and benzotrichloride) are at a minimum. The resultant chlorinated mixture is mixed with dilute alkali and the benzyl chloride recovered by distillation.

Alternatively, toluene can be photochlorinated at a temperature between 65 and 90°C until a 25% increase in weight is obtained. By-product hydrogen chloride is absorbed in water and benzyl chloride is recovered by distillation.

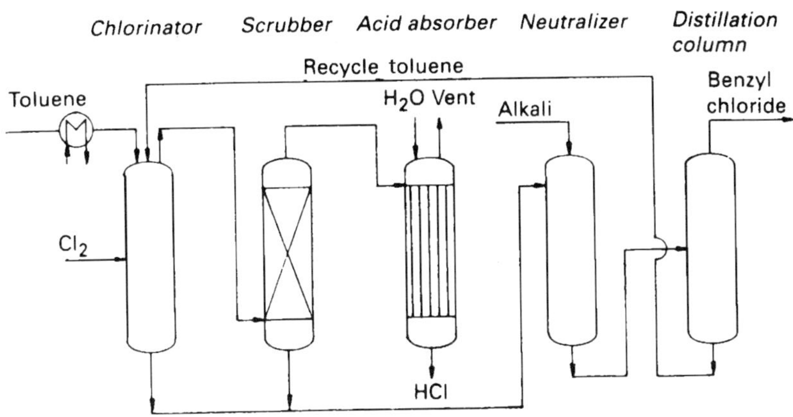

FIGURE 28 Benzyl chloride from toluene by chlorination

Reaction

$C_6H_5CH_3 + Cl_2 \rightarrow C_6H_5CH_2Cl + HCl$

Raw material requirements and yield

Raw materials required per tonne of benzyl chloride:

Toluene	780kg
Chlorine	640kg

Yield 95%

PROPERTIES

Colourless liquid with a pungent odour which fumes in moist air. Insoluble in cold water but decomposes slowly in hot water to give benzyl alcohol. Soluble in most organic solvents.

Molecular Weight	126.58
Density at 20°C	1.100
Melting Point	–39.2°C
Boiling Point	179.4°C
Autoignition Temperature	585°C
Explosive Limits in air	
lower	1.1vol%
upper	14vol%
Flash Point Closed Cup	67°C
Vapour Density (air = 1)	4.88
Exposure Limit COSHH	1ppm TWA
Exposure Limit ACGIH	1ppm TLV-TWA

GRADES

Pure anhydrous >99%, pure and stabilized >99%.

INTERNATIONAL CLASSIFICATIONS

UN No.	1738
CAS Reg No.	100–44–7
EINECS No.	202–853–6
EC No.	602–037–00–3
Description	Toxic substance, corrosive
Packing Group	II
Emergency Action Code	2W
HI (Kemler Code)	68

APPLICATIONS

The importance of benzyl chloride is due to its role as a chemical intermediate. Its major outlet is for the manufacture of butyl benzal chloride, a plasticizer used in flexible vinyl resins, especially floor coverings.

Other outlets include the production of phenylacetic acid, the precursor of penicillin and amphetamines, and benzyl alcohol for aspartamine. Small volumes are consumed in quaternary ammonium salts, benzyl esters for the perfume and flavour industries, photographic developers, triphenyl dyes, and used as disinfectants and algacides.

HEALTH AND HANDLING

Benzyl chloride is a powerful lacrimator. Its vapour is irritating to the eyes and mucous membranes and can be absorbed into the lungs and gastrointestinal tract. Care must be exercised when handling benzyl chloride as it is a suspected carcinogen.

Glass-lined tanks are the preferred choice for storage, but nickel or ceramic-lined vessels can be used. Benzyl chloride must be stabilized except where nickel containers are employed. The storage area must be well ventilated.

Because benzyl chloride reacts with heavy metals, it is shipped in glass, enamel, lined steel or nickel containers. Suitable liners are lead or stable synthetic resins. The transportation of benzyl chloride is governed by many regulations owing to its toxicity, and local advice should be sought.

Spills should be collected and placed in containers which are vented to allow any hydrochloric acid to escape. Dry absorbents can be used to clean up the area prior to neutralization with soda ash. Care must be taken to avoid discharge into sewers.

Fires can be extinguished with carbon dioxide or dry chemical, but water must not be used because dangerous gases would be produced. Clean-up and firefighting personnel must wear full protective clothing and self-contained breathing equipment.

MAJOR PLANTS

Plants with capacities greater than 20 000 tonnes per year:

Monsanto Europe	Antwerp	Belgium
Bayer	Krefeld	Germany
Tessenderlo Chemie	Tessenderlo	Netherlands
Monsanto Chemicals	Bridgeport	US
Kawaguchi Chemical	Tokoyama	Japan

MAJOR LICENSORS

Monsanto

Bisphenol A

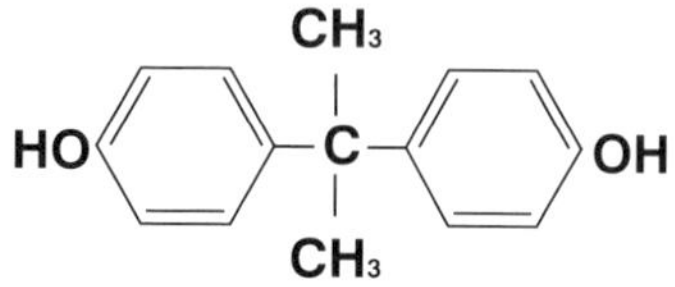

SYNONYMS

BISPHENOL A 4,4′-isopropylidenediphenol, BPA, 2,2-bis(4-hydroxyphenyl) propane, *p,p*-dihydroxydiphenyl-dimethylmethane, 4,4′-isopropylidene-bis-phenol, 4,4′-(1-methylethylidene)bi-phenol

The interest in bisphenol A is as a raw material for the manufacture of polycarbonates and epoxy resins.

All commercial processes for the manufacture of bisphenol A involve the condensation of phenol and acetone. The reaction is catalysed by acids, but in new plants these have been replaced by cationic ion exchange resins.

The major differences in the processes used are in the recovery and purification stages. Two grades of bisphenol A are produced, one suitable for epoxy resin manufacture and a high-purity product used in the manufacture of polycarbonate resins.

Capacities range from 60 000 to 210 000 tonnes per year. Almost all production is used captively.

PROCESSES

From acetone and phenol by condensation

Acetone and excess phenol in a molar ratio of 1:20 together with 1:4wt% of hydrogen chloride are pumped into a glass-lined tank. A promoter, methyl mercaptan, is usually added to the acid catalyst solution. The temperature is kept at around 50°C by cooling to remove the heat generated by the reaction. (*See Figure 29*)

After 30 minutes, when almost all of the acetone has reacted, the slurry is distilled to separate water, acid and some phenol overhead, which separates into two layers on cooling. Acid is recovered from the aqueous layer and recycled, and the organic layer is also returned to the reactor.

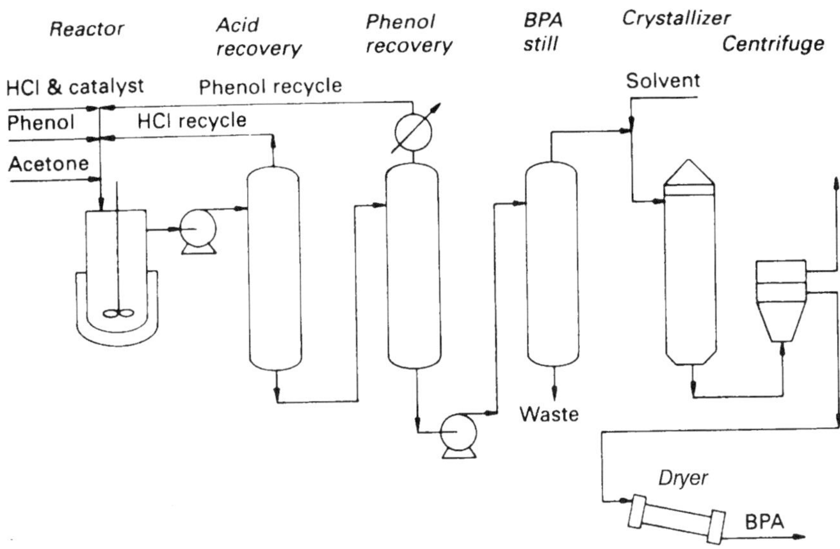

FIGURE 29 Bisphenol A from acetone and phenol by condensation

Phenol, removed from the reaction product by distillation under vacuum, is recycled. After steam stripping the bottoms to separate out any remaining impurities, the hot crude bisphenol A is mixed with benzene under pressure. In the crystallizer, the pressure is released and pure bisphenol A crystals are formed. These are separated in a centrifuge and dried. The crystals can be further purified by melt crystallization. The benzene filtrate is distilled to recover pure benzene for reuse. Any unreacted acetone is separated out by distillation and fed back to the reactor.

Union Carbide have developed a process which uses sulphonated styrene divinylbenzene cation exchange resins to catalyse the condensation reaction. The reaction takes place at 70–80°C with a 1 hour residence time. A 50% conversion of the acetone and 10% of the phenol per pass is achieved. Newer plants use cation exchange resins as the catalyst in preference to hydrochloric acid.

A new continuous ion exchange catalytic stripping process has been developed which removes the water formed in the condensation reaction by using nitrogen. The high phenol conversion to bisphenol A obtained permits the crystallization of the product to be achieved in one stage with resultant cost savings.

Reaction

$$2C_6H_5OH + CH_3COCH_3 \rightarrow HOC_6H_4C(CH_3)_2C_6H_4OH + H_2O$$

Raw material requirements and yield

Raw materials required per tonne of bisphenol A:

Acetone	260kg
Phenol	850kg
HCl catalyst	20kg

Yield 91%

PROPERTIES

White crystals or flakes with a slightly phenolic odour. Soluble in ether, ethyl alcohol and acetone. Slightly soluble in water.

Molecular Weight	228.28
Density at 20°C	1.195
Melting Point	157°C
Boiling Point	220°C
Flash Point Open Cup	207°C

Exposure Limit COSHH None allocated

Substance on the current Advisory Committee on Toxic Substances (ACTS) and the Working Group on the Assessment of Toxic Chemicals (WATCH) work programme, and being reviewed under the Existing Substances Regulations.

Exposure Limit ACGIH 10mg/m^3 TLV-TWA (total dust)

GRADES

Epoxy resin minimum melting point = 155°C.
Polymer-grade minimum melting point = 156.5°C.

INTERNATIONAL CLASSIFICATIONS

UN No.	Not allocated
CAS Reg No.	80–05–7
EINECS No.	201–245–8
EC No.	603–074–00–8

APPLICATIONS

Bisphenol A is the raw material for the manufacture of epoxy and polycarbonate resins. Together these account for 95% of total world demand for bisphenol A.

Epoxy resins are used in surface coatings, laminates, adhesives, composites, and for castings, and mouldings. The outlets for polycarbonates are in communications and electronics, compact disks and associated equipment, lighting and signs, building and glazing, and appliances.

Future growth will be dependent on the demand for epoxy and polycarbonate resins.

HEALTH AND HANDLING

Dust, vapour and solutions can cause irritation to the skin, lungs and eyes. Good ventilation, protective clothing and eye protection must be provided where risk of contact can occur.

Bisphenol A can be shipped in either bags, drums, boxes or in bulk. Aluminium is recommended for bulk storage, and handling. Bisphenol A should be stored under dry air or nitrogen at ambient temperature. Care must be taken on transfer as the dust can form explosive mixtures with air. The equipment should be earthed to prevent static build-up, and friction heat accumulation should be prevented. Local advice should be taken on handling spills.

Bisphenol A is a slight fire hazard but the dust can cause explosions. Carbon dioxide, dry chemical or foam can be used to blanket fires. Firefighters must wear protective clothing and self-contained breathing apparatus as toxic gases are given off during burning.

MAJOR PLANTS

Plants with capacities greater than 80 000 tonnes per year:

Bayer	Antwerp	Belgium
Bayer	Leverkusen	Germany
Dow Chemical	Stade	Germany
GE Plastics	Bergen op Zoom	Netherlands
Shell Nederland Chemie	Pernis	Netherlands
Aristech Chemical	Haverhill	US
Dow Chemical	Freeport	US
GE Plastics	Mount Vernon	US
Shell Chemical	Deer Park	US
Mitsui Toatsu Chemical	Nagoya	Japan

MAJOR LICENSORS

Chiyoda
Hooker Chemicals
Kellogg/Dow Chemical
Rhone-Poulenc
Sinopec/ABB Lummus Global
Union Carbide

Butadiene

$CH_2 = CHCH = CH_2$

SYNONYMS

BUTADIENE 1,3-butadiene, biethylene, buta-1,3-diene, vinylethylene, divinyl

Butadiene is a basic raw material used in the manufacture of synthetic rubbers and an intermediate for the production of styrene copolymers.

Originally produced from acetylene by the Adol or Reppe routes, these processes are no longer economic. Butadiene was also produced from ethyl alcohol but this process is obsolete.

The most important source of butadiene worldwide is C_4 fractions obtained as a by-product from the cracking of naphtha and gas oil for ethylene. Naphtha contains 4.3% butanes/butenes and 4.6% butadiene, while the figures for gas oils range from 4.9 to 6.2% and from 4.5 to 5.3% respectively (see Ethylene). Most manufacturers purchase C_4 streams for processing but many consume them captively as part of an integrated cracker complex. Details of butadiene content from the cracking of various feedstocks are listed under Ethylene.

In the US, butanes – a component of natural gas liquids and liquefied petroleum gas and butenes – are hydrogenated to butadiene on a large scale.

Recently, catalysts have been developed which will oxydehydrogenate butane/butene feeds. This could provide a useful source of butadiene from butane/butene residues remaining after the isolation of butadiene and isobutene. The increasing use of ethane for ethylene manufacture leading to a reduction in the availability of C_4 streams could give an impetus to its greater commercialization, and, in the longer term, this route is expected to become more important.

Worldwide, most butadiene is produced from C_4 steam cracker fractions, a by-product of ethylene manufacture, with the remainder from butane/butenes. Because of the butadiene over-supply situation, the C_4 stream from naphtha crackers is being hydroisomerized at some sites. The resultant 2-butenes-rich cut is metathesized to propylene which is in high demand (see Propylene).

Capacities range from 20 000 to 350 000 tonnes per year.

PROCESSES

1. From C_4 steam cracking fractions by extraction

C_4 fractions are produced primarily by the steam cracking of naphtha, gas oil, butanes and propanes. Although yields of C_4s are only marginally altered by cracking conditions, their composition is dependent on the cracking intensity. (*See Figure 30*).

Butadiene is isolated from other components by liquid extraction or extractive distillation. A wide range of extractive solvents have been suggested such as cuprous ammonium sulphate, furfural, acetonitrile and dimethylacetamide, but the ones in most current use are N-methylpyrrolidone or dimethylformamide.

The C_4 feed is extractively distilled with the solvent, and the more unsaturated hydrocarbons are dissolved. Butanes and butene are recovered overhead.

A second extractive distillation removes C_4 acetylenes and 1,2-butadiene. Butadiene is recovered from the top of the extraction tower and passes to a distillation column where the more volatile compounds are taken off at the top. In the second distillation column, 1,3-butadiene is separated from the lower boiling C_5s and 1,2-butadiene. C_4 acetylenes are hydrogenated before being recycled to the C_4 feed.

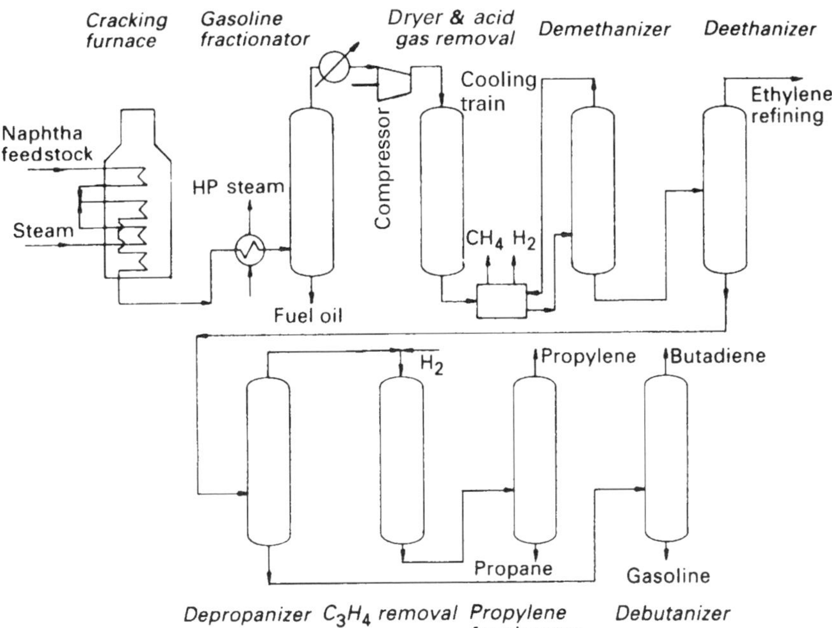

FIGURE 30 Butadiene from C_4 steam cracking fractions by extraction

2. From *n*-butene by dehydrogenation

Crude *n*-butene feed obtained by steam cracking or from C_4 refinery fractions is extractively distilled to remove butanes, pentanes and other by-products. Any isobutene is removed by acid extraction or converted to butyl ether. (*See Figure 31*)

The purified *n*-butene stream is preheated to 550°C and mixed with superheated steam in a steam/butene molar ratio that varies from 8:1 to 18–20:1 depending on the catalyst used. The mixed gases are passed over a fixed catalyst consisting of iron oxide with bauxite or chromium oxide, or calcium–nickel phosphate stabilized with chromium oxide. Reaction conditions are 620–670°C with a residence time of 0.2–0.25 seconds. The reaction is endothermic, with heat having to be added to maintain the required temperature.

Coke deposits build up on the catalyst and, in order to maintain its activity, the feedstock is alternated between two reactors so that while one is in operation, the second is regenerated by steam which burns off the carbon residues.

The hot reactor effluent is quenched, compressed and cooled. The condensate is fractionated to remove volatiles overhead and crude butadiene is separated by extractive distillation. Selectivities of 70–90% and conversion per pass of 28–45% are attained.

Phillips and Petro-Tex have developed processes which improve both the conversion and selectivity of the process by oxidative dehydrogenation. The addition of oxygen removes hydrogen released by the dehydrogenation reaction as water. Operating conditions are 480–590°C and a residence time of 0.25–0.5 seconds with an oxygen/butene molar ratio of 0.9–1.2:1. Conversions of 65–89% and selectivities of 93–98% are claimed.

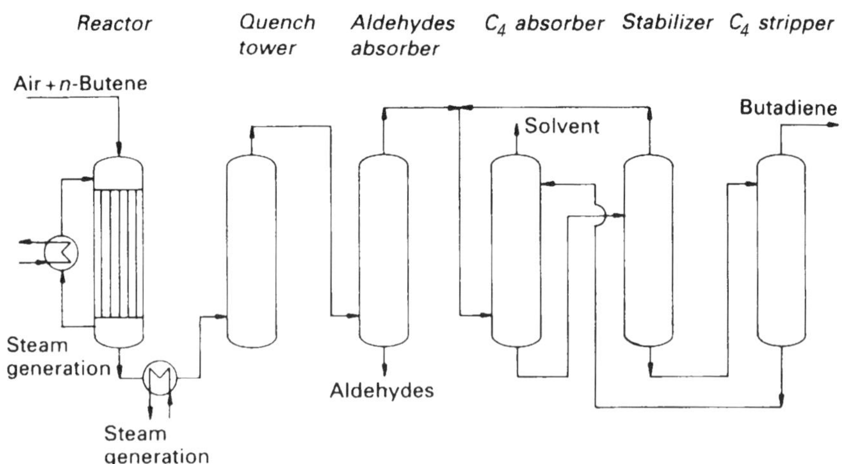

FIGURE 31 Butadiene from *n*-butene by dehydrogenation

Additional advantages are that oxygen reduces the carbon deposits on the catalyst giving it a longer life and lower energy consumption.

Reaction

$C_4H_8 \rightarrow CH_2{=}CHCH{=}CH_2 + H_2$ dehydrogenation

$C_4H_8 + \frac{1}{2}O_2 \rightarrow CH_2{=}CHCH{=}CH_2 + H_2O$ oxidative dehydrogenation

Raw material requirements and yield

Raw materials required per tonne of butadiene 100%:

n-Butene 1160–1370kg

Yield 75–90% (depending on the process used)

3. From *n*-butane by dehydrogenation

n-Butane feed is preheated before being fed into a fixed-bed reactor containing aluminium oxide mixed with chromium oxide as the catalyst. The endothermic reaction takes place at 550–650°C and 0.1–0.2 bar pressure. (*See Figure* 32)

After 5–15 minutes of operation, the incoming feed is switched to a second reactor while the first is regenerated. After purging to remove any residual butanes, heated air is passed through to burn off the carbon deposit. The heat produced during the regeneration heats the catalyst up to reaction temperature again before reuse. When regeneration is complete, any residual air is removed by purging with fuel gas before the reactor is brought back into service.

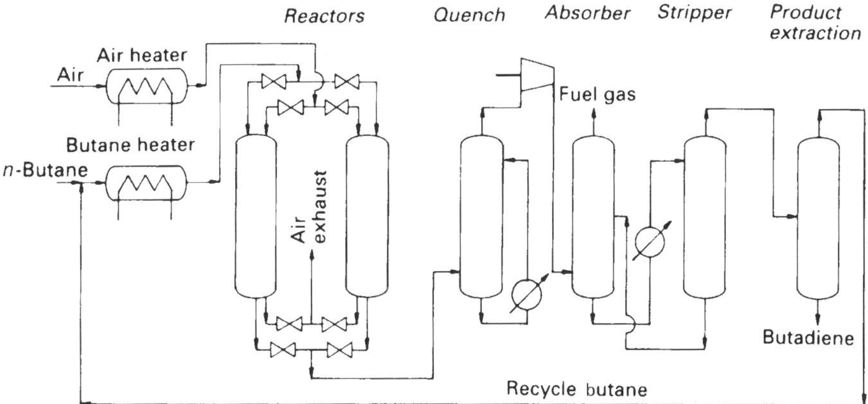

FIGURE 32 Butadiene from *n*-butane by dehydrogenation

The reacted gases are quenched in spray towers with gas oil. The gases are compressed to 10 bar, and light ends and high boilers are removed by scrubbing prior to fractionation. Butadiene is extracted with a solvent and purified.

Reaction

$$C_4H_{10} \rightarrow C_4H_8 + H_2$$

$$C_4H_8 \rightarrow CH_2{=}CHCH{=}CH_2 + H_2$$

Raw material requirements and yield

Raw materials required per tonne of butadiene 100%:

n-Butane	1820kg

Yield 82%

OTHER PROCESSES

A number of companies have developed one- or two-step catadiene dehydrogenation processes for the production of butadiene from mixed butane/butene streams. Several parallel fixed-bed reactors are used and their switching to on-stream or regeneration can be controlled automatically.

In the two-step Phillips route, *n*-butanes and *n*-butenes are catalytically dehydrogenated at 600°C and 1 bar pressure. For *n*-butane, the preferred catalyst system consists of chromium–sodium–aluminium oxides, and for *n*-butenes iron potassium–aluminium oxides.

PROPERTIES

Non corrosive, colourless, flammable gas with mild aromatic odour. Insoluble in water, soluble in ethyl alcohol.

Molecular Weight	54.092
Density at 20°C (liquid)	0.621
Melting Point	–108.9°C
Boiling Point (1.013 bar)	–4.411°C
Autoignition Temperature	420°C
Flammability Limits in air	
lower	2.0vol%
upper	11.5vol%
Flash Point Closed Cup	–70°C
Vapour Density (air = 1)	1.87

Exposure Limit COSHH (Schedule 1) 10ppm 8 hour TWA (Maximum Exposure Limit)
Defined as a carcinogen under COSHH regulations. Substance being reviewed under the Existing Substances Regulations.

Exposure Limit ACGIH 2ppm TLV-TWA
Classified A2, suspected carcinogen potential for humans.

GRADES

Technical 99.6% inhibited with *p-tert*-butyl pyrocatechol (TEC) 70–150ppm; 2,6-*tert*-butyl-*p*-cresol (TBK) can be used as an alternative.

INTERNATIONAL CLASSIFICATIONS

UN No.	1010
CAS Reg No.	106–99–0
EINECS No.	203–450–8
EC No.	601–013–00-X
Description	Flammable gas
Emergency Action Code	2WE
HI (Kemler Code)	239

APPLICATIONS

The major use for butadiene is in the manufacture of synthetic rubbers either as a monomer or comonomer. In the US, the most important product is styrene-butadiene rubber and latex which consume 43% of butadiene production. Styrene-butadiene rubber, 31% of the total, is used in the manufacture of tyres and the latex finds applications in carpet backing and paper. About 27% of the butadiene consumed is for the manufacture of polybutadiene also used in the tyre industry. Some 3% of butadiene usage is used for polychloroprene and another 3.5% goes into the manufacture of nitrile rubber.

Other important outlets for butadiene consuming 6.5% of US production, is adiponitrile, a chemical intermediate for the production of nylon and acrylonitrile-butadiene-styrene (ABS) which accounts for a further 10%.

Globally, demand is forecast to grow at 4% per year to 2000, with the Asia–Pacific region seeing the greatest increase.

HEALTH AND HANDLING

Butadiene has low toxicity but at high concentrations the vapours can lead to eye and lung irritation and finally unconsciousness. The liquid causes burns when in contact with the skin. Contact lenses must not be worn as they may absorb the vapours. Protective clothing, gloves, goggles and impervious boots must be worn when handling the product to avoid contact. Butadiene is listed as an A2 carcinogen.

Steel or aluminium closed containers are employed for storage as butadiene is non-corrosive, but copper and its alloys, which act as oxidizing agents, must not be used. It can be stored under its own pressure or under an inert atmosphere of carbon dioxide or nitrogen if higher pressures are required. Oxygen must be excluded as butadiene forms potentially explosive peroxides. To prevent the accumulation of peroxides only liquid must be drawn from containers. Inhibitors are added to block dangerous polymer formation and checks should be carried out to ensure adequate levels are maintained, especially if butadiene is stored for longer than one month. Additional inhibitor can be added if necessary to maintain satisfactory levels.

Butadiene is highly flammable and forms explosive mixtures with air over a wide concentration. It must be stored in a well-ventilated area away from sources of ignition and all electrical equipment must be flameproof, bonded and earthed to prevent static build-up.

Small fires may be extinguished with carbon dioxide, dry chemical or foam. The flow of gas should be stopped and containers cooled with water sprays. As butadiene vapour is heavier than air, it can travel long distances and lead to flashback. Breathing apparatus should be used by firefighters as butadiene gives off carbon dioxide and carbon monoxide gases on burning.

MAJOR PLANTS

Plants with capacities greater than 145 000 tonnes per year:

Erdoelchemie	Cologne	Germany
ROW	Wesseling	Germany
Dow Chemical	Terneusen	Netherlands
DSM	Geleen	Netherlands
Shell Nederland Chemie	Moerdijk	Netherlands
Complex	Nizhnekamsk	Russia
Complex	Tobolsk	Russia
Exxon Chemical	Baton Rouge	US
Huntsman Corp.	Port Neches	US
Lyondell	Channelview	US
Shell Chemical	Norco	US
Texas Petrochemicals	Houston	US
Copene	Camacari	Brazil
Chiba Butadiene	Chiba	Japan
Nippon Zeon	Tokuyama	Japan

MAJOR LICENSORS

Extractive distillation	*ABB Lummus Global* *BASF* *Exxon* *Japan Synthetic Rubber* *Lurgi* *Nippon Zeon* *Shell Development*
Dehydrogenation	*ABB Lummus Global* *Air Products & Chemicals* *Nippon Zeon* *Phillips Petroleum*
Oxidative dehydrogenation	*Petro-Tex Chemical* *Phillips Petroleum* *Shell Development* *UOP*

Butyl Acetate $CH_3COOC_4H_9$

SYNONYMS

BUTYL ACETATE butyl ethanoate, acetic acid butyl ester, butyl acetic ether

Four butyl acetates exist but only *n* and iso isomers are of any commercial importance due to their usage as solvents.

n-Butyl acetate is made by the esterification of acetic acid with butyl alcohol in the presence of sulphuric acid as a catalyst. Azeotropic removal of the water formed as a by-product pushes the reaction to completion.

Capacities range from 5000 to 77 000 tonnes per year.

PROCESSES

From butyl alcohol and acetic acid by esterification

Acetic acid, butyl alcohol and sulphuric acid in a 1–2% concentration overall are charged into a reactor. Most commonly, 80% acid is employed and butyl alcohol, free from other isomers, is present in 10% excess. Steam is used to heat the mixture until refluxing begins at a temperature of around 89°C. (*See Figure 33*)

Vapours from the top of the column containing butyl acetate, butyl alcohol and water as a ternary azeotrope, are removed and condensed in a separator. Part of the top layer in the separator is bled off and recycled to the reactor where it acts as a reflux and helps to maintain a constant temperature. The remainder of the top layer is fed to a low-boiler column where unreacted alcohol is flashed off and recycled to the reactor.

From the base of the column, the crude ester mixture undergoes a second distillation to separate off the high-boiling by-products and butyl acetate is collected overhead. Water, produced by the esterification reaction which forms the aqueous layer, is stripped of any remaining organics with steam prior to disposal.

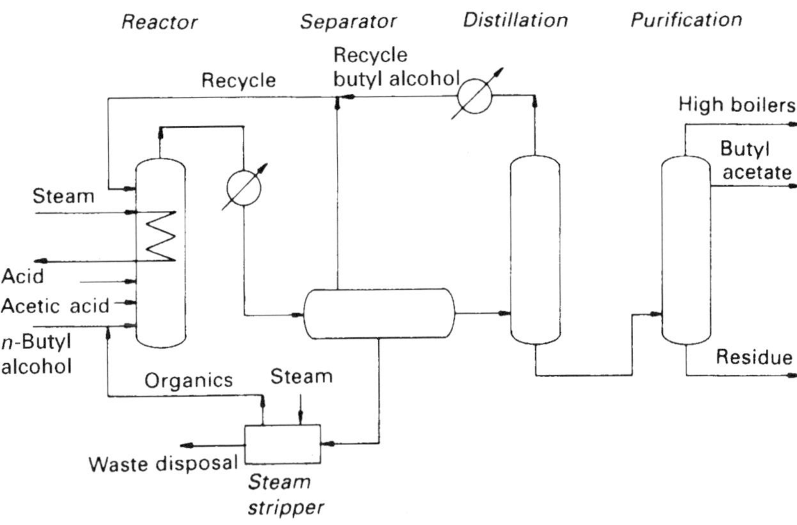

FIGURE 33 Butyl acetate from butyl alcohol and acetic acid by esterification

Reaction

$$C_4H_9OH + CH_3COOH \rightarrow CH_3COOC_4H_9 + H_2O$$

Raw material requirements and yield

Raw materials required per tonne of n-butyl acetate:

Butyl alcohol 95%	710kg
Acetic acid 90%	550kg

Yield 95%

PROPERTIES

Colourless, mobile liquid with a characteristic fruity odour of the simple acetate esters. Highly flammable. Soluble in ethyl alcohol and ether.

n-B*utyl acetate* – $CH_3COOCH_2CH_2CH_2CH_3$

Molecular Weight	116.16
Density at 20°C	0.884
Melting Point	–73.5°C
Boiling Point	126.5°C
Autoignition Temperature	421°C
Flammability Limits in air	
lower	1.7vol%
upper	7.6vol%
Flash Point Closed Cup	23°C
Vapour Density (air = 1)	4.0
Exposure Limits COSHH	200ppm 15 minutes 150ppm 8 hour TWA
Exposure Limit ACGIH	50ppm skin (Ceiling Exposure Limit)

iso-*Butyl acetate* – $CH_3COOCH_2CH(CH_3)_2$

Molecular Weight	116.16
Density at 20°C	0.873
Melting Point	–98.85°C
Boiling Point	118°C
Autoignition Temperature	422.8°C
Flammability Limits in air	
lower	2.4vol%
upper	10.5vol%
Flash Point Closed Cup	17°C
Vapour Density (air = 1)	4.0
Exposure Limit COSHH	187ppm 15 minutes 150ppm 8 hour TWA
Exposure Limit ACGIH	150ppm 8 hour TLV-TWA

sec-*Butyl acetate* – $CH_3COOCH(CH_3)CH_2CH_3$

Molecular Weight	116.6
Density at 20°C	0.865
Melting Point	–73.5°C
Boiling Point	112.2°C
Autoignition Temperature	420°C
Flammability Limits in air	
lower	1.7vol%
upper	9.8vol%
Flash Point Closed Cup	18.9°C
Vapour Density (air = 1)	4.0
Exposure Limits COSHH	250ppm 15 minutes 200ppm 8 hour TWA
Exposure Limit ACGIH	200ppm TLV-TWA

tert-*Butyl acetate* – $CH_3COOC(CH_3)_3$

Molecular Weight	116.16
Density at 20°C	0.884
Melting Point	–76.8°C
Boiling Point	126.5°C
Autoignition Temperature	421°C
Flammability Limits in air	
lower	1.7vol%
upper	15.0vol%
Flash Point Closed Cup	23°C
Exposure Limits COSHH	250ppm 15 minutes 200ppm 8 hour TWA
Exposure Limits ACGIH	200ppm TLV-TWA

GRADES

Technical >97%

INTERNATIONAL CLASSIFICATIONS

n-*Butyl acetate*

UN No.	1123
CAS Reg No.	123–86–4
EINECS No.	204–658–1
EC No.	607–025–00–1
Description	Flammable liquid
Packing Group	III
Emergency Action Code	3Y
HI (Kemler Code)	30

iso-*Butyl acetate*

UN No.	1213
CAS Reg No.	110–19–0
EINECS No.	203–745–1
EC No.	607–026–00–7
Description	Flammable liquid
Packing Group	II
Emergency Action Code	3YE
HI (Kemler Code)	33

sec-*Butyl acetate*

UN No.	1123
CAS Reg No.	105–46–4
EINECS No.	203–300–1
EC No.	607–026–00–7
Description	Flammable liquid
Packing Group	II
Emergency Action Code	3YE
HI (Kemler Code)	33

tert-*Butyl acetate*

UN No.	1123
CAS Reg No.	540–88–5
EINECS No.	208–760–7
EC No.	607–025–00–1
Description	Flammable liquid
Packing Group	III
Emergency Action Code	3YE
HI (Kemler Code)	30

APPLICATIONS

The major outlet for *n*-butyl acetate is as a solvent for lacquers because it imparts good flow and blush resistance. These properties are particularly good for film-coating resins such as cellulose nitrate, cellulose acetate butyrate, ethyl cellulose, polystyrene and methyl methacrylate resins.

Butyl acetate is also widely employed as a solvent for the preparation of artificial leathers, textiles and plastics, an extractant for oils and pharmaceuticals, and as an ingredient for perfumes and synthetic flavours.

Butyl acetate and butyl alcohol can be used as a dehydrating agent as together they form a ternary azeotrope with water.

Where problems exist with high concentrations of vapour, *n*-butyl acetate and methyl isobutyl ketone are frequently replaced by isobutyl acetate because of its less pronounced odour.

HEALTH AND HANDLING

High concentrations of *n*-butyl acetate cause irritation to the eyes and throat, leading to coughing, nausea and unconsciousness. The liquid in contact with the skin causes defatting. When handling butyl acetate, protective clothing, gloves, aprons, safety boots and goggles should be worn. Any contaminated clothing should be removed immediately, dried and laundered before reuse.

Store butyl acetate in closed containers in a well-ventilated, explosion-proof, cool place, away from any forms of ignition. Equipment should be spark-proof and earthed to prevent static build-up. Oxidizing agents should not be stored in proximity.

In the event of spillage, cleaning staff should wear protective clothing. Small spills can be absorbed onto sand or vermiculite, but large overflows should be collected and burnt by atomization in an approved incinerator. Care should be taken to keep butyl acetate away from waterways.

Carbon dioxide, dry chemical or foam can be used to fight fires. As butyl acetate is flammable and its vapour is heavier than air, flashback can occur and care should be taken with sumps or ducts where vapour could collect. Firefighting staff should wear protective clothing and breathing apparatus.

MAJOR PLANTS

Plants with capacities greater than 30 000 tonnes per year:

BP Chemicals	Antwerp	Belgium
BASF	Ludwigshafen	Germany
Huels	Marl	Germany
BP Chemicals	Hull	UK
Eastman Chemical	Kingsport	US
Hoechst Celanese	Bishop	US
Union Carbide	Seadrift	US
	Texas City	US
Kyowa Yuka	Yokkaichi	Japan
Chang Chun	Miaoli City	Taiwan

MAJOR LICENSORS

BP Chemicals
Hoechst

Butyl Alcohol

***n*-Butyl alcohol**	$CH_3(CH_2)_2CH_2OH$
Isobutyl alcohol	$(CH_3)_2CHCH_2OH$
***sec*-Butyl alcohol**	$CH_3CH_2CHOHCH_3$
***tert*-Butyl alcohol**	$(CH_3)_3COH$

SYNONYMS

N-BUTYL ALCOHOL *n*-butanol, 1-butyl alcohol, 1-hydroxybutane, butyl hydroxide, *n*-propylcarbinol

ISOBUTYL ALCOHOL iso-butanol, 2-methyl-1-propanol, iso-propylcarbinol,

SEC-BUTYL ALCOHOL *sec*-butanol, 2-butyl alcohol, butylene hydrate, 2-hydroxybutane, methylethylcarbinol

TERT-BUTYL ALCOHOL *tert*-butanol, 2-methyl-2-propanol, trimethylcarbinol

Butyl alcohol occurs as four isomeric alcohols consisting of two primary, one secondary and one tertiary.

Production was initially based on the fermentation of carbohydrates by bacteria which yielded a mixture of acetone and butyl alcohol. As demand grew, chemical methods were employed; although many processes are available, very few have been used commercially.

Originally, *n*-butyl alcohol was made by the hydrogenation of crotonaldehyde, obtained from acetaldehyde, but this route was overtaken by the Reppe process. In this method, propylene is reacted with carbon monoxide and water to yield a mixture of *n*- and iso-alcohols. Despite lower operating conditions and high *n*/iso product ratios, this route has never been widely used, because of its high operating costs.

The most important process in use at the present time is the catalytic hydroformylation of propylene – the oxo process – followed by the hydrogenation of the aldehydes formed. This reaction yields a mixture of *n*- and isobutyl alcohols, but as the former is in greater demand, research has been concentrated on maximizing its yield.

Initially, processes used catalysts based on cobalt carbonyls followed by the development of cobalt salts of organic acid, trialkyl phosphines and alkali. The latest technology employs the carbonyls of rhodium and ruthenium. Not only is the reaction carried out at lower temperatures and pressures, but the *n*/iso ratio is doubled or quadrupled by the use of rhodium and faster reaction rates are obtained. *n*-Butyl alcohol is the preferred product because it is a better solvent than the iso-isomer and because it can be converted to methyl ethyl ketone. Olefins can be hydrated directly to alcohols at high temperatures and pressures, but these routes are not yet competitive.

Secondary and tertiary alcohols cannot be made by the oxo reaction. *sec*-Butyl alcohol is produced by the hydration of *n*-butene, and *tert*-butyl alcohol from isobutylene. Some *tert*-butyl alcohol is also formed as a by-product of propylene oxide manufacture from 2-methylpropane, and the splitting of methyl *tert*-butyl ether, but neither route is of any commercial importance.

Currently, around 90% of butyl alcohol production is based on propylene. If oil prices rise and fermentation alcohol provides a cheaper source of acetaldehyde, then interest in the crotonaldehyde route could revive.

Capacities range from 10 000 to 180 000 tonnes per year, but individual capacities are flexible because the product mix can be varied to meet demand by changing process conditions and catalysts.

PROCESSES

1. *n*-Butyl alcohol: From propylene by hydroformylation (oxo process)

Carbon monoxide and hydrogen are added to propylene in the presence of a catalyst; the aldehydes formed are hydrogenated to *n*-butyl alcohol and isobutyl alcohol. Since sulphur compounds are catalyst poisons, both propylene and carbon monoxide have to be desulphurized before use. (*See Figure 34*)

Propylene, carbon monoxide, hydrogen and a dissolved catalyst consisting of a rhodium complex, hydrodocarbonyltris (triphenylphosphine) rhodium, are fed into a reactor. The liquid-phase reaction takes place at 100°C and 10–50 bar. If a cobalt catalyst is used, a temperature of 145–180°C and a pressure in the range 200–300 bar are employed.

The catalyst is separated from the reaction mixture and recycled and any unreacted gases are withdrawn and used as fuel. The crude oxo mixture is distilled and the isomeric butyraldehydes removed overhead. The bottoms are separated into butyl alcohols and the esters by further distillation, and the residues are burnt as fuel.

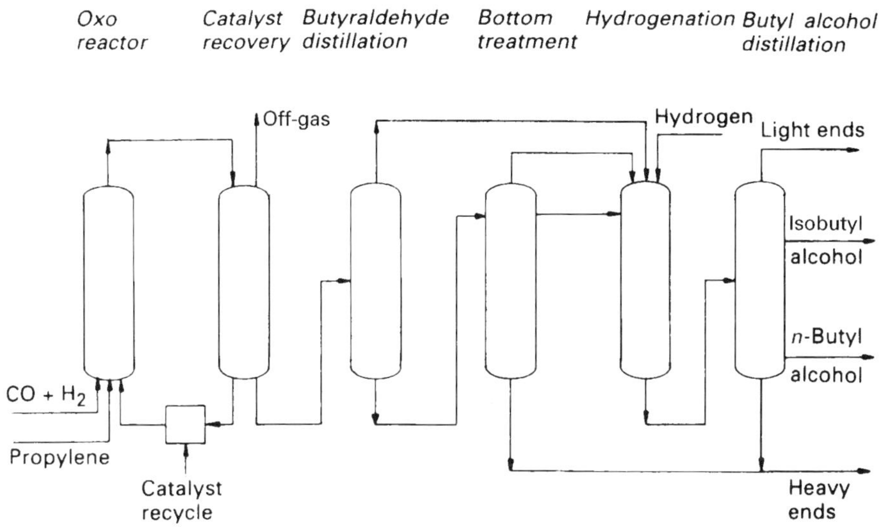

FIGURE 34 *n*-Butyl alcohol from propylene by hydroformylation (oxo process)

As the oxo process can produce a range of products, yields of butyl alcohol can be increased by the hydrogenation of the butyraldehyde formed and its integration into the crude butyl alcohol stream.

Butyraldehyde is hydrogenated over a fixed-bed catalyst, at 130–160°C and 3–5 bar pressure. A wide range of catalysts can be used including chromium and copper oxides on a silica support, or cobalt or nickel on silica or alumina. Up to 10% of water is added to the reaction mixture to suppress the formation of ethers.

The mixture containing crude butyl alcohol is combined with the butyl alcohols stream obtained from the first reactor and distilled. Isobutyl alcohol, water and low-boiling impurities are separated overhead in the first column. After removal of high-boiling impurities overhead, pure *n*-butyl alcohol is recovered by rectification of the bottoms in a second column.

If rhodium catalysts are used, then around 85–95% *n*-butyl alcohol is produced compared to 75% for cobalt catalysts. By modifying the operating conditions, the amount of isobutyl alcohol formed can be increased up to 50%.

Reaction

$$C_3H_6 + CO + H_2 \rightarrow C_3H_7CHO$$

$$C_3H_7CHO + H_2 \rightarrow C_4H_9OH$$

Raw material requirements and yield

Raw materials required per tonne of *n*-butyl alcohol:

Propylene	920kg

Yield 95%

2. *n*-Butyl alcohol: From propylene by carbonylation (Reppe process)

Propylene, carbon monoxide and water are reacted in the aqueous phase in the presence of a catalyst (consisting of at least a 10% solution of carbonyl triferrate in a solvent such as N-alkylpyrrolidine or butyl alcohol). The catalyst is formed in the reactor by the interaction of iron pentacarbonyl, butyl pyrrolidine and water to give the butyl pyrrolidine salt of carbonyl triferrate as well as iron pentacarbonyl. *(See Figure 35)*

The reaction is carried out at 100°C and a pressure of 5–20 bar. Gases from the top of the reactor pass to a recycle column where butyl alcohol, unreacted feed and by-products formed during the reaction are separated from butyl pyrrolidine. The pressure is reduced and butyl alcohol, water and iron pentacarbonyl are condensed. In a separator, the condensate forms three phases. The bottom two containing water and catalyst are returned to the recycle column. The top phase of butyl alcohol with some by-product is fed into the carbonyl column where, under pressure, the remaining traces of catalyst are removed. Spent catalyst is removed from the recycle stream as iron carbonate, with loss made up by addition of iron pentacarbonyl.

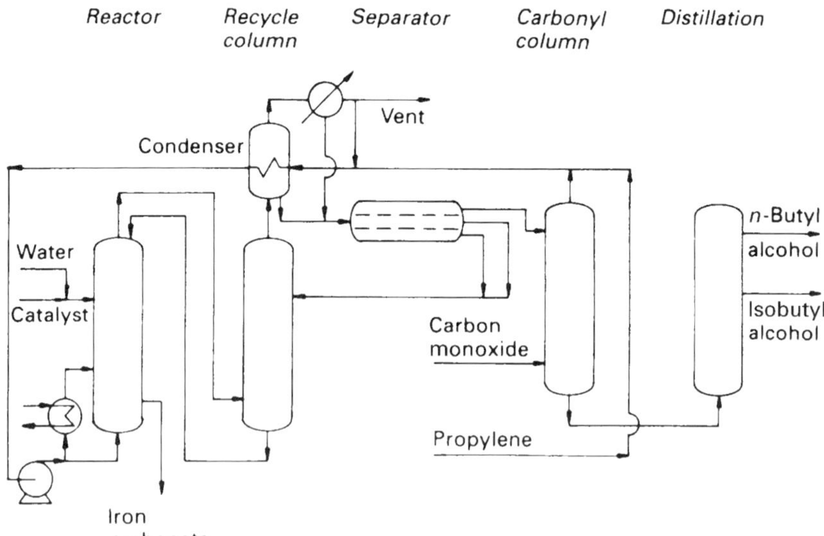

FIGURE 35 *n*-Butyl alcohol from propylene by carbonylation (Reppe process)

Crude butyl alcohol from the base of the column is dehydrated and separated into its isomers by fractionation. The ratio of *n*-butyl alcohol and isobutyl alcohol produced is 6:1.

Propylene and carbon monoxide (separated from the recycle gases from the recycle column) are returned to the reactor, while by-product carbon dioxide and other impurities are continuously vented.

Reaction

$$CH_3CH{=}CH_2 + 3CO + 2H_2O \rightarrow C_4H_9OH + 2CO_2$$

Raw material requirements and yield

Raw materials required per tonne of butyl alcohol:

Propylene	920kg

Yield 90%

3. *sec*-Butyl alcohol: From *n*-butene by hydration

sec-Butyl alcohol can be produced from a *n*-butene/butane mixture by esterification with sulphuric acid in a double reactor. (*See Figure 36*)

Butene feed, 75–80% sulphuric acid and recycled ester reaction product are fed countercurrently into the first reactor. The mixture is stirred and the reaction takes

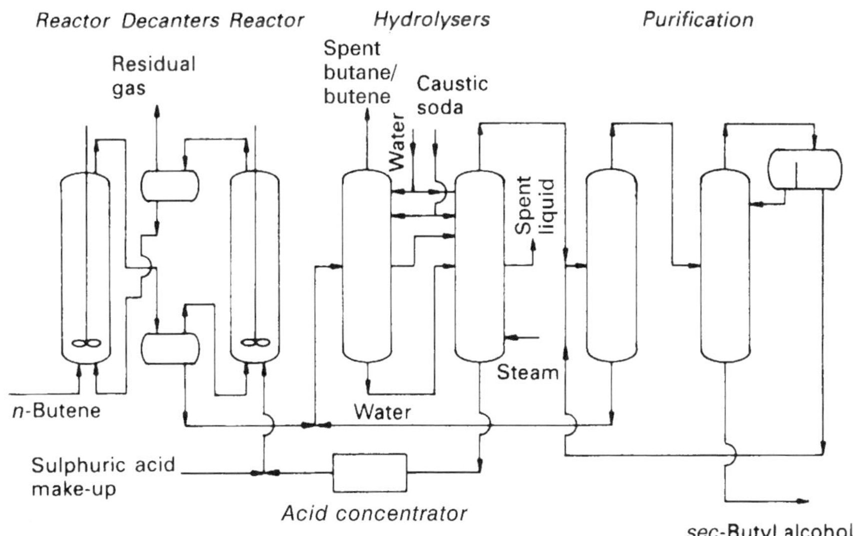

FIGURE 36 *sec*-Butyl alcohol from *n*-butene by hydration

place at around 40°C. Heat from the exothermic reaction is removed by water cooling. The mixture from the first reactor goes to a decanter, where the upper phase (consisting of unreacted butane/butene) is charged into a second stirred reactor together with further 75% concentrated sulphuric acid. The additional acid esterifies any unconverted butene from the first reactor.

The reaction products from the second reactor are fed into a second decanter, where the lower ester phase is separated out and recycled to the first reactor.

The ester mixture from the first decanter is hydrolysed, and vapours from the top of the hydrolysor containing crude butyl alcohol, are neutralized with caustic soda and cooled. Dilute sulphuric acid from the bottom of the hydrolysor is concentrated to 75% acid and recycled.

The crude alcohol stream is separated by azeotropic distillation, and the dry *sec*-butyl alcohol bottoms are purified by further distillation.

Reaction

$$CH_3CH{=}CHCH_3 + H_2SO_4 \rightarrow CH_3CH(OSO_3H)CH_2CH_3$$

$$CH_3CH(OSO_3H)CH_2CH_3 + H_2O \rightarrow CH_3CH(OH)CH_2CH_3 + H_2SO_4$$

Raw material requirements and yield

Raw materials required per tonne of *sec*-butyl alcohol:

Butene	893kg

Yield 85%

sec-Butyl alcohol: Other processes

A direct hydration process has been developed whereby *n*-butane and water react in the presence of a cationic exchange resin catalyst consisting of sulphonated polystyrene resin cross-linked with divinyl benzene. The optimum degree of cross-linking is in the range of 8–12%.

The reactants, preheated to 150–170°C and under a pressure of 50–70 bar, are passed through several catalyst beds, forming *sec*-butyl alcohol. After cooling, water is separated off and the *sec*-butyl alcohol remaining is purified by further distillation. Although less by-products are obtained by this route, it has not yet obtained wide acceptance as a competitor to the sulphuric acid method.

4. *tert*-Butyl alcohol: From isobutylene

tert-Butyl alcohol can be obtained from isobutylene containing streams with yields of 80–85%.

Isobutylene and water react in the presence of a catalyst consisting of a 65% solution of sulphuric acid. Some *tert*-butyl alcohol is added as a solvent. *tert*-Butyl alcohol can be produced by direct hydration using either tungsten oxide and an aqueous aluminium hydroxide gel suspension or acidic ion exchange sulphonic acid resin as the catalyst at a temperature of 190–245°C.

Unreacted olefins are separated by distillation overhead. *tert*-Butyl alcohol is recovered from the remaining mixture by azeotropic distillation.

Reaction

$$CH_2{=}CHCH_2CH_3 + H_2O \rightarrow CH_3CH(OH)CH_2CH_3$$

Raw material requirements and yield

Raw materials required per tonne of *tert*-butyl alcohol:

Isobutylene	920kg

Yield 82–85%

PROPERTIES

n-*Butyl alcohol*

Colourless, mobile liquid with a mild odour. Slightly hygroscopic; oxidizes slowly in contact with air.

Molecular Weight	74.12
Density at 20°C	0.809
Melting Point	–89.7°C
Boiling Point	117.6°C
Autoignition Temperature	380°C
Flammability Limits in air	
lower	1.4vol%
upper	11.3vol%
Flash Point Closed Cup	28.9°C
Vapour Density (air = 1)	2.6
Exposure Limit COSHH	50ppm 15 minutes (can be absorbed through skin)
Exposure Limit ACGIH	50ppm TLV-STEL (skin) (Ceiling Limit)
	75ppm TLV-TWA

Isobutyl alcohol

Colourless, mobile liquid with slightly sweetish odour. Soluble in ether and ethyl alcohol. Slightly soluble in water.

Molecular Weight	74.12
Density at 20°C	0.802
Melting Point	–108°C
Boiling Point	108°C
Autoignition Temperature	430°C
Flammability Limits in air	
lower	1.7vol%
upper	10 9vol%
Flash Point Closed Cup	28°C

Vapour Density (air = 1)	2.55
Exposure Limit COSHH	75ppm 15 minutes 50ppm 8 hour TWA
Exposure Limit ACGIH	75ppm TLV-STEL 50 ppm TLV-TWA

sec-*Butyl alcohol*

Colourless, mobile liquid with a slight peppermint odour. Soluble in ethyl alcohol.

Molecular Weight	74.12
Density at 20°C	0.807
Melting Point	–114.7°C
Boiling Point	99.5°C
Autoignition Temperature	390°C
Flammability Limits in air	
lower	1.7vol%
upper	10 9vol%
Flash Point Closed Cup	24°C
Vapour Density (air = 1)	2.55
Exposure Limits COSHH	150ppm 150 minutes 100 ppm 8 hour TWA
Exposure Limit ACGIH	150ppm TLV-STEL 100ppm TLV-TWA

tert-*Butyl alcohol*

White, hygroscopic, crystalline solid. Soluble in water and ethyl alcohol.

Molecular Weight	74.12
Density at 20°C	0.787
Melting Point	25.8°C
Boiling Point	82.5°C
Autoignition Temperature	470°C
Flammability Limits in air	
lower	2.4vol%
upper	11.1vol%
Flash Point Closed Cup	11°C
Vapour Density (air = 1)	2.55
Exposure Limit COSHH	150ppm 10 minutes 100ppm 8 hour TWA
Exposure Limit ACGIH	150ppm TLV-STEL 100ppm TLV-TWA

Classified A4, not classified as a human carcinogen.

GRADES

Technical 99%.

INTERNATIONAL CLASSIFICATIONS

n-*Butyl alcohol*

UN No.	1120
CAS Reg No.	71–36–3
EINECS No.	200–751–6

EC No.	603–004–00–6
Description	Flammable liquid
Packing Group	
– flash point less than 23°C	II
Emergency Action Code	3YE
HI (Kemler Code)	33
Packing Group	
– flash point 23°C to 61°C	III
Emergency Action Code	3Y
HI (Kemler Code)	30

isobutyl alcohol

UN No.	1212
CAS Reg No.	78–83–1
EINECS No.	201–148–0
EC No.	603–004–00–6
Description	Flammable liquid
Packing Group	III
Emergency Action Code	3Y
HI (Kemler Code)	30

sec-*Butyl alcohol*

UN No.	1121
CAS Reg No.	78–92–2
EINECS No.	201–158–5
EC No.	603–004–00–6
Description	Flammable liquid
Packing Group	III
Emergency Action Code	3Y
HI (Kemler Code)	30

tert-*Butyl alcohol*

UN No.	1122
CAS Reg No	75–65–0
EINECS No.	200–889–7
EC No.	603–005–00–1
Description	Flammable liquid
Packing Group	II
Emergency Action Code	3YE
HI (Kemler Code)	33

APPLICATIONS

The principal use for *n*-butyl alcohol is as a solvent either directly or in mixtures with esters, toluene, or ethyl alcohol. Butyl alcohol/butyl acetate mixtures are excellent solvents for nitrocellulose lacquers and coatings. They are frequently used as diluents or thinners because of the improvement in flow properties of varnishes and the reduction of streaking in paints.

Butyl glycol ethers, produced by the reaction of *n*-butyl alcohol with ethylene oxide, are used in vinyl and acrylic paints and lacquers, and to solubilize organic surfactants in surface cleaners.

Butyl acrylate and methacrylate are the most important commercial derivatives of *n*-butyl alcohol. They are used in emulsion polymers for latex paints, in textile manufacturing and in impact modifiers for rigid polyvinyl chloride (PVC).

Butyl esters of higher acids (such as phthalic, adipic and stearic) are employed as plasticizers and surface-coating additives. Other outlets for *n*-butyl alcohol include a diluent for brake fluids and in acrylic fibre spinning.

Isobutyl alcohol can be used as a substitute for *n*-butyl alcohol and its lower price and less pronounced odour makes it an attractive alternative additive for nitrocellulose lacquers, paints, printing inks and cleaning fluids. It is an excellent process solvent widely employed in the pharmaceutical, fragrance and pesticide industries.

It forms esters with higher alcohols which are employed as plasticizers, especially for PVC and its copolymers. Isobutyl alcohol can be used as an antifreeze in gasoline.

Almost all *sec*-butyl alcohol is converted to methyl ethyl ketone which has excellent solvent properties. Some *sec*-butyl alcohol is used in hydraulic fluids, flotation solutions, industrial cleaning compounds and paint removers. It is a good extractant for oils, essences, perfumes and dyes.

tert-Butyl alcohol is used in the preparation of artificial musk – a constituent of some perfumes. It can be dehydrated to isobutylene, a feedstock utilized in the manufacture of methyl methacrylate and methyl *tert*-butyl ether (MTBE). Glycol ethers, obtained by the interaction of *tert*-butyl alcohol with propylene oxide, are employed in industrial coatings and in aqueous cleaners containing organic surfactants.

HEALTH AND HANDLING

Butyl alcohols are irritating to the eyes leading to conjunctivitis; they defat skin causing cracking. They are toxic on inhalation. Neoprene gloves, aprons and protective clothing to avoid splashing should be worn. Contact lenses should be avoided.

The product should be stored in closed mild steel or enamelled steel containers in approved safety cabinets or storage areas which are cool, explosion-proof and well ventilated. All equipment used must be earthed to avoid static build-up. Oxidizing agents must be avoided as butyl alcohols will explode if brought into contact with hypochlorous acid, chlorine or peroxides. Special transport and shipping regulations apply to the movement of butyl alcohols, and normally stainless steel containers are used.

Spills should be absorbed with sand or vermiculite, placed in containers and disposed of in accordance with local regulations.

Carbon dioxide, alcohol foam or water fog can be used to fight fires. Because butyl alcohol vapour is heavier than air, flashback can occur. Firefighting and clean-up staff must wear protective clothing and respirators.

MAJOR PLANTS

n-*Butyl alcohol*

Capacities greater than 80 000 tonnes per year:

BASF	Ludwigshafen	Germany
Hoechst	Oberhausen	Germany
Huels	Marl	Germany
Eastman Chemical	Longview	US
Hoechst Celanese	Bay City	US
Union Carbide	Taft	US
	Texas City	US
Kyowa Yuka	Yokkaichi	Japan

Isobutyl alcohol

Capacities greater than 30 000 tonnes per year:

Oxochimie	Lavera	France
BASF	Ludwigshafen	Germany
Hoechst	Oberhausen	Germany
Huels	Marl	Germany
Eastman Chemical	Longview	US
Union Carbide	Texas City	US

All capacities are flexible depending on the ratio of *n*- and iso- aldehydes in the feedstock used.

sec-*Butyl alcohol*

Capacities greater than 45 000 tonnes per year:

Elf Atochem	Notre Dame de Gravenchon	France
CONDEA Chemical	Moers	Germany
Shell Nederland Chemie	Pernis	Netherlands
Exxon Chemical	Fawley	UK
Exxon Chemical	Bayway	US
Shell Chemical	Norco	US
Maruzen Petrochemical	Ichihara	Japan
Tonen Chemical	Kawasaki	Japan
TASCO Chemical	Linyuan	Taiwan

tert-*Butyl alcohol*

Capacities greater than 45 000 tonnes per year:

ARCO Chimie	Fos	France
ARCO Chemie Nederland	Rozenburg	Netherlands
Lyondell	Bayport	US
Texaco Refining	Port Neches	US
Kuraray	Kashima	Japan
Mitsui Petrochemical Industries	Ichihara	Japan

Capacities are flexible as the oxo process produces a wide range of products.

MAJOR LICENSORS

n-, isobutyl alcohol	*BASF*
	Hoechst
	Huels
	Rhone-Poulenc
	Shell Development
	Union Carbide
sec-Butyl alcohol	*Edeleanu*
	Huels/UOP
	Idemitsu Petrochemical
tert-Butyl alcohol	*Arco Technology*
	Huels/UOP

Caprolactam

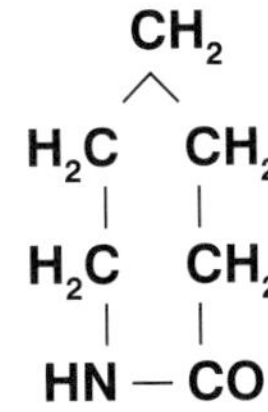

SYNONYMS

CAPROLACTAM aminocaproic lactam, 1,6-hexanolactam, hexahydro-2H-azepin-2-one, 2-oxohexamethyleneimine, nylon 6 monomer, ketohexamethylen-imine

All commercial processes for the manufacture of caprolactam start with cyclohexane, phenol or toluene. They are multi-stage processes in which ammonium sulphate and other by-products are formed.

Whichever route to cyclohexanone is chosen – the hydrogenation of phenol followed by dehydrogenation of the cyclohexanol formed, or the catalytic oxidation of cyclohexane – cyclohexanone is the key intermediate. The addition of hydroxylamine sulphate to cyclohexane gives cyclohexanone oxide which, followed by a Beckmann rearrangement, yields caprolactam.

The major problem with these processes has been the formation of substantial quantities of ammonium sulphate by-product, over four tonnes for each tonne of caprolactam produced. Since the cost-efficiency of the process depends on the value of the by-products to a large extent, recent efforts have concentrated on finding alternative routes which reduce or avoid this problem.

Various methods have been proposed and the most important include:

- reaction of cyclohexane with hydroxylammonium nitrate;
- Beckmann rearrangement using polyphosphoric acid;
- neutralization of the products of the Beckmann rearrangement with metal oxides and subsequent decomposition to recover sulphur dioxide;
- extraction of caprolactam with a solvent;
- decomposition of ammonium sulphate to ammonia and sulphur dioxide.

Caprolactam can be recovered from nylon 6 waste by depolymerization using superheated steam in the presence of an aluminium oxide catalyst.

A new process to manufacture cyclohexane from butadiene via adiponitrile is being developed by Du Pont and BASF. It has not as yet been commercialized.

Around 80% of caprolactam is produced from cyclohexane, 10% from phenol, 7% from toluene, and the remainder from nylon waste.

Plant sizes range from 20 000 to 320 000 tonnes per year.

PROCESSES

1. From cyclohexanone

Hydroxylamine sulphate is produced by the oxidation of ammonia to nitrous oxide, followed by hydrogenation over a platinum catalyst in the presence of sulphuric acid. The addition of hydroxylamine sulphate to cyclohexanone in a weak acidic solution at 85–90°C produces cyclohexanone oxime. (*See Figure 37*)

After separation from the aqueous solution, anhydrous oxime is converted to caprolactam by heating to 100–120°C with 27% oleum. In order to control the exothermic reaction, it is carried out in the presence of excess already-rearranged product. The reaction mixture is diluted with water and then neutralized with ammonia.

The caprolactam formed is separated from the ammonium sulphate solution and extracted with toluene or benzene. Pure caprolactam is obtained by distillation of the solvent under reduced pressure. Ammonium sulphate is recovered from the reaction mixture by evaporation and crystallization.

Ammonium sulphate by-product production has been reduced in the newer processes developed by DSM and BASF. In the DSM process, the oxime reaction is carried out in a hydroxylamine–phosphoric acid buffer solution. This solution is reduced with hydrogen in the presence of palladium on graphite or alumina as the

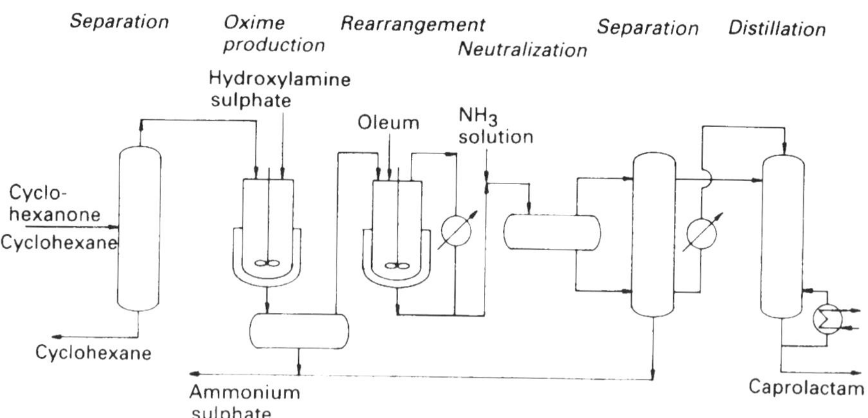

FIGURE 37 Caprolactam from cyclohexanone

catalyst. After oxime formation with cyclohexanone and its separation, the excess ammonium ions are replaced by the oxides of nitrogen.

In the BASF process, the oxime is formed by reacting cyclohexane with ammonium hydroxyl ammonium sulphate. The ammonium hydrogen sulphate formed is recycled to the hydroxylamine unit where it reacts with nitric acid over a platinum or graphite catalyst to form ammonium hydroxyl ammonium sulphate.

Reaction

$$C_6H_{10}O + NH_2OH \rightarrow C_6H_{10}{=}NOH + H_2O$$

$$C_6H_{10}{=}NOH \xrightarrow[\text{oleum}]{} C_6H_{11}NO.H_2SO_4$$

$$C_6H_{11}.H_2SO_4 + 2NH_3 \rightarrow NH(CH_2)_5{=}O + (NH_4)_2SO_4$$

Raw material requirements and yield

Raw materials required per tonne of caprolactam:

Cyclohexanone	890kg
Oleum (100%)	1100kg
Ammonia	900kg

Yield 90%

2. From toluene

Toluene and air are fed into a reactor, where the liquid-phase oxidation takes place at 165–170°C and 8–10 bar pressure using a cobalt catalyst. Gases from the top of the reactor are cooled to 7–8°C and any unreacted toluene is recovered and recycled. (*See Figure 38*)

The reaction mixture (containing 30% benzoic acid in toluene, catalyst, various intermediates and by-products) is fractionated. Recovered toluene and reaction intermediates collected overhead are returned to the reactor.

Benzoic acid is obtained from a side stream, and the bottom fraction of heavy by-products are left as a residue from which the cobalt catalyst is reprocessed. The hydrogenation of benzoic acid is carried out in a series of stirred tank reactors at 170°C and a pressure of 10–17 bar in the presence of a palladium-on-graphite catalyst.

Conversion is almost complete. The catalyst is separated by centrifuging from the liquid reaction product and reused. The cyclohexane carboxylic acid formed is purified by distillation under reduced pressure and blended with oleum before being introduced into a multi-stage nitrozation reactor. A 73% solution of nitrosyl-

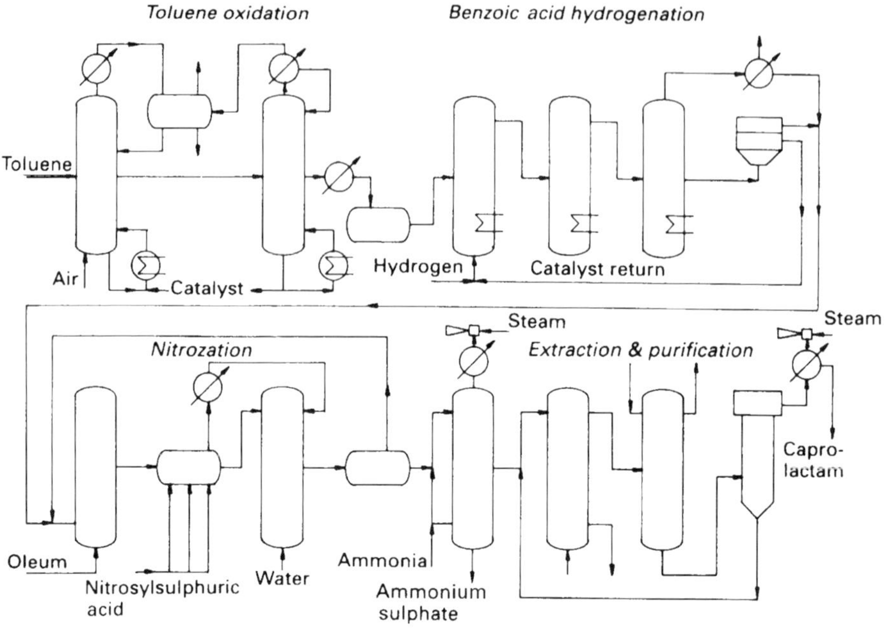

FIGURE 38 Caprolactam from toluene

sulphuric acid in sulphuric acid is added, the quantity being sufficient to carry out the nitrozation.

Heat generated by the exothermic reaction is removed by carrying out the nitrozation in boiling cyclohexane at atmospheric pressure. The reaction product is hydrolysed with water. Any unreacted cyclohexane carboxylic acid is extracted by the cyclohexane present for subsequent recycle.

The caprolactam solution, diluted with water to give a 50% acid concentration, is extracted with a countercurrent of alkylphenol. Any residual acid is neutralized by treatment with water and alkali prior to fractionation to separate out the caprolactam.

Sulphuric acid in the aqueous extract is thermally cracked and the sulphur dioxide gases are recycled. No ammonium sulphate and no impurities are formed in this process, thus avoiding any pollution problems.

Reaction

$$C_6H_5CH_3 + 1\tfrac{1}{2}O_2 \rightarrow C_6H_5COOH + H_2O$$

$$C_6H_5COOH + 3H_2 \rightarrow C_6H_{11}COOH$$

$$N_2O_3 + H_2SO_4 + SO_3 \rightarrow 2NOHSO_4$$

$$C_6H_{11}COOH + NOHSO_4 \rightarrow NH(CH_2)_5CO + H_2SO_4 + CO_2$$

Raw material requirements and yield

Raw materials required per tonne of caprolactam:

Toluene	960kg
Ammonia	200kg
Hydrogen	800Nm^3

Yield 90%

3. From cyclohexane by photooximation

Nitrosyl chloride, prepared by reacting nitrosylsulphuric acid (produced from nitrous gases obtained by the combustion of sulphuric acid and ammonia) with hydrogen chloride, is bubbled through cyclohexane liquid in a photoreactor. The reaction takes place in the presence of actinic light from 60kW mercury lamps immersed in the mixture. Heat produced by the lamps is removed using cooling water. Unreacted nitrosyl chloride is withdrawn from the photoreactor and recycled. (*See Figure 39*)

The crude cyclohexane oxime dichloride formed is separated out as a heavy oil at the bottom of the reactor. Following treatment with oleum and the Beckmann rearrangement, the product is cooled and neutralized with ammonia to yield an aqueous solution of caprolactam and ammonium sulphate.

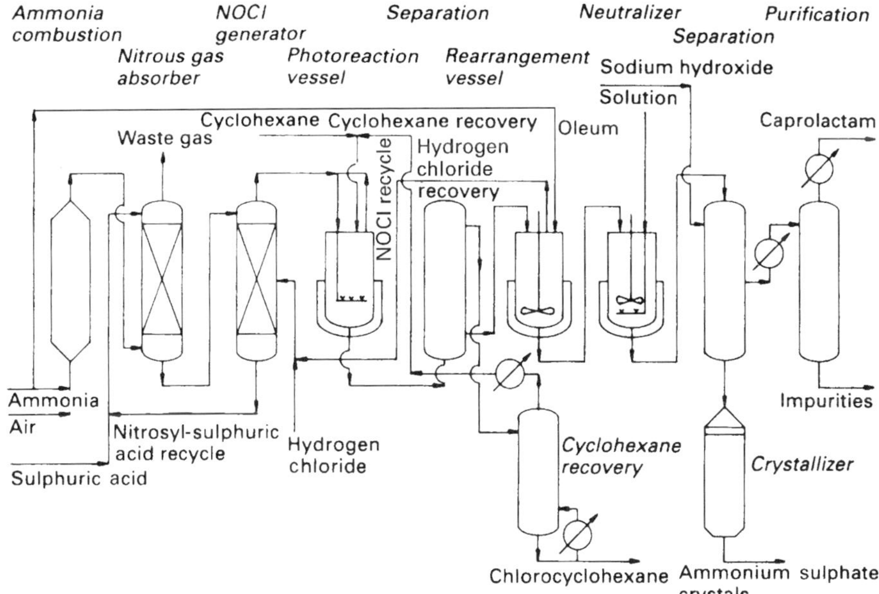

FIGURE 39 Caprolactam from cyclohexane by photooximation

Reaction

$2H_2SO_4 + NO + NO_2 \rightarrow 2NHOSO_4 + H_2O$ nitrosyl chloride

$NHOSO_4 + HCl \rightarrow NOCl + H_2SO_4$ preparation

$$C_6H_{10} + NOCl \xrightarrow{HCl} C_6H_{10}{=}NOH.2HCl$$

$$C_6H_{10}{=}NOH.2HCl + H_2SO_4 \rightarrow C_6H_{11}{=}NO.H_2SO_4 + 2HCl$$

$$C_6H_{10}{=}NO.H_2SO_4 + 2NH_3 \rightarrow NH(CH_2)_5CO + (NH_4)_2SO_4$$

Raw material requirements and yield

Raw materials required per tonne of caprolactam:

Cyclohexane	910kg
Yield 80%	

1550kg ammonium sulphate is produced as by-product.

PROPERTIES

White, crystalline, hygroscopic solid with distinctive odour. Soluble in water, cyclohexane, and chlorinated hydrocarbons.

Molecular Weight	113.2
Density at 20°C	0.998
Melting Point	69.2°C
Boiling Point	174°C
Autoignition Temperature	375°C
Explosive Limits in air	
lower	1.4vol% at 135°C
upper	8.0vol% at 180.5°C
Flash Point Closed Cup	125°C
Vapour Density (air = 1)	3.91
Exposure Limit COSHH	3mg/m³ 15 minutes STEL (dust)
	1mg/m³ 8 hour TWA (dust)
	10ppm 15 minutes STEL (vapour)
	5ppm 8 hour TWA (vapour)
Exposure Limit ACGIH	3mg/m³ TLV-STEL (dust)
	1mg/m³ TLV-TWA (dust)
	10ppm TLV-STEL (vapour)
	5ppm TLV-TWA (vapour)

GRADES

Flake, molten.

INTERNATIONAL CLASSIFICATIONS

UN No.	Not listed
CAS Reg No.	105–60–2
EINECS No.	203–313–2
EC No.	613–069–00–2

APPLICATIONS

Around 80% of caprolactam is used for the manufacture of nylon 6 fibres, and the remainder for nylon 6 resins and other copolymers. Nylon 6 fibres are used for carpets, textiles and tyres.

The market for caprolactam is growing at 2–3% per year, and its future is entirely dependent on the demand for nylon 6, its only outlet.

The fastest growth in capacity has taken place in the Asia–Pacific region where, by 1997, production accounted for over 30% of world capacity.

HEALTH AND HANDLING

Dust can cause irritation to the eyes, lungs and skin. Good ventilation and protective clothing and goggles must be provided.

Caprolactam is stable at room temperature and can be shipped either in drums or in the molten state. Steel is recommended for storage. Caprolactam must be stored in a well-ventilated, cool area away from strong oxidizing agents.

Spills should be covered with sand or similar absorbent, avoiding dust generation. Waste can be collected into containers for disposal or mixed with a flammable solvent and incinerated. Care must be taken not to generate dust. Wash the contaminated area with soap and water. Waste material must not be allowed to get into waterways or sewers.

Caprolactam is a slight fire hazard when exposed to heat or flame. In the event of fire, evacuate the area and extinguish all sources of ignition. Fires can be put out with carbon dioxide, dry chemical, alcohol foam or water fog but not water jets which would disperse the material. Firefighters must wear protective clothing and self-contained breathing apparatus as toxic fumes are given off during burning.

MAJOR PLANTS

Plants with capacities greater than 95 000 tonnes per year:

BASF	Antwerp	Belgium
Bayer	Antwerp	Belgium
BASF	Ludwigshafen	Germany
DOMO Group	Leuna	Germany
DSM	Geleen	Netherlands
EniChem	Manfredonia	Italy
	Porto Marghera	Italy
Complex	Kemerova	Russia
AlliedSignal	Hopewell	US
BASF	Freeport	US
DSM	Augusta	US
Mitsubishi Chemical	Kurosaki	Japan
Toray Industries	Nagoya	Japan
Ube Industries	Sukai	Japan
	Ube	Japan
Chung Tal Chemical	Taipei	Taiwan
Formosa Chemical & Fibre	Shin Yuan	Taiwan

MAJOR LICENSORS

Toluene	*Snia Engineering*
Cyclohexane	*BASF*
	Stamicarbon
	Inventa
	Toray Industries
Phenol	*Allied Signal*
Butadiene	*Du Pont/BASF*

Carbon Tetrachloride CCl_4

SYNONYMS

CARBON TETRACHLORIDE perchloromethane, tetrachloromethane, benzinoform, methane tetrachloride

The original route for the manufacture of carbon tetrachloride was by the chlorination of carbon disulphide. Because no by-product hydrogen chloride is formed, it is still employed in a few plants. This route was superseded in the 1950s by the high-temperature chlorination of methyl alcohol or methane which gives a mixture of chloromethanes. High yields of carbon tetrachloride can be achieved by recycling and using an excess of chlorine. In an alternative route, carbon tetrachloride is obtained as a by-product by the chlorination of propane.

Residues from chlorinated processes and chlorinated hydrocarbons are being increasing used as a source of carbon tetrachloride in order to recover the expensive chlorine molecule.

However, concern over the depletion of the ozone layer has led to uncertainty over future demand and a consequent decline in process development and new plant construction.

Plants range in size from 5000 to 110 000 tonnes per year.

PROCESSES

1. From methane by chlorination

Methane of high purity and excess chlorine are mixed and preheated before entering the reactor. In the presence of a nickel catalyst, a range of chlorinated methanes is produced. If carbon tetrachloride is the desired end product, then the lower-boiling methyl chloride, methylene chloride and chloroform pass to a second reactor where more chlorine is added. By employing a number of reactors in series and an excess of chlorine, almost all the methane can be fully chlorinated. The mixture of

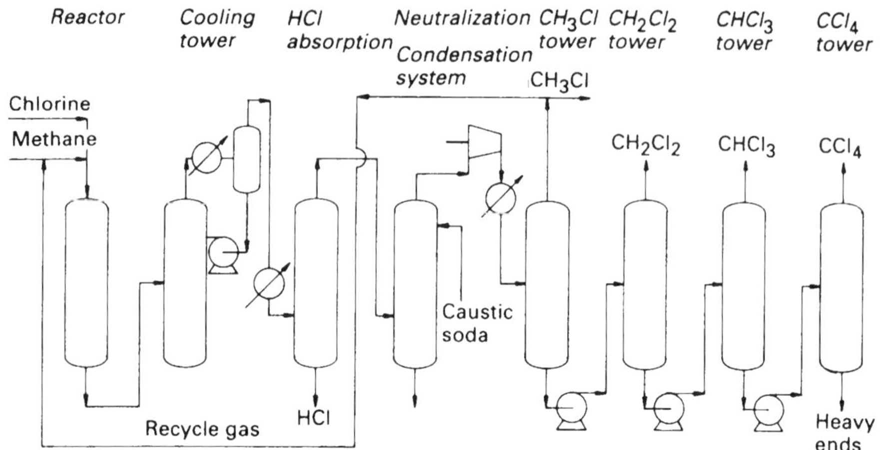

FIGURE 40 Carbon tetrachloride from methane by chlorination

chloromethanes produced is separated by distillation. In many plants a range of products is marketed (see Chloroform). (*See Figure 40*)

Reaction

$CH_4 + 4Cl_2 \rightarrow CCl_4 + 4HCl$

Raw material requirements and yield

Raw materials required per tonne of carbon tetrachloride:

Methane	110kg

Yield 95%

2. From carbon disulphide and chlorine

Carbon disulphide is dissolved in a solution of carbon tetrachloride containing ferrous or ferric chloride as a catalyst and fed into a reactor maintained at a temperature of 30°C by a cooling jacket. Chlorine gas is bubbled into the liquid and the reaction products pass overhead and are distilled. Crude carbon tetrachloride containing small amounts of disulphur dichloride is recovered. Chlorine is present in excess to ensure complete conversion. The carbon tetrachloride formed is purified by treatment with dilute alkali to hydrolyse any sulphur compounds and dried by azeotropic distillation. Any residual sulphur compounds are removed by air stripping. (*See Figure 41*)

Residue by-product disulphur dichloride is cycled to a second reactor where carbon disulphide is present in excess. The liquids are agitated and kept at a temperature of 60°C. Sulphur formed during the reaction separates out as a solid on cool-

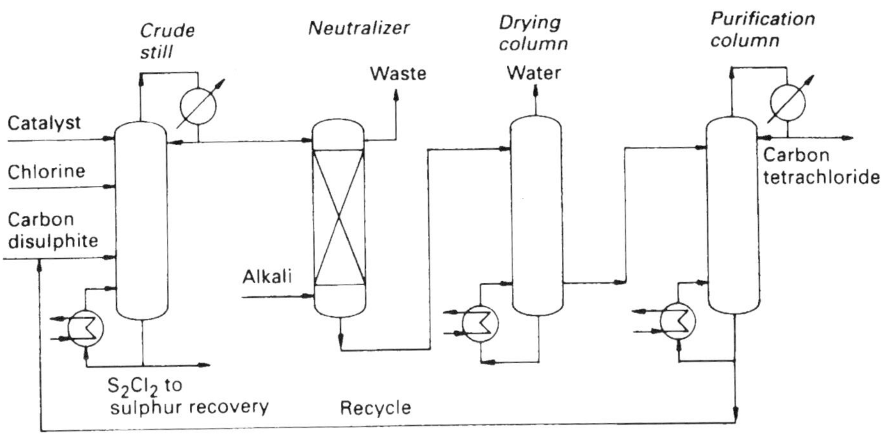

FIGURE 41 Carbon tetrachloride from carbon disulphide and chlorine

ing and is removed. Carbon tetrachloride and excess carbon disulphide are obtained as a distillate overhead. The carbon tetrachloride is separated from excess carbon disulphide by distillation. Carbon disulphide is recycled to the chlorinator while the carbon tetrachloride is purified by treatment with dilute alkali before being dried and air stripped to remove any residual sulphur compounds. The sulphur can be recovered and converted to carbon disulphide by burning with coke.

Reaction

$2CS_2 + 6Cl_2 \rightarrow 2CCl_4 + 2S_2Cl_2$

$CS_2 + 2S_2Cl_2 \rightarrow CCl_4 + 6S$

Raw material requirements and yield

Raw materials required per tonne of carbon tetrachloride:

Carbon disulphide	550kg
Chlorine	1145kg

Yield

on carbon disulphide	90%
on chlorine	80%

3. From methyl chloride by chlorination

Methyl alcohol and hydrochloric acid are reacted over an aluminium oxide catalyst in the vapour phase to produce methyl chloride (see Methyl Chloride). Additional chlorine is added to the methyl chloride formed and, at a temperature of 500–700°C under high pressure, carbon tetrachloride is formed (see Chloroform).

Reaction

$CH_3Cl + 3Cl_2 \rightarrow CCl_4 + 3HCl$

OTHER PROCESSES

From chlorinated wastes by chlorinolysis

Carbon tetrachloride is produced as a by-product from the low-pressure chlorinolysis of hydrocarbons or chlorinated wastes to perchloroethylene. Chlorinated aliphatic hydrocarbons obtained as by-products from the chlorination of hydrocarbons such as methane, methyl alcohol, acetylene and ethylene can be used as the feedstock.

Liquid chlorine compressed to 240 bar is heated by high-pressure steam and mixed with the chlorinated hydrocarbon feedstock. The mixed gases are fed into a reactor which is heated electrically to start the chlorination. Excess chlorine is used to maintain the reaction temperature around 620°C. Cold carbon tetrachloride is used to quench the exit gases, initially to 500°C, and further cooling is carried out by reducing the pressure to around 20 bar. The reaction mixture (containing hydrogen chloride, chlorine, carbon tetrachloride, perchloroethylene and other chlorinated hydrocarbons) is separated by distillation in four separate columns.

PROPERTIES

Colourless, mobile liquid with a sweetish odour. Non-flammable.

Molecular Weight	153.8
Specific Gravity at 20°C	1.585
Melting Point	–22.6°C
Boiling Point	76.78°C
Vapour Density (air = 1)	5.32
Exposure Limit COSHH	2ppm 8 hour TWA
Exposure Limit ACGIH	10ppm TLV-STEL
	5ppm 8 hour TLV-TWA

Classified A3, animal carcinogen.

GRADES

Technical 99%, pure 99.9%.

INTERNATIONAL CLASSIFICATIONS

UN No.	1846
CAS Reg No.	56–23–5
EINECS No.	200–262–8
EC No.	602–008–00–5
Description	Toxic substance
Packing Group	II
Emergency Action Code	2Z
HI (Kemler Code)	60

APPLICATIONS

No longer used as a dry cleaning agent for clothes, carbon tetrachloride's major outlet is for the manufacture of trichlorofluoromethane (Fluorocarbon 11), and dichlorodifluoromethane (Fluorocarbon 12), which are used as refrigerants and aerosol propellants. Because of concern over the pollution of the atmosphere and the 'greenhouse effect', they have been replaced by less toxic materials in many countries. Despite a ban by the European Union on the use of CFCs in the domestic market, ample supplies are still available.

Carbon tetrachloride is used as a speciality solvent and foam-blowing agent. It can be used as a fungicide for grain. Future demand is expected to either remain static or decline as alternative less toxic products are employed.

HEALTH AND HANDLING

Carbon tetrachloride vapour is toxic and at low concentrations is irritating to the eyes, nose and throat. At moderate concentrations, the vapour rapidly leads to unconsciousness and injury to the liver and kidneys. Carbon tetrachloride is absorbed by the skin so that protective clothing and goggles must be worn by personnel handling the product. Sensitivity can develop with prolonged exposure and carbon tetrachloride is also a suspected carcinogen in the US. The product's toxicity is increased by the synergistic effects of alcohol.

Store in closed containers made of iron or steel; aluminium, copper and lead should be avoided. The area should be well ventilated and away from sources of heat and sunlight. On exposure to moisture and light, hydrochloric acid is liberated which can cause corrosion.

Spills should be contained, the liquid absorbed with paper or vermiculite and then disposed of according to local regulations. Carbon tetrachloride is not flammable, but at high temperatures, phosgene and other toxic gases can be produced. It will explode if brought into contact with sodium, potassium, lithium, powdered magnesium, aluminium or fluorine. Firefighters and clean-up staff must wear protection against skin contact and inhalation of the vapour. Strict conditions apply to the filling, storage, handling, transportation and labelling of carbon tetrachloride.

MAJOR PLANTS

Plants with capacities greater than 30 000 tonnes per year:

Solvay	Jemeppe	Belgium
Elf Atochem	St Auban	France
Dow Chemical	Stade	Germany
EniChem	Porto Marghera	Italy
ERKIMIA	Flix	Spain
Akzo Chemicals	Le Moyne	US
Dow Chemical	Pittsburg	US
	Plaquemine	US
Vulcan Chemicals	Geismar	US
Shin-Etsu Chemical	Naoetsu	Japan

Most plants produce a mixture of chloromethanes.

MAJOR LICENSORS

ABB Lummus Global
Asahi Glass
Braun
Halcon-Scientific Design
Hoechst
Huels
Rhone-Poulenc
Solvay
Stauffer Chemical
Vulcan Materials

Chlorobenzene

SYNONYMS

CHLOROBENZENE monochlorobenzene, chlorobenzol, phenyl chloride

Most monochlorobenzene is produced from benzene by chlorination in the vapour or liquid phase. Chlorination can be either direct with chlorine in the presence of ferric chloride or by oxychlorination with hydrogen chloride using promoted copper oxide or copper chloride as the catalyst.

Monochlorobenzene, ortho-dichlorobenzene and para-dichlorobenzene are formed during the process, the proportions of each depend on the ratio of raw materials, process conditions and catalyst used.

Other benzene chlorination processes which have been developed include the electrolysis of benzene with hydrogen chloride, the oxychlorination of benzene with hydrogen chloride in the aqueous phase, and the use of chlorine-containing compounds. None are in commercial operation.

Most monochlorobenzene is produced by the continuous chlorination of benzene in the liquid phase.

Capacities range from 15 000 to 145 000 tonnes per year.

PROCESSES

1. From benzene by chlorination

In the liquid-phase process, dry benzene is introduced into a glass-lined reactor containing anhydrous ferric chloride as a catalyst and chlorine is bubbled into the mixture. In order to favour the production of monochlorobenzene, the ratio of benzene to chlorine is kept low. The reaction is allowed to proceed with vigorous agitation at a temperature of 80–100°C and atmospheric pressure. The monochlorobenzene formed is removed as quickly as possible to reduce the quantities

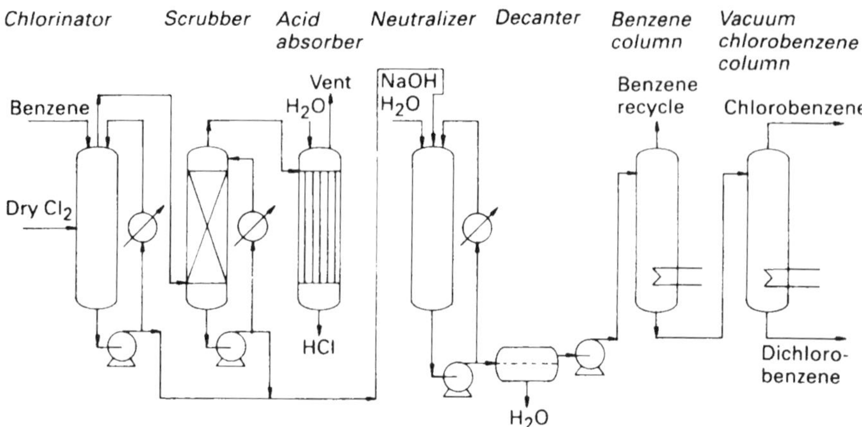

FIGURE 42 Chlorobenzene from benzene by chlorination

of dichlorinated by-products. Heat from the highly exothermic reaction is removed by external heat exchangers. (*See Figure 42*)

During the reaction hydrogen chloride is evolved. The vapour is scrubbed with benzene and then absorbed with water or oil, from which the acid can be recovered if desired.

The liquid phase can either be neutralized with dilute sodium hydroxide and the chlorinated benzenes allowed to separate out from the aqueous layer, or it can be distilled directly. The chlorobenzene layer or liquid phase is fractionated and monochlorobenzene is isolated from the dichlorobenzenes. Any unreacted benzene is dried and recycled to the reactor. Residues containing the iron catalyst are removed as waste.

The chlorides of aluminium, tin or manganese have been used as alternative catalysts for the reaction. As well as causing severe corrosion, the presence of water inactivates the catalyst and excess chlorine can lead to local overheating.

Reaction

$C_6H_6 + Cl_2 \rightarrow C_6H_5Cl + HCl$

Raw material requirements and yield

Raw materials required per tonne of monochlorobenzene:

Benzene	960kg
Chlorine	860kg

Yield 80%

15% of para-dichlorobenzene and 5% of ortho-dichlorobenzene are additionally formed.

OTHER PROCESSES

Monochlorobenzene can be produced by the reaction of benzene with hydrogen chloride in the presence of air. The vapour-phase process is carried out at a temperature of 220–300°C and a pressure between atmospheric and 6 bar in the presence of a catalyst, which can be copper oxide promoted by cobalt oxide and aluminium oxide, or copper chloride promoted by ferric chloride and aluminium oxide. About 6% of dichlorobenzenes are produced as by-products.

The process is uneconomic both because of the high cost of corrosion-resistant plant, and the low conversion rate required in order to control the temperature of the solid catalyst bed. Yields of 95% are obtained.

Reaction

$$2C_6H_6 + 2HCl + O_2 \rightarrow 2C_6H_5Cl + 2H_2O$$

PROPERTIES

Colourless, mobile, volatile liquid with an almond odour. Soluble in benzene and ethyl alcohol but insoluble in water.

Molecular Weight	112.56
Density at 20°C	1.106
Melting Point	–45.6°C
Boiling Point	132.1°C
Autoignition Temperature	638°C
Flammability Limits in air	
lower	1.3vol%
upper	7.1vol%
Flash Point Closed Cup	29°C
Vapour Density (air = 1)	3.9
Exposure Limit COSHH	50ppm 8 hour TWA
Exposure Limit ACGIH	10ppm TLV-TWA

GRADES

Technical 99%.

INTERNATIONAL CLASSIFICATIONS

UN No.	1134
CAS Reg No.	108–90–7
EINECS No.	203–628–5
EC No.	602–033–00–1

Description	Flammable liquid
Packing Group	III
Emergency Action Code	2Y
HI (Kemler Code)	30

APPLICATIONS

The major outlet for monochlorobenzene, which consumes around 65% of total demand, is for nitrobenzene production. The other market for monochlorobenzene – phenol manufacture – has declined due to its replacement by cumene. Other uses include the manufacture of dyes, pharmaceuticals and as a solvent.

Monochlorobenzene is a mature product and only modest growth can be expected to 2002.

HEALTH AND HANDLING

Monochlorobenzene is moderately toxic by inhalation as well as being absorbed by the skin. It is irritating to the nose, throat and eyes and is a strong narcotic. Protective clothing and eye protection must be worn when handling the product.

Store in closed containers in a separate, cool, well-ventilated area, away from sunlight, sources of heat and oxidizing agents. Monochlorobenzene is stable to air, light and moisture at room temperature. Because of its flammability, all equipment must be non-sparking and earthed to prevent the build-up of static.

Spills should be absorbed with paper or sawdust and placed in a container for disposal using spark-proof tools. Operatives should wear protective clothing and goggles to avoid skin contact.

Fires can be extinguished with carbon dioxide, dry chemical, foam or water spray. Because of the dense vapour, flashback is a hazard. Hydrochloric acid and carbon monoxide gases are given off on burning and firefighters must wear full protective clothing, eye protection and self-contained breathing apparatus.

MAJOR PLANTS

Plants with capacities greater than 22 000 tonnes per year:

Elf Atochem	Jarrie	France
Bayer	Leverkusen	Germany
EniChem	Pieve Vergonte	Italy
Monsanto Chemicals	Sauget	US
PPG Industries	Natrium	US

Standard Chlorine	Delaware	US
Mitsui Toatsu Chemical	Omuta	Japan

MAJOR LICENSORS

Dow Chemical
Gulf Oil Chemicals
Nippon Shokubai Chemical
PPG Industries
Rhone-Poulenc
Union Carbide

Chloroform

$CHCl_3$

SYNONYMS

CHLOROFORM trichloromethane, formyl trichloride, methyl trichloride

The first industrial processes for the production of chloroform were the reaction of chlorine on ethyl alcohol or acetaldehyde followed by treatment of the chloral formed with calcium hydroxide, and the reaction of bleaching powder on acetone.

Newer processes employing methane or methyl alcohol as the raw material have superseded these earlier routes. Both processes result in a mixture of chloromethanes, but the ratio of products can be regulated by the molar ratio of methane or methyl chloride to chlorine employed. By-product formation can be minimized by the choice of reaction conditions.

Although methane can be chlorinated thermally, catalytically or photochemically, the first is the preferred commercial procedure. The methane route does, however, suffer from the problem of disposing of the by-product, hydrogen chloride.

In an attempt to avoid this difficulty, a two-step process was introduced in which methane is chlorinated and then oxychlorinated. The major advantage of this method is the lack of waste streams as all by-products can be recycled.

Chloromethanes can be obtained from chlorine residues and waste chlorinated hydrocarbons from vinyl chloride manufacture by treatment with chlorine.

Most chloroform is produced by the chlorination of methane and methyl chloride.

Capacities range from 6000 to 65 000 tonnes per year.

PROCESSES

1. From methane by chlorination

Chlorine is introduced into a mixture of methane and recycle methyl chloride and the vapours are fed into a nickel-coated loop reactor. In order to obtain an optimum yield of chloroform, the concentration of the mixture is adjusted to give a molar ratio of 2.6:1 chlorine to methane. Internal circulation of the gases is maintained by a valve system to prevent the formation of explosive mixtures. (*See Figure 43*)

The methane must be pure with the volume of other hydrocarbons kept to a minimum, to prevent the formation of a range of chlorinated hydrocarbons which would complicate the separation of the chloromethanes formed. Chlorine with a minimum purity of 97% is also used in order to reduce the need for the removal of off-gases from the recycle streams, with the resultant loss of chlorinated products.

The reaction temperature is controlled at 350–400°C at a pressure slightly above atmospheric. When all the chlorine has been consumed, the gases are cooled and passed through an absorber where by-product hydrogen chloride is removed by passing through a dilute acid–water solution. Any remaining acid and chlorine in the gases are extracted by washing with caustic soda.

The gases are compressed to a pressure of 8 bar, dried and cooled to about −14°C. The resultant liquid is distilled under pressure to separate the chloromethanes. The principal products are methylene chloride (70wt%) and chloroform. Uncondensed gases, consisting of methane and some methyl chloride, are combined with overheads from the first column and recycled to the reactor.

Various designs of reactors have been proposed which permit partial to full mixing of the gases in order to optimize the process. Newer plants use a dry process to remove by-product hydrogen chloride, thus avoiding the problems associated with the disposal of aqueous acid wastes. Several companies have developed a hydrogen chloride removal system whereby the acid is absorbed by methyl chloride or a mixture of chloromethanes.

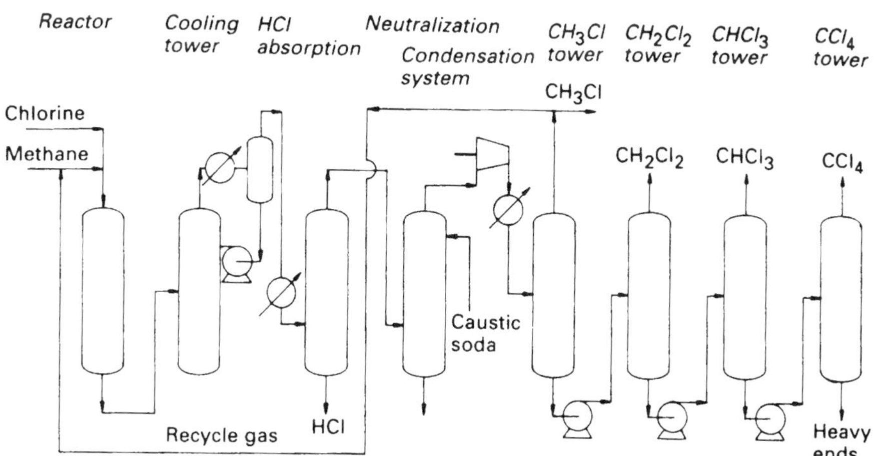

FIGURE 43 Chloroform from methane by chlorination

Reaction

$CH_4 + Cl_2 \rightarrow CH_3Cl + HCl$

$CH_3Cl + Cl_2 \rightarrow CH_2Cl_2 + HCl$

$CH_2Cl_2 + Cl_2 \rightarrow CHCl_3 + HCl$

Raw material requirements and yield

Raw materials required per tonne of chloroform (theoretical):

Methane	140kg
Chlorine	1780kg

Yield

on methane	70–85%
on chlorine	95–99%

2. From methyl chloride by chlorination

Methyl chloride (produced by the hydrochlorination of methyl chloride with hydrogen chloride) and chlorine are premixed before entering the chlorinator, operating at a temperature of 350–400°C and a pressure of 8–15 bar. (*See Figure 44*)

The reaction gases are cooled in a quench boiler, dried and distilled to separate the chlorinated methanes formed. Methyl chloride is recycled to the chlorinator. By-product hydrogen chloride is mixed with methyl alcohol before entering a hydrochlorination reactor where methyl chloride is produced (see Methyl Chloride).

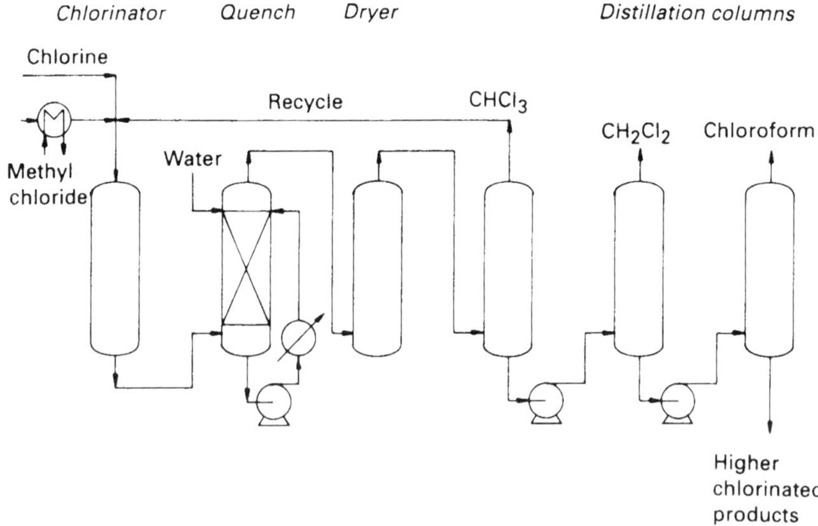

FIGURE 44 Chloroform from methyl chloride by chlorination

Reaction

$CH_3Cl + Cl_2 \rightarrow CHCl_3 + H_2$

Raw material requirements and yield

Raw materials required per tonne of chloroform:

Methyl chloride	430kg

Yield 98%

PROPERTIES

Colourless, volatile, non-flammable liquid with a sweetish odour and taste. Decomposes in a flame giving off toxic gases. Miscible with many organic solvents.

Molecular Weight	119.4
Specific Gravity at 20°C	1.489
Melting Point	–63.8°C
Boiling Point	61.3°C
Vapour Density (air = 1)	4.13

Exposure Limit COSHH	2ppm 8 hour TWA
Exposure Limit ACGIH	10ppm TLV-TWA

Classified A2, suspected human carcinogen.

GRADES

Technical 99%, stabilized with 0.1–0.2% ethyl alcohol or methyl alcohol.

INTERNATIONAL CLASSIFICATIONS

UN No.	1888
CAS Reg No.	67–66–3
EINECS No.	200–663–8
EC No.	602–006–00–4
Description	Toxic substance
Packing Group	III
Emergency Action Code	2Z
HI (Kemler Code)	60

APPLICATIONS

The market for chloroform is the smallest of all the chlorinated methanes. Its largest outlet is in the manufacture of chlorodifluoromethane (Fluorocarbon 22), which is the precursor for the production of polytetrafluoroethylene (PTFE), as well as being used as a refrigerant and propellant.

Chloroform is also used in the pharmaceutical industry and for the preparation of explosive- and shock-sensitive products. Another minor outlet is its use in the manufacture of ortho-formic esters. It is no longer employed as an anaesthetic, having been replaced by less toxic compounds. Although chloroform is an excellent solvent, it has been replaced by other more acceptable substitutes due to environmental concerns about chlorinated compounds.

HEALTH AND HANDLING

Chloroform vapour affects the central nervous system leading to dizziness, nausea, headache and anaesthesia. Long-term exposure affects the kidneys, liver, heart and eyes. Skin contact should be avoided and protective clothing and goggles worn. Contact lenses must not be used as they can absorb and concentrate the vapour which leads to irritation.

Chloroform is stable if stored in closed containers protected from air and sunlight in a well-ventilated area. It is incompatible with strong alkalis. When stabilized with ethyl alcohol, chloroform develops an acidity on prolonged exposure to air and sunlight. Chloroform must not be transferred through plastic or rubber pipes or hoses.

Spills should be contained and absorbed with an inert material such as sand, earth or vermiculite which is then placed in containers for disposal. Care must be taken to avoid leakage into sewers or watercourses. Clean-up personnel should wear protective clothing which is laundered before reuse. Contaminated footwear should be discarded.

Although chloroform is non-flammable, it will produce toxic and corrosive fumes – chlorine, hydrogen chloride and carbon monoxide – during fires. Its vapour is heavier than air and can lie along the ground and collect in hollows. Firefighters should wear protective clothing and self-contained breathing apparatus.

MAJOR PLANTS

Plants with capacities greater than 25 000 tonnes per year:

Elf-Atochem	Lavera	France
Celanese	Frankfurt	Germany
Dow Chemical	Stade	Germany
Akzo Chemicals	Delfzijl	Netherlands
Ausimont	Bussi sul	Italy
	Tirino	Italy
ICI	Runcorn	UK
Dow Chemical	Freeport	US
	Plaquemine	US
Vulcan Chemicals	Geismar	US
	Wichita	US
Tokuyama Soda	Tokuyama	Japan

Plants produce a mixture of chloromethanes.

MAJOR LICENSORS

ABB Lummus Global
Asahi Glass
Halcon-Scientific Design
Hoechst
Huels
Solvay
Vulcan Materials

Cumene

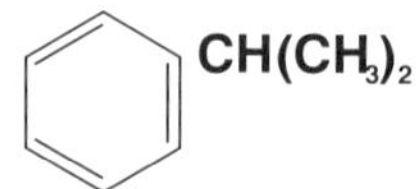

SYNONYMS

CUMENE isopropyl benzene, methyl ethyl benzene, cumol, 2-phenylpropane

Interest in cumene began during World War II when it was used in high-octane aviation gasoline. The original route for its manufacture was the propyl alkylation of benzene in the liquid phase using sulphuric acid as catalyst. Although higher temperatures and pressures increase the speed of the reaction, the yield can decline if the excess of benzene falls.

Because of the complicated neutralization and recycle steps required, together with corrosion problems, this process has been largely replaced. The major process now used is the vapour-phase cumene hydroperoxide route based on propylene and benzene.

The older phosphoric acid based processes are being replaced with zeolite catalyst technology introduced by Mobil, Dow Chemical and ABB Lummus Global. It is more flexible regarding feedstocks and produces lower quantities of waste products. An added advantage is that the capacity of existing plants can be increased without debottlenecking due to increased catalyst conversion efficiency which makes it economically attractive.

A recent development has been the combination of a catalytic reaction with distillation in a single column. The heat generated by the exothermic reaction is utilized in the distillation process thus reducing the amount of energy required.

Capacities range from 30 000 to 700 000 tonnes per year.

PROCESS

1. From benzene and propylene

Chemical-grade propylene, or a propylene cut containing up to 60% of propane, and benzene are mixed and fed into a reactor where they are brought into contact with a catalyst consisting of phosphoric acid supported on kieselguhr or pumice. (*See Figure 45*)

The reactor temperature is kept at 200–250°C with a pressure range of 15–35 bar. An excess of benzene in the molecular ratio of 5:1 benzene to propylene is maintained in order to suppress dialkylation, oligomerization and other side reactions and to attain a high conversion rate.

The gases from the reactor are used to heat incoming feed before entering the recycle column where any unreacted benzene is recovered and recycled. The remainder of the liquid stream is fed into the cumene distillation column where heavy by-products such as di- and tri-isopropylbenzene are recovered as bottoms and pure cumene passes overhead.

Reaction

$C_6H_6 + CH_2{=}CHCH_3 \rightarrow C_6H_5CH(CH_3)_2$

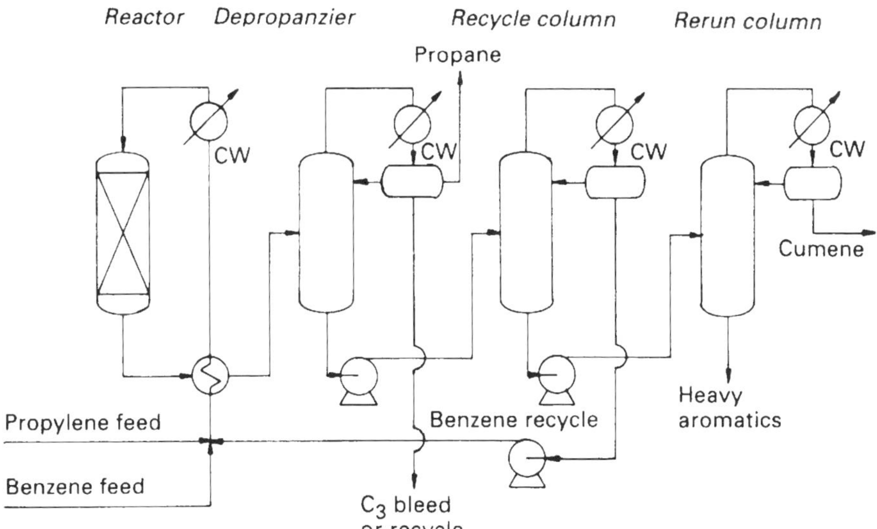

FIGURE 45 Cumene from benzene and propylene

Material requirements and yields

Materials required per tonne of cumene:

Propylene	780kg
Benzene	410kg

Yield

on propylene	91–93%
on benzene	96–97%

OTHER PROCESSES

In the combined catalytic-distillation process, unreacted feed is held in the reactor zone which promotes an equilibrium towards cumene.

Propylene vapour enters between the distillation and reaction section of the column while benzene is fed via a reflux drum into the top of the reactor. The vapour condenses overhead and returns to the reactor as reflux.

The liquid product leaving the reactor goes through a stripper where any unreacted benzene is vaporized, condensed overhead and returned as reflux. Cumene is recovered from the liquid product by distillation.

In the exothermic reaction, enough heat is released during the formation of cumene to vaporize three times its weight of benzene.

In the liquid-phase process, the reaction takes place at 35–70°C and 5–15 bar pressure, with a residence time of 20–30 minutes. Benzene is present in excess in the reactor feed, a molar ratio of benzene to propylene of 6:1 being recommended. The conversion rate on propylene is 98%.

In Germany, the Huels liquid-phase process employs hydrogen fluoride as the catalyst, while Monsanto's catalyst is aluminium chloride.

A wide range of promoted and unpromoted catalysts have been mentioned in the literature, for example, boron trifluoride with phosphoric and sulphuric acids. The range of products produced is very sensitive to the catalyst used, the purity and composition of the reactor feed and the operating conditions.

PROPERTIES

Colourless liquid, insoluble in water. Soluble in ethyl alcohol, ether and benzene.

Molecular Weight	120.19
Density at 20°C	0.86
Melting Point	–96°C
Boiling Point	152.6°C

Flash Point Open Cup	40.6°C
Explosive Limits in air	
lower	1.1vol%
upper	8.0vol%
Exposure Limit COSHH	75ppm 15 minutes 25ppm 8 hour TLV-TWA
Exposure ACGIH	50ppm TLV-TWA (skin)

GRADES

Technical, pure >99%

INTERNATIONAL CLASSIFICATION

UN No.	1918
CAS Reg No.	98–82–8
EINECS No.	202–704–5
EC No.	601–024–00-X
Description	Flammable liquid
Packing Group	III
Emergency Action Code	3Y
HI (Kemler) Code	30

APPLICATIONS

Cumene has essentially only one outlet and that is for the manufacture of phenol with acetone as a valuable by-product. Future growth will depend therefore on the demand for synthetic phenol since the two products are so closely linked. There is still a strong commercial interest in finding alternative routes to phenol to avoid by-product acetone which is in oversupply.

Very small amounts of cumene are used for α-methyl styrene which can be recovered from the process.

HEALTH AND HANDLING

Cumene vapour is irritating to the eyes and nose, and excessive exposure can lead to headaches and narcosis. Skin contact should be avoided and protective clothing, goggles, gloves, aprons, boots and a respirator should be worn when handling the product. Any contaminated clothing should be removed immediately and laundered before reuse.

Store in a well-ventilated area. All equipment must be spark-proof and earthed to prevent explosions. Cumene is stable at room temperature but must be kept well away from oxidizing agents.

In the event of leaks, extinguish any sources of ignition, evacuate personnel and absorb with sand or vermiculite. Collect with a non-sparking scoop, and containerize for disposal by burning in an approved incinerator. Cumene must not be allowed to get into sewers or watercourses. Clean-up staff must wear full protective clothing.

Cumene is a fire and explosion hazard when heated. Carbon dioxide, dry chemical or foam can be used to fight fires. Water will scatter the flames and is best used to cool storage containers. There is a risk of flashback from heavy vapours. As toxic gases are evolved on burning, firefighters must wear protective clothing and self-contained breathing apparatus.

MAJOR PLANTS

Plants with capacities greater than 200 000 tonnes per year:

Vestolen	Gelsenkirchen	Germany
Dow Chemical	Terneuzen	Netherlands
Agip Petroli	Priolo	Italy
	Sarroch	Italy
EniChem	Porto Torres	Sardinia
Ashland Petrochemical	Catlettsburg	US
Chevron Chemicals	Port Arthur	US
Citgo Petroleum	Corpus Christi	US
Georgia Gulf	Pasadena	US
Gulf Oil	Philadelphia	US
Koch Refining	Corpus Christi	US
Shell Chemical	Deer Park	US
Sun	Philadelphia	US
Mitsubishi Chemical	Kashima	Japan
Mitsui Petrochemical Industries	Senboku	Japan
Tosoh Corp.	Yokkaichi	Japan

MAJOR LICENSORS

ABB Lummus Global
CDTECH
Dow Chemical/Kellogg
Englehard
Huels
Mitsubishi Chemical
Mitsui Petrochemical
Mobil Oil/Badger
Monsanto
Unocal
UOP

Cyclohexane

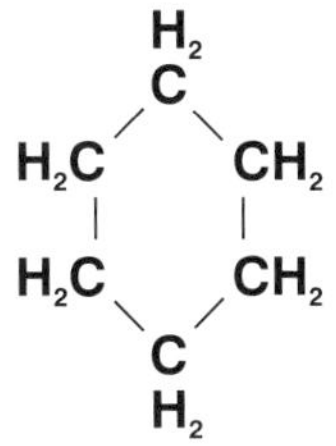

SYNONYMS

CYCLOHEXANE hexahydrobenzene, hexamethylene, naphthene, hexanaphthene, benzene hexahydride

Cyclohexane is present in very small quantities, ranging from 0.05 to 1.0%, in crude oil. Traditionally it was obtained from the naphtha cut by fractional distillation, but separation is difficult.

Although cyclohexane has been known for years, it was the discovery of nylon which led to its importance as a chemical intermediate. The routes to nylon 66 (from adipic acid and hexamethylenediamine) and nylon 6 (from caprolactam) both start from cyclohexane.

Nearly all cyclohexane is produced industrially by the hydrogenation of benzene because of its simplicity and high efficiency. The reaction can be carried out in the liquid or vapour phase in the presence of a highly dispersed catalyst or in a catalytic fixed bed. Because of the economic importance of cyclohexane, there are a large number of industrial licensed processes. Since temperature control is critical to obtain complete benzene conversion, the design of the reactor and method of dissipating the heat generated by the exothermic reaction have been addressed in several different ways. Recent developments have been aimed at improving energy utilization and reducing operating costs.

Over 90% of cyclohexane is produced by the hydrogenation of benzene. Because the price of cyclohexane is dependent on the cost of benzene and hydrogen, plants tend to be built adjacent to large refineries where cheap raw materials are readily available.

Capacities range from 55 000 to 420 000 tonnes per year.

PROCESSES

1. From benzene by hydrogenation

Benzene and hydrogen are fed cold into a reactor containing a catalyst consisting of finely divided Raney nickel in liquid cyclohexane. As the reaction is highly exothermic, the temperature in the reactor is maintained at 180–200°C by passing the catalyst slurry through an external heat exchanger at high speed using a circulating pump. The vigorous agitation caused by the pump ensures that the catalyst remains in suspension; the heat produced by the reaction is sufficient to generate low-pressure steam from the cooling water. (*See Figure 46*)

The vapour from the reactor containing cyclohexane and inert gases, and a small excess of hydrogen, goes to a fixed-bed reactor where conversion of any remaining benzene is completed with a minimal rise in temperature.

The effluent is condensed and flashed in a high-pressure separator and hydrogen-rich gas is returned to the reactor. The liquid from the separator goes to the stabilizer which removes any remaining hydrogen and other dissolved light gases. Hydrogen can be recovered before the gases are burnt as fuel gas. Cyclohexane is taken off from the bottom, and a portion is recycled as make-up to the first reactor.

The purity of the resultant cyclohexane depends on the purity of the benzene feed. Purity in excess of 99% can be achieved if the hydrogen feedstock contains 75% or

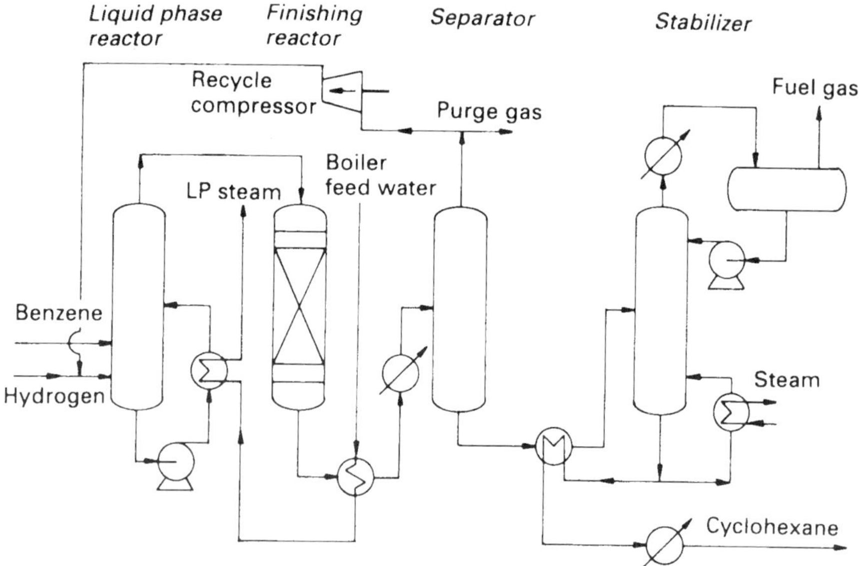

FIGURE 46 Cyclohexane from benzene by hydrogenation

more of hydrogen, and once-through operation at 27 bar ensures its optimal consumption.

Alternative processes take place in the vapour phase using palladium, platinum or nickel catalysts. The reaction is conducted in multi-stage tubular reactors containing the catalyst. Benzene, hydrogen and recycled cyclohexane are fed into the first reactor and the heat of the reaction is removed by heat exchangers between the two reactors. The exit temperature from the final reactor is kept below 300°C to ensure the complete conversion of benzene to cyclohexane. The reaction mixture is separated in a similar manner to that described previously.

Reaction

$C_6H_6 + 3H_2 \rightarrow C_6H_{12}$

Raw material requirements and yield

Raw materials required per tonne of cyclohexane:

Benzene	930kg
Hydrogen	935cm^3
Catalyst	small

Yield 99%

2. From naphtha

Recovery of cyclohexane from naphtha is difficult because of the number of other hydrocarbons, hexanes, methylcyclopentane and benzene which have similar boiling points.

Naphtha is fractionated to remove methylcyclopentane and to concentrate the cyclohexane to give a purity of 85%. After separation of the benzene and methylcyclopentane, they are hydrogenated and isomerized to increase the yield and purity of the resultant cyclohexane.

PROPERTIES

Colourless liquid with a mild ethereal odour. Immiscible with water. Flammable.

Molecular Weight	84.16
Density at 20°C	0.779
Melting Point	6.55°C
Boiling Point	80.8°C
Autoignition Temperature	245°C
Explosive Limits in air	
lower	1.35 vol%
upper	8.35 vol%
Flash Point Closed Cup	–18.4°C

Vapour Density (air = 1)	2.91
Exposure Limit COSHH	300ppm 15 minutes 100ppm 8 hour TWA
Exposure Limit ACGIH	300ppm TLV-TWA

GRADES

Industrial 95% and 99%, solvent-grade 85%.

INTERNATIONAL CLASSIFICATIONS

UN No.	1145
CAS Reg No.	110–82–7
EINECS No.	203–806–2
EC No.	601–017–00–1
Description	Flammable liquid
Packing Group	II
Emergency Action Code	3YE
HI (Kemler Code)	33

APPLICATIONS

The major outlet for cyclohexane is in the manufacture of nylon intermediates. Adipic acid and caprolactam used for the production of nylon 66 and nylon 6 respectively, together account for about 95% of cyclohexane consumption. Hexamethylenediamine, the other intermediate used in nylon 66 manufacture, can be made from cyclohexane.

Other minor uses of cyclohexane are as a solvent, reaction diluent, and a starting material for the production of cyclohexanol–cyclohexanone, 'KA oil'.

With the recent developments in alternative routes to nylon intermediates using butadiene, phenol and acrylonitrile, future growth in demand for cyclohexane is expected to be around 1% per year.

Cyclohexane of 85% purity, containing methylcyclopentane and paraffins, finds use as a solvent.

HEALTH AND HANDLING

Cyclohexane can cause irritation to the skin, eyes, nose and throat, but it is less toxic than benzene. In high concentrations, dizziness and unconsciousness occur. Dermatitis is a risk with prolonged skin exposure.

Cyclohexane is stable at room temperature and as it is non-corrosive to metals, iron, steel, copper, aluminium or lead can be used for containers. Rubber gaskets and seals should be avoided. The storage area should be well ventilated and containers must be kept away from oxidizing agents.

Spills should be absorbed with paper, vermiculite or dry sand and placed in containers using non-sparking tools for disposal by burning. All sources of ignition must be extinguished and personnel evacuated from the contaminated area. The liquid must be kept away from sewers because of the high explosion risk.

Cyclohexane vapour readily forms explosive mixtures in air and flashback is a great hazard to firefighters. Carbon dioxide, dry chemical or foam should be used to extinguish fires but water is ineffective as it scatters the flames. Firefighting and clear-up staff must wear protective clothing and self-contained breathing apparatus to prevent inhalation of vapour and skin contact.

Handling cyclohexane involves the same flammability risks as for other similarly volatile saturated hydrocarbons. All equipment must be earthed to prevent static build-up. Special regulations govern its storage and transport.

MAJOR PLANTS

Plants with capacities greater than 100 000 tonnes per year:

Fina Antwerp Olefins	Antwerp	Belgium
Vestolen	Gelsenkirchen	Germany
Wintershall	Lingen	Germany
Exxon Chemical	Botlek	Netherlands
ICI	Middlesbrough	UK
Citgo Petroleum	Corpus Christi	US
Clark Oil & Refining	Port Arthur	US
Huntsman Corp.	Port Arthur	US
Phillips Petroleum	Sweeny	US
Phillips Petroleum	Guayama	Puerto Rico
Gujarat State Fertilizer	Baroda	India
Idemitsu Petrochemical	Tokuyama	Japan
Mitsubishi Chemical	Mizushima	Japan
Ube Industries	Sakai	Japan
Ukishima Aromatic Chemicals	Kawasaki	Japan
Yukong Oxichemical	Ulsan	South Korea
Mobil Oil	Jurong	Singapore

MAJOR LICENSORS

Air Products & Chemicals/Houdry
Arco Technology
ABB Lummus Global
BP Chemicals
CDTECH
Haines & Associates
IFP
Stamicarbon
Texaco Development
Toray Industries
Uhde
UOP/Hydrar
Zimmer

Cyclohexanol & Cyclohexanone

```
      H2                 H2
      C                  C
    /   \              /   \
 H2C     CH2       H2C     CH2
  |       |         |       |
 H2C     CH2       H2C     CH2
    \   /              \   /
     CH                  C
     OH                  ||
                         O
```

SYNONYMS

CYCLOHEXANOL hexahydraphenol, hexalin, Anol, hydroxycyclohexane, Adronal, cyclohexyl alcohol

CYCLOHEXANONE ketohexamethylene, pimelic ketone, cyclohexyl ketone, Anone, Nadone, Sextone

Historically, cyclohexanol and cyclohexanone were produced commercially by the hydrogenation of phenol. When high-purity cyclohexane became available, it led to the development of oxidation processes for the production of cyclohexanol and cyclohexanone. The ratio of the two products, known as a 'mixed oil' or ketone–alcohol (KA) oil, can be controlled by the choice of metal catalyst. The reaction is usually carried out in the liquid phase, although vapour-phase oxidation can be used. The cyclohexanol–cyclohexanone mixture can be separated by fractional distillation. Asahi Chemical are developing a new process in which benzene is partially dehydrogenated to cyclohexane which can be hydrolysed to cyclohexanol.

Although phenol is still used as the starting material, cyclohexane oxidation is the preferred route because of its lower cost. Approximately 95% of KA oil is manufactured from cyclohexane, while the remainder is produced almost equally from phenol and cyclohexanol.

Capacities range from 90 000 to 360 000 tonnes per year.

PROCESSES

1. From cyclohexane

Cyclohexane of 98% purity is oxidized to a mixture of cyclohexanol and cyclohexanone with air in a reactor at 90–120°C and a pressure of 10 bar. The catalyst used consists of a cobalt salt (usually naphthenate or oleate) which is soluble in hydrocarbons, at a concentration of around 100ppm if cyclohexanone is desired.

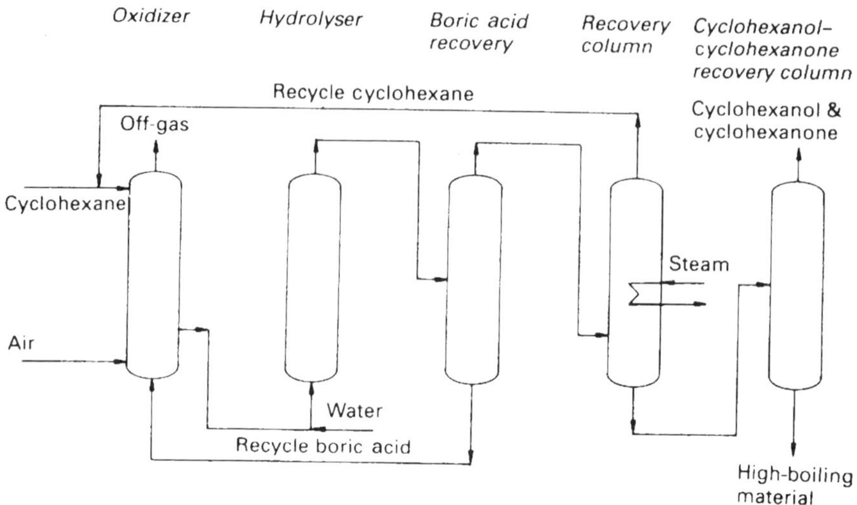

FIGURE 47 Cyclohexanol–Cyclohexanone from cyclohexane

For cyclohexanol, boric acid is used as the catalyst. In the latter case the speed of the oxidation must be limited by controlling the amount of oxygen to the reactor. This can be achieved by mixing air with a quantity of inert gas to reduce the oxygen content to about 8–10%. (*See Figure 47*)

The reaction takes place at a temperature of 160–175°C and 8–10 bar. The resultant products are treated with water with the boric esters rapidly hydrolysed to cyclohexanol and orthoboric acid. It is necessary to remove water formed during the reaction to obtain the desired oxidation selectivity. Boric acid is recovered from the aqueous phase and recycled.

The organic phase containing the crude cyclohexanol–cyclohexanone mixture is treated with alkali to remove any remaining acids. Any unreacted cyclohexane is removed by fractionation at atmospheric pressure and recycled. The liquid from the base of the column is purified by distillation under vacuum at a temperature not exceeding 140°C to give a pure cyclohexanol–cyclohexanone mixture. If cyclohexanone is required, the mixture is dehydrogenated in the liquid phase in the presence of a cupric chromate catalyst at 200°C.

Reaction

$$2C_6H_{12} + O_2 \rightarrow C_6H_{11}OH + C_6H_{10}O + H_2$$

Raw material requirements and yield

Raw materials required per tonne of cyclohexanone:

Cyclohexane	925kg
Steam at 25atm	1400kg
Steam at 10atm	4000kg
Boric acid	5kg
Sodium hydroxide (100%)	33kg
Zinc oxide	0.75kg

Yield 90–95%

2. From phenol

The hydrogenation of phenol can be carried out in the liquid or vapour phase, the reaction products being determined by the catalyst used. Of the two processes, the liquid-phase route is more selective and requires less catalyst, making it economically attractive. (*See Figure 48*)

In the vapour-phase process, phenol and hydrogen are fed into a reactor containing a supported nickel catalyst (or nickel containing copper, cobalt or manganese supported on alumina or silic acid) if cyclohexanol is desired. Palladium, platinum, osmium or iridium are used as the catalyst if cyclohexanone is the required end product.

The reaction takes place at 140–200°C and 1–4 bar. By adjusting the operating conditions, a mixture of cyclohexanol and cyclohexanone can be obtained which finds many outlets as a solvent.

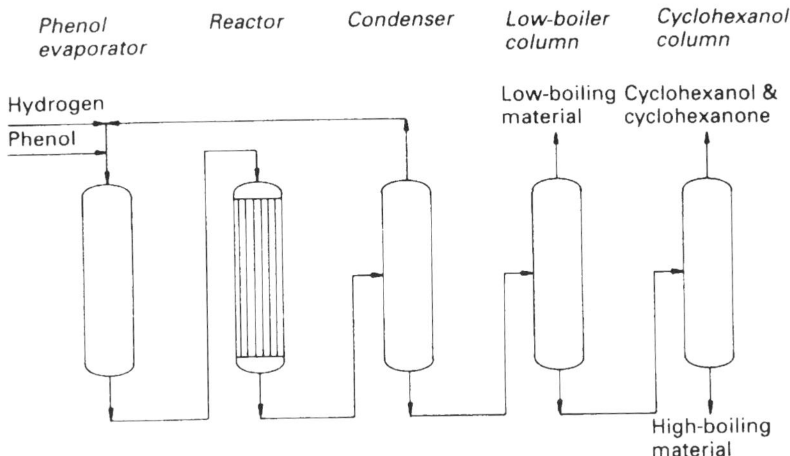

FIGURE 48 Cyclohexanol–Cyclohexanone from phenol

The liquid-phase reaction is carried out at 140°C and 2–18 bar using a palladium catalyst on a carbon or alumina carrier if cyclohexanone is required, or finely divided nickel on silica if cyclohexanol is the desired product.

Reaction

$C_6H_5OH + 2H_2 \rightarrow C_6H_{10}O$

Raw material requirements and yield

Raw materials required per tonne of mixed product:

Phenol	950kg
Hydrogen	50kg

Yield 99%

PROPERTIES

Cyclohexanol

Colourless, oily liquid with a slight camphor-like odour.

Molecular Weight	100.16
Density at 20°C	0.949
Melting Point	23.6°C
Boiling Point	160.65°C
Autoignition Temperature	300°C
Explosive Limits in air	
lower	2.4vol%
Flash Point Open Cup	67.2°C
Vapour Density (air = 1)	3.45
Exposure Limits COSHH	50ppm 8 hour TWA
Exposure Limit ACGIH	50ppm TLV-TWA

Cyclohexanone

Colourless liquid with a characteristic odour which is similar to acetone.

Molecular Weight	98.14
Density at 20°C	0.9478
Melting Point	–32°C
Boiling Point	156.7°C
Autoignition Temperature	420°C
Flammable Limits in air	
lower	1.1 vol%
upper	9.4 vol%
Flash Point Closed Cup	42°C
Vapour Density (air = 1)	3.38
Exposure Limit COSHH	100ppm 15 minutes STEL 25ppm 8 hour TLV-TWA
Exposure Limit ACGIH	25ppm TLV-TWA, (skin)

GRADES

Cyclohexanol technical 95%, pure 100% (can be inhibited with up to 2.5% methyl alcohol).

Cyclohexanone commercial 95% and 100%.

INTERNATIONAL CLASSIFICATIONS

Cyclohexanol

UN No.	Not allocated
CAS Reg No.	108–93–0
EINECS No.	203–603–6
EC No.	603–009–00–3

Cyclohexanone

UN No.	1915
CAS Reg No.	108–94–1
EINECS No.	203–631–1
EC No.	606–010–00–7
Description	Flammable liquid
Packing Group	III
Emergency Action Code	3Y
HI (Kemler Code)	30

APPLICATIONS

Cyclohexanol

The two major uses of cyclohexanol are as an intermediate for the production of adipic acid which is used in the manufacture of nylon 66, and for cyclohexanone which is converted to caprolactam for nylon 6.

Cyclohexanol is an excellent solvent for lacquers, varnishes, gums, oils, dyes, and alkyd resins. It is also used in plasticizer preparation. Other outlets include paint and varnish removers, dry cleaning of textiles, and as a stabilizer for synthetic detergents and soaps.

Cyclohexanone

Over 95% of cyclohexanone is used either in the manufacture of caprolactam, or oxidized to adipic acid. Other minor outlets include its use as a solvent for synthetic resins and lacquers, and as a chemical intermediate in the pharmaceutical and pesticide industries. Future consumption will depend on the demand for nylon.

HEALTH AND HANDLING

Cyclohexanol

Exposure to cyclohexanol vapour may cause eye irritation leading to conjunctivitis, headaches and mild intoxication even at low concentrations. As the vapour pressure is low at normal temperatures, eye and nasal irritation will give sufficient warning before dangerous levels are reached.

Cyclohexanol can be absorbed through the skin; evidence exists that regular exposure can have a cumulative effect with the risk of dermatitis. Care should be taken when handling the product and protective clothing must be worn. Cyclohexanol is flammable but because of its low flash point it is not a high risk.

Iron, mild steel, or aluminium are usually used for storage containers as cyclohexanol has no appreciable corrosive effect on these metals. If water is present, which is often the case in some of the technical grades, then corrosion can occur. Cyclohexanol must be kept away from oxidizing agents which can cause it to ignite.

If spills occur the contaminated area should cleared of personnel and all sources of ignition extinguished. Absorb with dry sand or vermiculite and scoop up with non-sparking tools into containers for disposal by burning. Contaminated clothing should be laundered before reuse.

Fires should be blanketed with carbon dioxide, dry chemical or foam as water can scatter the flames. Firefighting and clean-up staff should wear protective clothing and self-contained breathing apparatus.

Cyclohexanone

Cyclohexanone vapour is irritating to the eyes, nose and throat, which usually provides adequate warning of harmful levels. High concentrations lead to headaches, dizziness and unconsciousness and regular exposure can have a cumulative effect on the body. Contact lenses must not be worn as they concentrate the vapours leading to eye injury. The liquid degreases the skin and cracking and dermatitis may result. Care must be taken when handling cyclohexanone and protective clothing should be worn at all times.

Cyclohexanone is not corrosive to metals, so plant and storage containers can be made of iron, steel, copper or aluminium. Contact with oxidizing agents must be avoided.

Spills should be contained, absorbed with vermiculite or dry sand, and collected with non-sparking tools for containerization. Dispose of the waste in accordance with local regulations.

Cyclohexanone is a moderate fire and explosion hazard. Carbon dioxide, dry chemicals and foam are used to extinguish flames. Flashback is a hazard. Protective clothing and self-contained breathing apparatus must be worn by clean-up and firefighting staff. Launder contaminated clothing before reuse.

MAJOR PLANTS

Plants with capacities greater than 130 000 tonnes per year:

Cyclohexanol

Bayer	Antwerp	Belgium
Rhone-Poulenc	Roussillon	France
BASF	Ludwigshafen	Germany
Bayer	Leverkusen	Germany
Henkel	Dusseldorf	Germany
EniChem	Mantua	Italy
Allied/Signal	Hopewell	US
BASF	Freeport	US
Honshu Chemical	Wakayama	Japan
Japan Lactam	Niihama	Japan
Kanto Denka	Mizushima	Japan
Mitsubishi Kasei	Kurosaki	Japan
Toagosei Chemical	Nagoga	Japan

Cyclohexanone

Bayer	Antwerp	Belgium
BASF	Ludwigshafen	Germany
Bayer	Leverkusen	Germany
DSM	Geleen	Netherlands
EniChem	Mantua	Italy
Allied/Signal	Hopewell	US
BASF	Freeport	US
DSM	Augusta	US
Du Pont	Orange	US
	Victoria	US
Monsanto Chemicals	Pensacola	US
Honshu Chemical	Wakayama	Japan
Japan Lactam	Niihama	Japan
Kanto Denka	Mizushima	Japan
Mitsubishi Kasei	Kurosaki	Japan

MAJOR LICENSORS

BASF
IFP
Rhone-Poulenc
Scientific Design
Stamicarbon

Epichlorohydrin

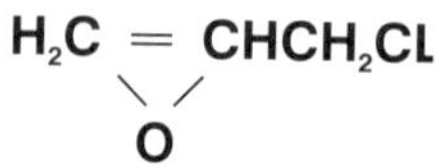

SYNONYMS

EPICHLOROHYDRIN 1-chloro-2,3-epoxypropane, chloromethyloxirane, 3-chloro-1,2-propylene oxide, 1,2-epoxy-3-chloropropane, γ-chloropropylene oxide

The development of a commercial process for the manufacture of synthetic glycerine provided the initial interest in epichlorohydrin, which is the intermediate product. Subsequently, further growth in demand for the product was linked to its use as one of the most important raw materials used for the manufacture of epoxy resins.

The original industrial process, the chlorohydrination of allyl chloride produced from propylene oxide, is now obsolete. Today all epichlorohydrin is produced by the chlorohydrination of allyl chloride. Two routes to allyl chloride exist: the chlorination of propylene and the newer oxychlorination of propylene. Corrosion is a problem and plants must be constructed with acid-resistant materials.

Capacities range from 21 000 to 250 000 tonnes per year.

PROCESSES

From propylene

Dry propylene is preheated before being mixed with chlorine and fed into a reactor. The volume of the gases is propylene:chlorine in a molar ratio of 4:1. The reaction conditions are 500°C and up to 1 bar pressure, with a contact time of approximately 2 seconds. (*See Figure 49*)

After cooling, the reaction products are fractionated. Crude allyl chloride from the bottom of the column is purified by further distillation. By-products hydrogen chloride and 2-chloropropane are removed overhead. Other major by-products of the allyl chloride process are 1,2-dichloropropane and 1,3-dichloropropene.

Allyl chloride is reacted with hypochlorous acid (formed by the interaction of water with chlorine) in the aqueous phase at a temperature of 28°C. The liquid

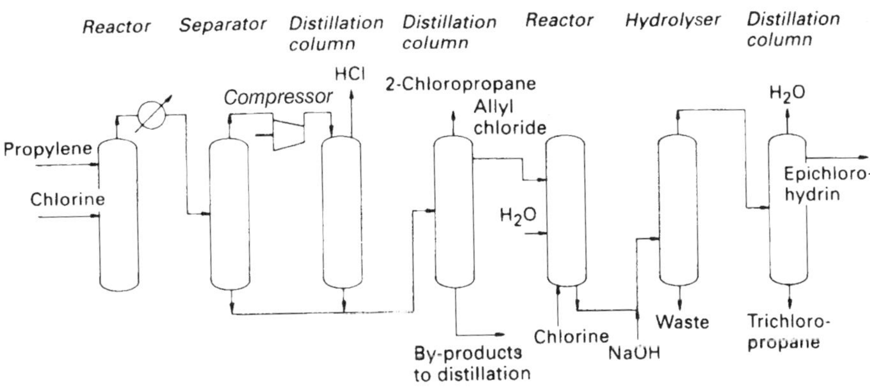

FIGURE 49 Epichlorohydrin from propylene

products pass to a separator where the aqueous phase is recycled to the hypochlorous generator, and the hydrocarbon phase is fed into a second reactor. Here the dichlorohydrins are converted to epichlorohydrin by treatment with alkali, either sodium hydroxide or calcium hydroxide, at 60–70°C.

Epichlorohydrin is separated as an azeotrope from by-products by steam distillation and concentrated by further distillation.

Reaction

$$C_2H_4=CH_2 + Cl_2 \rightarrow ClC_2H_3=CH_2 + HCl$$

$$2ClC_2H_3=CH_2 + 2HOCl \rightarrow ClCH_2CHClCH_2OH + ClCH_2CHOHCH_2Cl$$

$$ClCH_2CHOHCH_2Cl + NaOH \rightarrow \underset{\diagdown\ \diagup \atop O}{H_2C-CHCH_2Cl} + NaCl + H_2O$$

Raw material requirements and yield

Raw materials required per tonne of epichlorohydrin:

Propylene	760kg
Chlorine	2350kg
Caustic soda	1150kg

Yield 65%

OTHER PROCESSES

Hoechst have developed an alternative route to allyl chloride based on the oxychlorination of propylene. In this process, propylene, aqueous hydrogen chloride and oxygen are mixed and reacted at 240°C and atmospheric pressure in the presence of a tellurium catalyst. The allyl chloride formed is separated and converted to epichlorohydrin as in the previous process.

Reaction

$$2C_2H_4{=}CH_2 + 2HCl + O_2 \rightarrow 2\underset{\diagdown\;O\;\diagup}{CH_2 - CHCH_2Cl} + 2H_2O$$

PROPERTIES

Colourless, mobile liquid with an irritating, sweet, garlic-like odour. Flammable. Miscible with alcohols, esters, ethers, ketones and aromatic hydrocarbons. Sparingly soluble in water.

Molecular Weight	92.53
Density at 20°C	1.181
Melting Point	–58.1°C
Boiling Point	115.2°C
Autoignition Temperature	411°C
Flammability Limits in air	
lower	3.8vol%
upper	21.1vol%
Flash Point Open Cup	40.6°C
Vapour Density (air = 1)	3.29

Exposure Limit COSHH (Schedule 1)	1.5ppm 15 minutes (Maximum Exposure Limit)
	0.5ppm 8 hour TWA (Maximum Exposure Limit)
Exposure Limit ACGIH	0.1ppm TLV-TWA

Classified A2, suspected human carcinogen.

GRADES

Technical, refined 99.0%.

INTERNATIONAL CLASSIFICATIONS

UN No.	2023
CAS Reg No.	106–89–8
EINECS No.	203–439–8
EC No.	603–026–00–6
Description	Toxic substance, flammable liquid
Packing Group	II
Emergency Action Code	2W
HI (Kemler Code)	63

APPLICATIONS

Epichlorohydrin is the most important material for the production of unmodified epoxy resins with bisphenol A accounting for over half of epichlorohydrin demand in the US and West Europe and three-quarters in Japan. The production of glycerine is the second largest consumer of epichlorohydrin but this market is in decline with the growth of natural glycerine production. Minor uses include elastomers, resins for the paper industry and in dyes and textile additives.

Current global growth rate is estimated at 4–5% per year.

HEALTH AND HANDLING

Epichlorohydrin is very irritating to the eyes, nose and throat leading to headache, nausea, vomiting and depression to the central nervous system. It is poisonous if inhaled. The liquid will burn skin and eyes. Contact with both liquid and vapour must be avoided and rubber protective clothing, gloves, goggles and self-contained breathing apparatus must be worn. Any contaminated clothing must be removed as quickly as possible and affected areas washed well with water. Epichlorohydrin has caused cancer in experimental animals.

Store in a well-ventilated area away from strong acids and bases, and sources of ignition. Spills should be contained and disposed of in accordance with local regulations. Flush the area with water, taking care to keep the waste away from sewers or watercourses.

Combat fires from a safe distance, and extinguish with water, dry chemical, alcohol foam or carbon dioxide. Poisonous gases are produced on burning and explosions can occur if the vapour is ignited in a confined area. Full protective clothing and self-contained breathing apparatus must be worn by firefighting and clean-up personnel who should stay upwind of the vapour if possible.

MAJOR PLANTS

Plants with capacities greater than 30 000 tonnes per year:

Solvay	Tavaux	France
Dow Chemical	Stade	Germany
Solvay	Rheinberg	Germany
Shell Nederland Chemie	Pernis	Netherlands
Complex	Angarsk	Russia
	Sterlitansk	Russia
Dow Chemical	Freeport	US
Shell Chemical	Deer Park	US
	Norco*	US
Daiso	Mitzushima	Japan
Kashima Chemical	Kashima	Japan

* Crude product made at Norco is refined at Deer Park.

MAJOR LICENSORS

ABB Lummus Global
Hoechst
Shell Development
Showa Denko
Solvay

Ethanolamines

Monoethanolamine	**$CH_2OHCH_2NH_2$**
Diethanolamine	**$(CH_2OHCH_2)_2NH$**
Triethanolamine	**$(CH_2OHCH_2)_3N$**

SYNONYMS

MONOETHANOLAMINE MEA, 2-aminoethanol

DIETHANOLAMINE DEA, 2,2-iminodiethanol

TRIETHANOLAMINE TEA, 2,2,2-nitrilotriethanol

Ethanolamines were initially prepared from ethylene chlorohydrin and ammonia, but this route was replaced when ethylene oxide became readily available after World War II.

The only commercial process in use today for their manufacture is the reaction of ethylene oxide with ammonia in the liquid phase. The ratio of end products formed depends on the molar excess of ammonia present. All plants produce a mixture of ethanolamines, the ratio depending upon the product composition required.

Monoethanolamine is the most important of the ethanolamines accounting for around half of total production, followed by diethanolamine with 30–35%.

Capacities range from 10 000 to 180 000 tonnes per year.

PROCESSES

From ethylene oxide and ammonia

Liquid ammonia and water, which act as a catalyst, are mixed to give an ammonia concentration of 50–100%. Ethylene oxide is metered into the ammonia–water mixture to avoid the risk of explosive polymerization; the feed then passes into a tubular reactor containing an ion exchange resin. Reaction temperatures up to 150°C and pressures of 160 bar are employed to keep the ammonia in the liquid phase. An excess of around 40 moles of ammonia to ethylene oxide is used depending on the product range required. (*See Figure 50*)

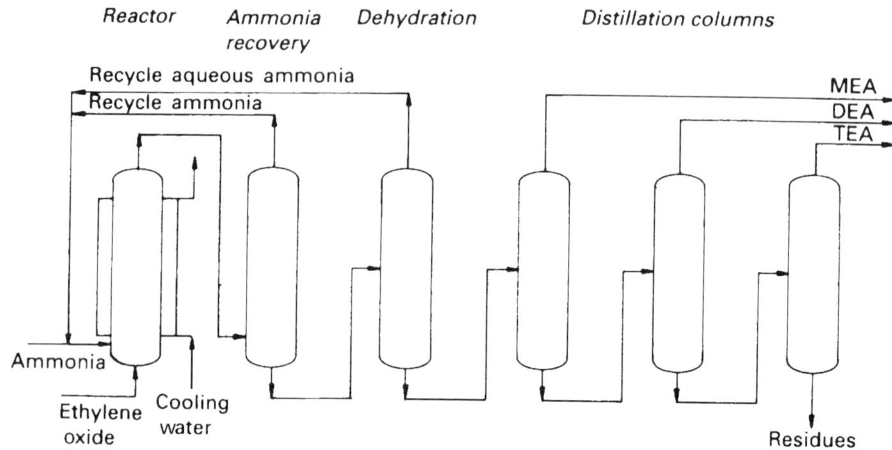

FIGURE 50 Ethanolamines from ethylene oxide and ammonia

Because the reaction is exothermic, the reaction mixture is cooled before passing to the first of the distillation columns where any unconsumed ammonia is removed overhead and recycled. In the second column, ammonia and water are removed from the end products. The mixture of ethanolamines is separated in a series of vacuum distillation columns.

Reaction

$$NH_3 + H_2COCH_2 \rightarrow HOCH_2CH_2NH_2$$

$$NH_2CH_2CH_2OH + H_2COCH_2 \rightarrow (HOCH_2CH_2)_2NH$$

$$NH(CH_2CH_2OH)_2 + H_2COCH_2 \rightarrow (HOCH_2CH_2)_3N$$

Raw material requirements and yield

Raw materials required per tonne of mixed ethanolamines:

Ethylene oxide	800kg
Ammonia	280kg

Yield 95%

PROPERTIES

Mono- and tri-ethanolamines are clear, colourless, viscous liquids with a mild ammonia-like odour. They are hygroscopic, absorbing water and carbon dioxide from air. Di-ethanolamine is crystalline or a viscous liquid. All are soluble in water and ethyl alcohol.

	Ethanolamine			
	Mono	*Di*	*Tri*	
Molecular Weight	61.08	105.1	149.2	
Density at 20°C	1.02	1.1	1.12	
Melting Point	10.3°C	27.4°C	21.6°C	
Boiling Point	170.3°C	286.5°C	336.1°C	
Autoignition Temperature	410°C	365°C	325°C	
Flammability Limits in air				
lower	5.5°C			
upper	17°C			
Flash Point Closed Cup	94.5°C	176°C	192°C	
Vapour Density (air = 1)	2.1	3.7	5.1	
Exposure Limit COSHH	Not listed			
Exposure Limit ACGIH ppm	6			TLV-STEL
ppm	3	0.46	0.5mg/m^3	TLV-TWA

GRADES

Monoethanolamine	99%
Diethanolamine	98%
Triethanolamine	98%

INTERNATIONAL CLASSIFICATIONS

UN No.	2491
CAS Reg No.	141–43–5
EINECS No.	205–483–3
EC No.	603–030–00–8
Description	Corrosive substance
Packing Group	III
Emergency Action Code	2X
HI (Kemler Code)	80

APPLICATIONS

The major outlet for ethanolamines is as an intermediate in the production of surfactants, textile and leather chemicals and emulsifiers for drilling and cutting oils.

Di- and tri-ethanolamines react with oleic, stearic, caprylic or lauric acids to form products used in medicinal soaps and toiletries. These soap derivatives find outlets as emulsifiers for cosmetics, shoe creams, ointments, car polishes, and drilling oils. Fatty ethanolamines are used in washing products, such as foam baths and shampoos, where they confer foam stability.

Monoethanolamine is used extensively in scrubbers to purify refinery and natural gases because of its solubility in hydrocarbons, and is used for the production of ethylenediamine and ethylenimine.

Other outlets for ethanolamines include dyeing of leather, paint removers, surface coatings, intermediates for pharmaceuticals, plasticizers, metal cleaning, chemical plating, corrosion inhibitors and lubricants for engines.

In US and West Europe, current growth is estimated at 2% per year.

HEALTH AND HANDLING

Ethanolamines can be absorbed by the skin in toxic amounts, and repeated contact will lead to redness and swelling. Their vapour is irritating to the eyes, skin and respiratory tract, and depression of the central nervous system can occur. Contact lenses should be avoided due to absorption of the vapour. When handling the product, protective clothing, safety goggles, gloves and boots must be worn. Any contaminated clothing must be laundered before re-use.

Ethanolamines are corrosive and will attack some plastics and rubbers. Stainless steel storage is normally used. In closed containers, ethanolamines are stable at temperatures below 35°C. A good ventilated storage area is required away from oxidizing agents or strong acids. Spills should be contained and absorbed with vermiculite, dry sand or clay and disposed of according to local regulations.

Ethanolamines are a moderate fire hazard. Carbon dioxide, or chemical foam must be used in firefighting as water will cause the liquid to spread. Firefighters should wear self-contained breathing apparatus.

MAJOR PLANTS

Plants with capacities greater than 30 000 tonnes per year:

BASF	Antwerp	Belgium
BASF	Ludwigshafen	Germany
Akzo Chemicals	Stenungsund	Sweden
Dow Chemical	Plaquemine	US
Huntsman Corp.	Port Neches	US
Union Carbide	Seadrift	US
Mitsui Toatsu Chemical	Osaka	Japan
Nippon Shokubai	Kawasaki	Japan

MAJOR LICENSORS

Leonard
Scientific Design
Union Carbide

Ethyl Acetate $C_2H_5COOCH_3$

SYNONYMS

ETHYL ACETATE ethyl acetic ester, acetic ether, acetidin, acetoxyethane, ethyl ethanoate

Ethyl acetate is used primarily as a lacquer solvent, but methyl ethyl ketone has largely replaced it as a constituent of low-boiling-point solvents.

Most of the ethyl acetate manufactured is prepared by the esterification of ethyl alcohol with acetic acid, but some is produced by the catalytic condensation of acetaldehyde with alkoxides.

BP Chemicals has developed a new direct process to ethyl acetate starting from ethylene and acetic acid. It is claimed to be simpler and cleaner than the route via acetaldehyde.

Capacities range from 5000 to 160 000 tonnes per year, but the data published frequently includes other aliphatic acetates.

PROCESSES

From ethyl alcohol and acetic acid by esterification

An aqueous solution of acetic acid is mixed with an excess of 95% ethyl alcohol; then 1% sulphuric acid is added. The mixture is preheated before passing into an esterifying column where it is refluxed and some of the distillate removed overhead. The mixture, containing about 20% of the ester, goes into a second refluxing column where a ternary azeotrope (consisting of 83% ethyl acetate, 9% ethyl alcohol and 8% of water) is removed overhead at a temperature of about 70°C. (*See Figure 51*)

An equal volume of water is mixed with the distillate after which it separates into two layers. The bottom layer is fed into a separator column and any ester is recovered as a ternary azeotrope. Any alcohol remaining in the bottom layer is vaporized and recycled to the esterification column.

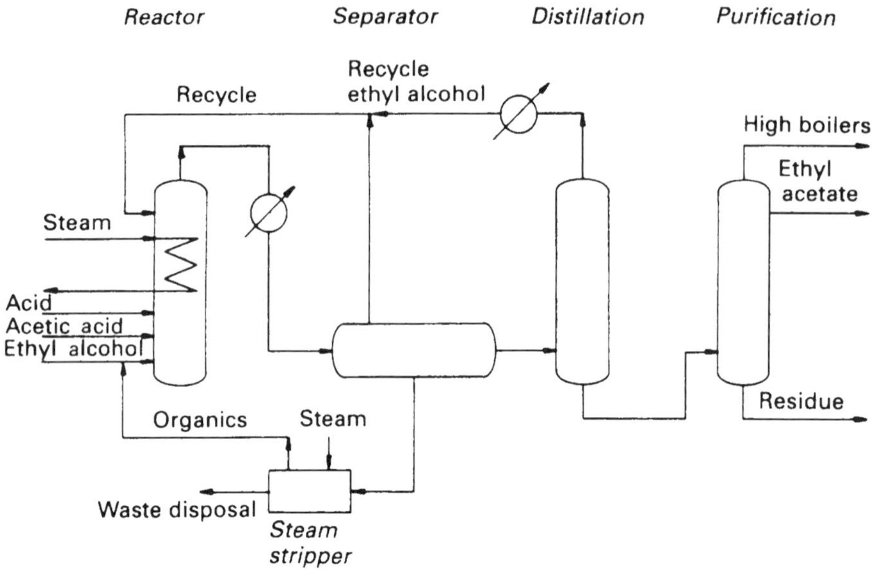

FIGURE 51 Ethyl acetate from ethyl alcohol and acetic acid by esterification

The upper layer, containing around 93% ethyl acetate together with some water and ethyl alcohol, passes to a refluxing column where it is mixed with sufficient ester to ensure that the water and ethyl alcohol are removed as a ternary azeotrope overhead. The residue in the column, consisting of 95% ethyl acetate, is redistilled to removed any impurities.

Reaction

$$C_2H_5OH + C_2H_5COOH \rightarrow C_2H_5COOCH_3 + H_2O$$

Raw material requirements and yield

Raw materials required per tonne of ethyl acetate:

Ethyl alcohol (95%)	620kg
Acetic acid (100%)	690kg

Yield 99%

PROPERTIES

Colourless, mobile, volatile liquid with a pleasant fruity odour. Highly flammable. Soluble in ethyl alcohol and ether and slightly soluble in water.

Molecular Weight	88.1
Density at 20°C	0.902
Melting Point	–83.6°C
Boiling Point	77.15°C
Autoignition Temperature	427°C

Explosive Limits in air	
lower	2.25vol%
upper	11.0vol%
Flash Point Closed Cup	–4.4°C
Vapour Density (air = 1)	3.04

Exposure Limit COSHH 400ppm 8 hour TWA
Substance on the current Advisory Committee on Toxic Substances (ACTS) and the Working Group on the Assessment of Toxic Chemicals (WATCH) work programme.
Exposure Limit ACGIH 400ppm TLV-TWA

GRADES

Commercial 100%; anhydrous, 85–88% with 12–15% ethyl alcohol for use as a lacquer solvent denatured with 1% methyl isobutyl ketone.

INTERNATIONAL CLASSIFICATIONS

UN No.	1173
CAS Reg No.	141–78–6
EINECS No.	205–500–4
EC No.	607–022–00–5
Description	Flammable liquid
Packing Group	II
Emergency Action Code	3YE
HI (Kemler Code)	33

APPLICATIONS

Ethyl acetate is a low-boiling-point solvent used alone or in a mixture with ethyl alcohol, primarily in lacquers and surface-coating resins.

HEALTH AND HANDLING

Ethyl acetate is moderately toxic by inhalation. High concentrations of the vapour will cause irritation to the eyes, nose, throat and lungs leading to lacrimation, headaches, nausea, conjunctivitis and finally damage to the liver and kidneys. It has a defatting effect on the skin leading to dermatitis.

Storage should be in closed steel or aluminium containers but copper should be avoided due to traces of acetic acid which can develop in the presence of moisture. Ethyl acetate is highly flammable and can form explosive mixtures with air. The product should be stored in a well-ventilated area at room temperature and away from sources of ignition. Non-sparking tools must be used.

Spills should be contained and absorbed with earth or sand. Care must be taken to prevent discharge into sewers or inland waterways.

Ethyl acetate is a dangerous fire hazard and, as its vapours are heavier than air, flashback can occur. Carbon dioxide, dry chemical or water fog can be used to fight fires, but a stream of water must be avoided as it will tend to spread the flames.

MAJOR PLANTS

Plants with capacities greater than 35 000 tonnes per year:

Huels	Marl	Germany
EniChem	Porto Marghera	Italy
BP Chemicals	Hull	UK
Eastman Chemical	Kingsport	US
	Longview	US
Hoechst Celanese	Pampa	US
Rhodia	Santo Andre	Brazil
Chiba Ethyl Acetate	Ichihara	Japan
Tokuyama Petrochemical	Shin-Nanyo	Japan
	Yamaguchi	Japan
InterEster	Ulsan	South Korea
Lee Chan Yung Chemical	Linyuan	Taiwan

MAJOR LICENSORS

BP Chemicals
Chisso
Hoechst

Ethyl Alcohol

C_2H_5COH

SYNONYMS

ETHYL ALCOHOL ethanol, ethyl hydroxide, methyl carbinol, ethylic alcohol, absolute alcohol, rectified spirit, spirits of wine

Ethyl alcohol has been produced by fermentation of carbohydrates for many thousands of years, but economic industrial manufacture of synthetic ethyl alcohol began in the 1930s.

The first process used was the indirect catalytic hydration of ethylene, but this route has several disadvantages: the large volumes of dilute sulphuric acid to be handled, the energy required for its concentration, and corrosion caused by the acid. It has been superseded by the direct catalytic gaseous-phase hydration of ethylene.

Recently, alternative routes based on methyl alcohol or synthesis gas (a mixture of carbon monoxide and hydrogen) as starting materials have been developed. Synthesis gas is converted to methyl alcohol which is then carbonylated to acetic acid. Although the acid formed can be hydrolysed directly to ethyl alcohol, the preferred route is by esterification with methyl alcohol followed by hydrolysis of the acetate.

Waste liquor from paper pulp manufacture, converted to ethyl alcohol by the sulphate or sulphite process, has provided a useful source in those countries having such raw materials.

In countries with large volumes of fermentation alcohol, such as Brazil and India, ethyl alcohol is being increasingly used as a chemical feedstock instead of as a fuel. Around 30% of world ethyl alcohol capacity is synthetic, most being produced by the direct hydration route from ethylene.

Capacities range from 5000 to 210 000 tonnes per year.

PROCESSES

1. From ethylene by direct hydration

Polymerization-grade ethylene and demineralized water in a molar ratio of 1:0.3–0.8 are compressed to 60–80 bar and heated to 230–300°C in a series of heat exchangers. The gases pass into a reactor containing a fixed-bed catalyst consisting of 77% phosphoric acid absorbed onto a carrier such as silica gel or a diatomaceous earth fused with aluminium oxide. The carrier must be resistant to phosphoric acid and have mechanical stability to provide a reasonable process life. Conversion per pass is 4–5%. (*See Figure 52*)

The gaseous mixture leaving the reactor is cooled and washed with a dilute alkali solution to neutralize any vaporized phosphoric acid that may be entrained with the gases. Excess heat from the exothermic reaction is used in the heat exchangers to raise the temperature of the incoming feed. After further cooling, the condensate passes into a separator. Gases are flashed off, washed with water to remove any residual alcohol, compressed and recycled to the heat exchangers. Part of the recycle gas is fed to the ethylene purification section to remove impurities and inhibit their build-up.

Crude ethyl alcohol solution is fed into a purification column where light-boiling products are removed overhead and recycled or sent to the ether recovery plant. About 2% of diethyl ether is produced as by-product. It is usually recovered and sold, but it can be recycled to the reactor for conversion to ethyl alcohol.

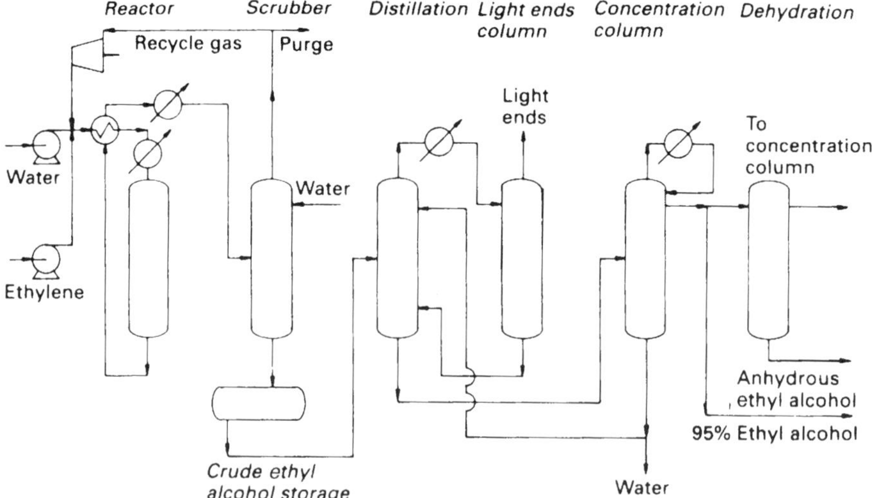

FIGURE 52 Ethyl alcohol from ethylene by direct hydration

The ethyl alcohol remaining is purified as its azeotrope by extractive distillation with water. The azeotrope containing 95vol% alcohol is dehydrated by further distillation to the anhydrous product.

To maintain catalyst activity, small amounts of phosphoric acid are added to the inlet stream at regular intervals to replace that lost during the reaction. Catalyst life is about 3 years. Other catalysts have been tried such as tungsten trioxide on silica, zeolites, phosphates, and magnesium or cobalt sulphate.

Recent improvements have concentrated on reducing process steam requirements by multi-stage high-pressure systems or by using overhead compressed vapour as the boiler heat source instead of steam. Considerable energy savings are claimed.

Reaction

$$CH_2{=}CH_2 + H_2O \rightarrow C_2H_5OH$$

Raw material requirements and yield

Raw materials required per tonne of ethyl alcohol:

Ethylene	627kg
Phosphoric acid (100%)	0.7kg

Yield

if ether is recovered	94–95%
if ether is recycled	96–97%

2. From ethylene by indirect hydration

Purified ethylene feedstock (containing only inert gases such as ethane of at least 35vol% ethylene) is passed through a tower in which 94–98wt% sulphuric acid flows in the opposite direction. Pressure is maintained at 10–35 bar with a temperature of 65–80°C and any unabsorbed gases pass from the top of the tower. The liquid is agitated for several hours to increase the rate of absorption. (*See Figure 53*)

The mixed esters consisting of mono- and di-ethyl sulphates, are hydrolysed with water in two stages. The first stage is carried out at 70°C in the presence of a limited amount of water, then further water is added until an acid concentration of 40–50wt% is reached.

Ethyl alcohol, diethyl ether by-product and some acid are stripped from the aqueous acid solution which is removed from the base of the column. The acid solution is concentrated and recycled.

The product mixture is washed with dilute alkali to remove the residual acid and any ether is removed by distillation prior to ethyl alcohol recovery as a 95% water azeotrope. The azeotrope is further distilled to give anhydrous ethyl alcohol. The

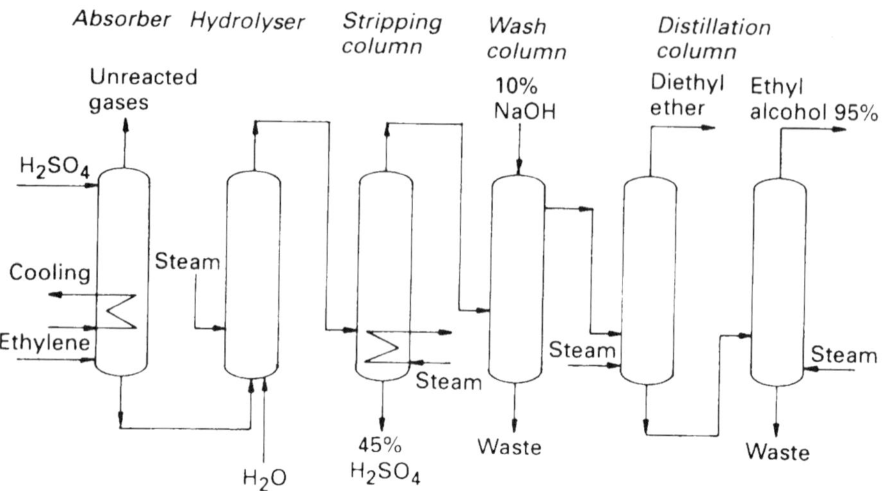

FIGURE 53 Ethyl alcohol from ethylene by indirect hydration

amount of by-product ethyl ether, normally in the range 5–10%, can be varied by controlling the ethylene:sulphuric acid ratio and hydrolysis conditions.

In order to reduce the amount of energy required to regenerate the sulphuric acid used, pressure in the absorption tower is reduced to 5 bar. The resultant lower absorption of ethylene has been improved by the addition of 7% silver sulphates. In this modified route, acid can be recovered by flash evaporation instead of expensive reconcentration.

Reaction

$$CH_2{=}CH_2 + H_2SO_4 \rightarrow C_2H_5OSO_2OH$$

$$2CH_2{=}CH_2 + H_2SO_4 \rightarrow C_2H_5OSO_2OC_2H_5$$

$$C_2H_5OSO_2OH + C_2H_5OSO_2OC_2H_5 + 3H_2O \rightarrow 3C_2H_5OH + 2H_2SO_4$$

$$2C_2H_5OH \rightarrow C_2H_5OC_2H_5 + H_2O$$

Raw material requirements and yield

Raw materials required per tonne of ethyl alcohol:

Ethylene	630kg

Yield 92%

3. From methyl alcohol and methyl acetate by carbonylation

The process is carried out in three stages. In the first stage methyl alcohol, produced from synthesis gas (see Methyl Alcohol), is combined with carbon monoxide in the liquid phase in the presence of carbonyls of non-noble metals such as tungsten, molybdenum or chromium. The acetic acid formed is esterified with methyl alcohol to methyl acetate in a tower reactor. (*See Figure 54*)

The reaction mixture is distilled and overheads are recycled to the reactor, while the crude acetic acid stream is dried before passing to the ethyl alcohol unit. The methyl acetate is dried and hydrolysed to ethyl alcohol and methyl alcohol.

This process has been modified so that the methyl acetate formed is carbonylated to acetic anhydride which is then reacted with methyl alcohol and ethyl alcohol to yield their respective acetates. These are separated by distillation and ethyl acetate is hydrolysed in the presence of sulphuric acid to ethyl alcohol. The methyl acetate is carbonylated to ethyl alcohol.

Reaction

$CH_3OH + CO + 2H_2 \rightarrow C_2H_5OH + H_2O$

Raw material requirements and yield

Raw materials required per tonne of ethyl alcohol:

Methyl alcohol	730kg

Yield 95%

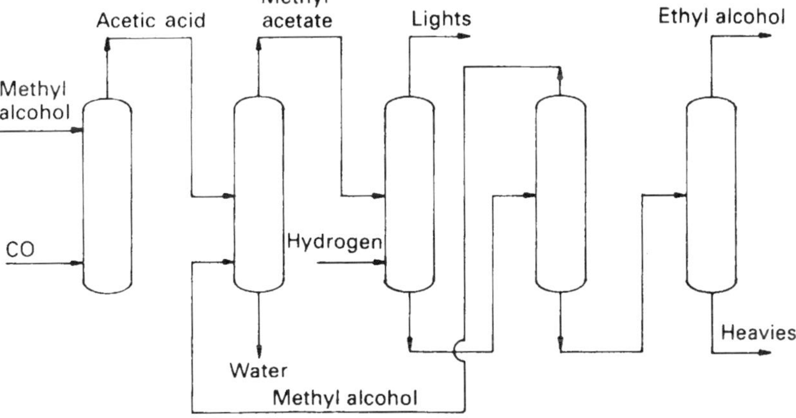

FIGURE 54 Ethyl alcohol from methyl alcohol and methyl acetate by carbonylation

PROPERTIES

Colourless, flammable, volatile liquid with a burning taste. Miscible with water, ether, acetone, benzene and a wide range of organic products. Can form explosive mixtures with air.

Molecular Weight	46.07
Density at 20°C	0.789
Melting Point	–114.1°C
Boiling Point	78.3°C
Autoignition Temperature	423°C
Flammability Limits in air	
lower	3.3vol%
upper	19.0vol%
Flash Point Closed Cup	13°C
Vapour Density (air = 1)	1.59
Exposure Limit COSHH	1000ppm 8 hour TWA
Exposure Limit ACGIH	1000ppm 8 hour TLV-TWA

GRADES

Absolute 99.7%, fine 96.5%, denatured spirit 88%. The denatured product, known as methylated spirit, or industrial spirit, has a wide range of additives or colorants added, such as 0.5–1.0wt% of crude pyridene coloured with methyl violet, which must be officially approved, to make the product unfit for human consumption.

INTERNATIONAL CLASSIFICATIONS

Ethyl alcohol >70%

UN No.	1170
CAS Reg No.	64–17–5
EINECS No.	200–578–6
EC No.	603–002–00–5
Description	Flammable liquid
Packing Group	II
Emergency Action Code	2YE
HI (Kemler Code)	33

Ethyl alcohol >24% <70%

UN No.	1170
CAS Reg No.	64–17–5
EINECS No.	200–578–6
EC No.	603–002–00–5
Description	Flammable liquid
Packing Group	III
Emergency Action Code	2Y
HI (Kemler Code)	30

APPLICATIONS

The two major outlets for ethyl alcohol are as a solvent and in chemical synthesis. Its solvent applications in pharmaceuticals, toiletries, cosmetics, detergents, flavours and surface coatings account for around 45% of total demand and have grown rapidly in recent years.

Ethyl alcohol is used as a chemical intermediate for the manufacture of esters, glycol ethers, acetic acid, acetaldehyde and ethyl chloride but several of the traditional routes have declined due the introduction of new processes which convert ethylene directly. Ethyl alcohol derivatives consume another 35% of ethyl alcohol production. World growth rate is 1–2% per year.

HEALTH AND HANDLING

Ethyl alcohol vapour is not hazardous under normal usage, but prolonged exposure can cause irritation to the lungs, dizziness, headaches, and in high concentrations nausea and narcosis. Concentrations of ethyl alcohol in excess of 70% can cause severe gastric damage (if taken internally). The liquid defats the skin leading to cracking and a risk of dermatitis. When handling the product, protective clothing, goggles, impervious gloves and rubber boots should be worn. Contact lenses, which can concentrate the vapours, should be avoided.

Ethyl alcohol is not corrosive to metals and storage containers can be made from mild steel, copper or aluminium. Tightly closed containers should be stored in a cool, well-ventilated area. Because of the risk of fire and explosion, all equipment must be earthed and any sources of ignition avoided.

Spills should be contained and absorbed with sand or vermiculite and disposed of according to local regulations. Fires should be extinguished using dry chemical or carbon dioxide, by staff wearing self-contained breathing apparatus.

All movements of ethyl alcohol, its storage and use are subject to governmental regulations.

MAJOR PLANTS

Plants with capacities greater than 60 000 tonnes per year:

SODES	Lillebonne	France
Erdolchemie	Cologne	Germany
Huels	Herne	Germany
BP Chemicals	Baglan Bay	UK
	Grangemouth	UK

Complex	Sumgait	Azerbaijan
Millenium Petrochemicals	Tuscola	US
Union Carbide	Texas City	US
Shell Oil/SABIC	Al Jubail	Saudi Arabia
Japan Ethanol	Yokkachi	Japan

MAJOR LICENSORS

Ethylene direct oxidation	*ABB Lummus Global*
	BP Chemicals
	Huels
	Millenium
	Shell Development
	Stone & Webster
Two-phase hydration	*Union Carbide*
Syngas	*IFP*
	Monsanto/BASF
Methyl alcohol	*Halcon-Scientific Design*
	Haldor Topsoe

Ethylbenzene

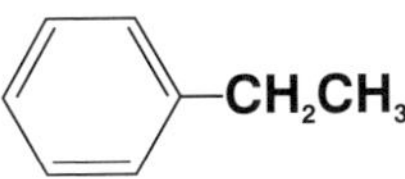

SYNONYMS

ETHYLBENZENE phenylethane, ethylbenzol

Over 90% of ethylbenzene is produced by the alkylation of benzene with ethylene. There are two commercial routes in use today:

- liquid-phase alkylation;
- vapour-phase alkylation.

Liquid-phase processes based on an aluminium chloride catalyst have been developed by many companies and achieved great importance. One of the major problems has been the disposal of waste aluminium chloride solution. In an attempt to reduce the quantity of catalyst required, ethyl chloride or hydrogen chloride have been used as promoters.

Monsanto's discovery – that at higher reaction temperatures, alkylation proceeds in a single homogeneous phase – led to a significant reduction in the quantity of aluminium chloride required.

Vapour-phase processes were unable to compete with aluminium chloride based processes until the introduction of the Alcar process by UOP. Although some plants still use this route, corrosion problems caused by traces of water on the boron trifluoride catalysts used have led to its demise. Mobil's vapour-phase process, developed in the 1970s, utilizes a synthetic zeolite catalyst to overcome the corrosion difficulties. An added advantage is the lack of waste disposal problems due to the environmental inertness of the catalyst. Both vapour-phase processes can use feedstocks containing as little as 10% of ethylene.

The latest development in alkylation technology, which was commercialized by Mobil in 1995, is based on a new zeolite catalyst. It overcomes oligomerization problems found in liquid-phase processes.

Ethylbenzene is also contained in C_8 aromatic streams arising from catalytic reforming, and absorption and distillation processes have been developed for its recovery. Superfractionation involving multi-column, multi-stage technology has been proposed, and several plants were built in the 1960s. Today's high energy costs have made this route uncompetitive, so that now little ethylbenzene production is made by this route.

In the non-communist world, the liquid-phase process accounts for around half of total ethylbenzene capacity. The Monsanto liquid-phase and Mobil vapour technologies are the dominant processes now in use.

Capacities range from 20 000 to 1 300 000 tonnes per year.

PROCESSES

1. From benzene and ethylene by liquid-phase alkylation

In the Monsanto process, 99% pure benzene is passed through a drying column to remove all traces of moisture before being fed into an alkylation reactor containing a catalyst. (*See Figure 55*)

The liquid catalyst complex consists of aluminium chloride promoted by hydrogen chloride. This is supplied by ethyl chloride in benzene as a diluent which breaks down to hydrogen chloride during the reaction. The presence of sulphur, toluene, xylenes and paraffins in the benzene feed are undesirable as they lead to unwanted by-products.

A moisture-free ethylene stream of 15–100mole% purity, containing no other unsaturated hydrocarbons, is sparged into the reactor at a carefully controlled rate. The

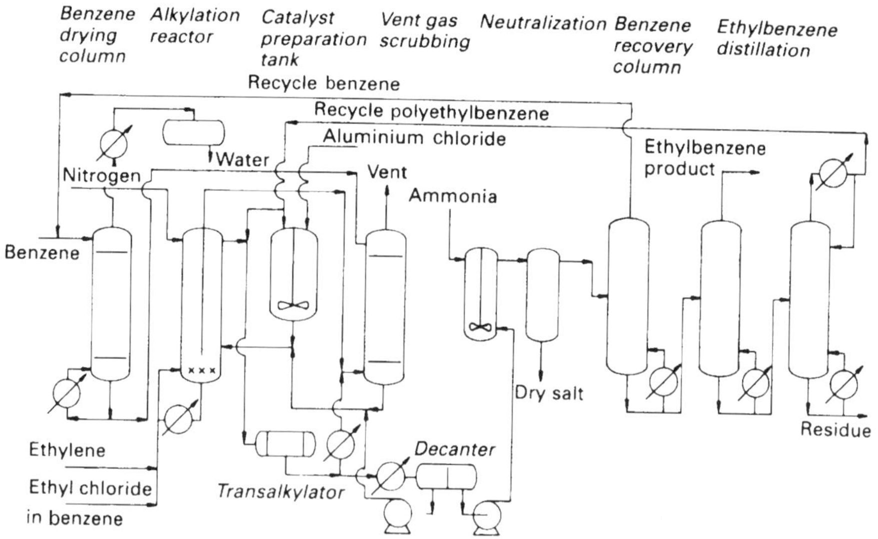

FIGURE 55 Ethylbenzene from benzene and ethylene by alkylation (liquid phase)

reaction temperature is maintained at 160–180°C and a pressure of 1 bar is used to maintain the reactants in the liquid phase. Benzene is present in excess to maximize the production of ethylbenzene and reduce the formation of higher alkyl benzenes.

The alkylated liquid leaving the reactor is cooled, washed with water to remove aluminium chloride and hydrogen chloride. The catalyst waste stream is not recovered but can be sold for use in water treatment. The organic ethylbenzene phase is washed with alkali to remove any remaining acid before being purified by fractionation.

The separation requires three columns. Unreacted benzene is recovered overhead from the first column, which is operated at slightly elevated pressure, and recycled. In the second column, which usually operates under reduced pressure, pure ethylbenzene is split from the heavy organic products. These are fed into the third column where diethylbenzene and polyethylbenzenes are separated. Residual organic compounds are burnt as fuel.

Because of the low catalyst concentrations employed, recycle polyethylbenzenes cannot be returned to the alkylation reactor, as in high concentrations they would terminate the reaction. Instead they pass to a separate reactor operating at a temperature below 130°C where transalkylation occurs. Aluminium chloride is removed from the reaction mixture before it joins the ethylbenzene stream in the separation section.

The advantages of the Monsanto process over other aluminium chloride processes are a decrease in the amount of catalyst used (because the reaction takes place in a single homogeneous phase instead of separation into a two-phase system) and higher yields. Waste by-product aluminium chloride is reduced and the higher heat of reaction can be used to generate low-pressure steam.

Reaction

$$C_6H_6 + CH_2{=}CH_2 \rightarrow C_6H_5C_2H_5$$

Raw material requirements and yield

Raw materials required per tonne of ethylbenzene:

Benzene	740kg
Ethylene	265kg
Catalyst	Small

Yield 99%

2. From benzene and ethylene by vapour-phase alkylation

In the Mobil process, dry benzene, ethylene and recycle polyethylbenzene are preheated and fed into a fixed multi-bed reactor containing a crystalline aluminosilicate zeolite catalyst. Usually two reactors are employed so that the catalyst in one

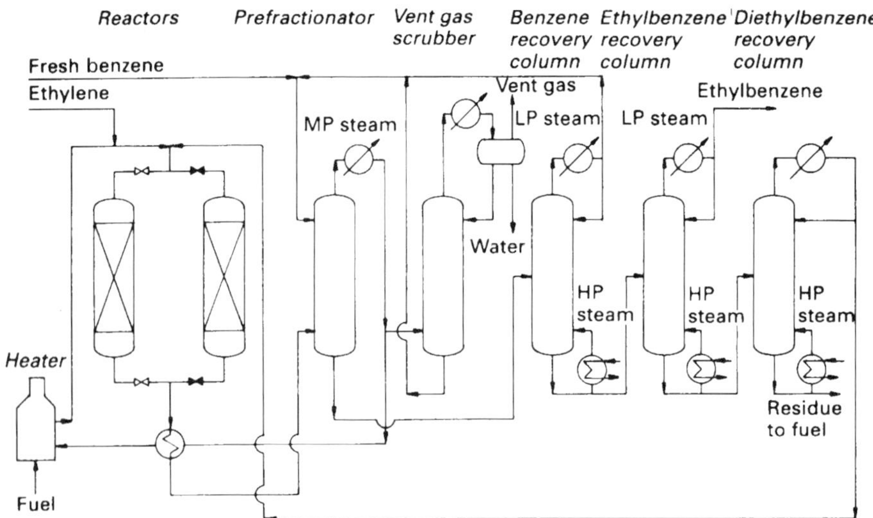

FIGURE 56 Ethylbenzene from benzene and ethylene by alkylation (vapour phase)

can be regenerated while the other is in production. The reaction conditions are 400–450°C and a pressure of 20–30 bar. (*See Figure 56*)

The composition of the feed is adjusted to give a molar ratio of benzene:ethylene of 7.5:1. Ethylene feedstocks of 15–100% purity can be used. Transalkylation and alkylation occur simultaneously and the heat generated by the exothermic reaction is used to generate steam.

Benzene is flashed off from the hot reactor vapours in the prefractionator and recycled. Any residual benzene is recovered in the first column, and ethylbenzene is separated from the heavy organic products in the second column. In the third column, the heavy organics are distilled to separate polyethylbenzenes for recycle.

The catalyst is regenerated every 6–8 weeks to remove any coke formed, an operation that takes approximately 36 hours. Usually two reactors are available so that one can be used while the other is being regenerated.

The major advantages of this process are that the catalyst is less sensitive to impurities, is non-hazardous and non-corrosive and no waste streams are produced. No catalyst recovery, waste treatment equipment, high-alloy materials and special linings for reactor construction are required, thus effecting considerable cost savings.

Yield 98–100%

3. From refinery gases

The Alcar vapour-phase process is designed to utilize refinery and coke-oven gas streams containing 8–10mole% of ethylene.

Dehydrated benzene and ethylene in a molar ratio around 7.5:1 are preheated and fed into a reactor. The ethylene must be free from oxygenates and sulphur compounds. The alkylation reaction is carried out at 100–150°C, a pressure of 25–35 bar and in the presence of a boron trifluoride catalyst on an inert base.

Transalkylation of recycled polyethylbenzene takes place in a separate reactor at a temperature of 180–230°C. Gases from both reactors are combined, flashed and fed into a benzene recovery column, where benzene is separated and recycled. The catalyst is removed and ethylbenzene is separated from the benzene column bottoms in the recovery column. Polyethylbenzene is recovered from the third column and recycled.

The major disadvantage of the Alcar process is the high maintenance cost which can be incurred due to severe corrosion caused by traces of water. Its advantage is that it can accommodate ethylene streams containing 8–10% ethylene providing catalyst poisons are absent.

4. From mixed xylene streams

Ethylbenzene can be recovered from xylene concentrates coming from catalytic reforming which may contain up to 25% of ethylbenzene. In order to separate out the ethylbenzene, close fractionation known as superfractionation has to be employed. This requires three 61m columns in series containing a large number of plates and high reflux ratios. This route is no longer competitive due to the high cost of energy required.

PROPERTIES

Colourless liquid with an odour of xylene. Highly flammable. Practically insoluble in water, but soluble in ethyl alcohol, ether, and benzene.

Molecular Weight	106.17
Density at 20°C	0.867
Freezing Point	–94.9°C
Boiling Point	136.2°C
Autoignition Temperature	460°C
Explosive Limits in air	
lower	0.99vol%
upper	6.7vol%
Flash Point Closed Cup	15°C
Vapour Density (air = 1)	3.7
Exposure Limit COSHH	125ppm 15 minutes 100ppm 8 hour TWA
Exposure Limit ACGIH	125ppm TLV-STEL 100ppm TLV-TWA

GRADES

Technical, styrene-grade >99.5% containing <40gm/kg diethylbenzene.

INTERNATIONAL CLASSIFICATIONS

UN No.	1175
CAS Reg No.	100–41–4
EINECS No.	202–849–4
EC No.	601–023–00–4
Description	Flammable liquid
Packing Group	II
Emergency Action Code	3YE
HI Kemler Code)	33

APPLICATIONS

The most important outlet for ethylbenzene, accounting for almost 99% of total demand, is as an intermediate for the manufacture of styrene.

Other minor uses are as a solvent in the paint industry, in the production of dyes, and as the raw material for the manufacture of diethylbenzene and acetophenone.

Future growth will depend on the demand for styrene, which is currently increasing at 4% per year. Because of their interdependence, the production of ethylbenzene and styrene has become fully integrated in many companies.

HEALTH AND HANDLING

Ethylbenzene is less toxic than benzene, but it has an irritating effect on the eyes, skin and respiratory tract. Care should be exercised when handling and protective clothing and goggles worn. Adequate ventilation must be provided as ethylbenzene absorption can result in chronic poisoning through inhalation leading to narcosis, kidney and liver disease.

It is normally stored in mild steel containers but iron, copper or aluminium can be used. The use of rubber valves should be avoided because ethylbenzene attacks rubber and some plastics. Care should be taken not to expose ethylbenzene to heat, flames or strong oxidizing agents.

Spills should be contained to avoid contamination of streams or waterways as the product is toxic to fish. Ethylbenzene is a dangerous fire hazard and the vapours being heavier than air can travel a distance with the risk of flashback. Carbon dioxide, dry chemical or alcohol foam can be used to fight fires. A water stream should be avoided because it could disperse the fire. Firefighters must wear self-contained breathing apparatus.

Ethylbenzene is very flammable and special regulations affect its transportation.

MAJOR PLANTS

Plants with capacities greater than 500 000 tonnes per year:

BASF	Antwerp	Belgium
Dow Chemical	Terneuzen	Netherlands
EniChem	Mantua	Italy
Lyondell	Channelview	US
Chevron	St. James	US
Cos-Mar	Carville	US
Dow Chemical	Freeport	US
Huntsman Corp.	Bayport	US
Sterling Chemicals	Texas City	US
SADAF	Al Jubail	Saudi Arabia
Asahi Chemical	Okayama	Japan

MAJOR LICENSORS

BASF
CDTECH
Chiyoda
Fina Technology
Halcon-Scientific Design
Mobil
Monsanto
Raytheon Engineering
Union Carbide/Badger
UOP

Ethyl Chloride C_2H_5Cl

SYNONYMS

ETHYL CHLORIDE chloroethane, muriatic ether, chlorethyl

The importance of ethyl chloride grew with the increasing use of the motor car, as it is the starting material for the production of tetraethyl lead, an anti-knock additive for gasoline. The current trend towards lead-free gasoline has severely affected demand for this product.

There are two commercial routes to ethyl chloride: the hydrochlorination of ethylene and the chlorination of ethane. Most ethyl chloride is made by the former process which can be carried out in the vapour or liquid phase. Shell have developed a balanced process which combines the hydrochlorination of ethylene with the chlorination of ethane. The hydrogen chloride produced by the ethane chlorination is consumed by the ethylene reaction leading to minimal hydrogen chloride by-product disposal problems.

An earlier route, the esterification of ethyl alcohol with hydrogen chloride, is no longer in use. Ethyl chloride is also obtained as a by-product of ethylene dichloride manufacture.

Capacities range from 3000 to 60 000 tonnes per year.

PROCESSES

From ethylene by hydrochlorination

In the vapour-phase process, ethylene and anhydrous hydrogen chloride in equal molecular proportions are mixed, preheated and introduced into a reactor containing a catalyst consisting of aluminium chloride present as either a fluid or fixed bed. Other catalyst systems used are thorium oxychloride on silica, or platinium on alumina. The reaction takes place at 250–400°C under a pressure of 5–15 bar. Alternatively, it can be carried out in the liquid phase with a mixture of aluminium chloride, ethyl chloride and ethylene dichloride at 35–40°C and a pressure below 5

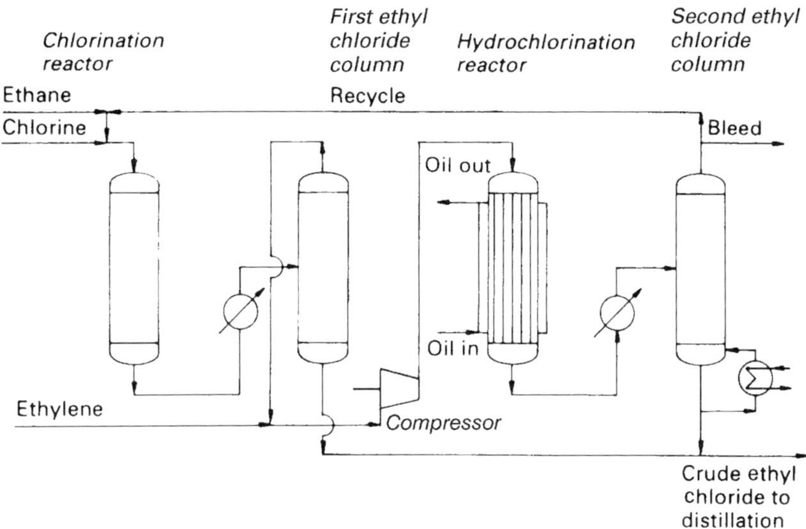

FIGURE 57 Ethyl chloride from ethylene by hydrochlorination

bar. In this case the temperature of the reaction is controlled by refluxing the ethyl chloride formed and varying the raw material feed rates. (*See Figure 57*)

The reaction products are fed into a column where ethyl chloride is flashed off from the less volatile by-products. Ethyl chloride is purified by distillation. Unreacted ethylene and hydrogen chloride are recycled. In the liquid-phase reaction, part of the remaining liquid is bled off to remove high-boiling by-products and additional catalyst is added to make up the loss.

Reaction

$C_2H_4 + HCl \rightarrow C_2H_5Cl$

Raw material requirements and yield

Raw materials required per tonne of ethyl chloride:

Ethylene	490kg
Hydrogen chloride	625kg

Yield

on ethylene	90%
on chlorine	95%

OTHER PROCESSES

In the Shell process, the hydrochlorination of ethylene and the chlorination of ethane are carried out in two separate reactors. Chlorine vapour is reacted non-catalytically with preheated ethane and a recycle gas stream containing ethylene at 400–500°C. Ethane is present in excess in order to achieve high selectivity for ethyl chloride. The reaction products are passed to a distillation column where unreacted ethane, inerts and hydrogen chloride by-product are removed overhead. Ethyl chloride and other by-products in the bottom stream are sent to storage.

Ethylene and the overhead stream are combined to give an ethylene–hydrogen chloride molar ratio of 1:1. The mixture is compressed and preheated before being fed into a second fixed-bed reactor where more ethyl chloride is formed. Any of the range of vapour-phase catalysts described in the ethylene-based process can be used. Unreacted materials are separated from the reaction mixture by distillation and collected overhead. The bottoms containing crude ethyl chloride are combined with the other ethyl chloride stream. Ethyl chloride is purified by distillation. A yield of 90–95% is claimed.

Reaction

$$C_2H_6 + Cl_2 \rightarrow C_2H_5Cl + HCl$$

$$C_2H_4 + HCl \rightarrow C_2H_5Cl$$

PROPERTIES

Colourless gas which can be compressed to a liquid with an ethereal odour. Highly flammable. Soluble in most organic hydrocarbons and slightly soluble in water.

Molecular Weight	64.52
Density at 20°C	0.921
Melting Point	–138.7°C
Boiling Point	12.2°C
Autoignition Temperature	518°C
Flammability Limits in air	
lower	3.6vol%
upper	14.8vol%
Flash Point Closed Cup	–43°C
Vapour Density (air = 1)	2.22
Exposure Limit COSHH	1250ppm 15 minutes 1000ppm 8 hour TWA
Exposure Limit ACGIH	1250ppm TLV-STEL 1000ppm TLV-TWA
Classified A3, animal carcinogen	

GRADES

Technical 99.5%.

INTERNATIONAL CLASSIFICATIONS

UN No.	1037
CAS Reg No.	75–00–3
EINECS No.	200–830–5
EC No.	602–009–00–0
Description	Flammable gas
Packing Group	Not allocated
Emergency Action Code	3WE
HI (Kemler Code)	23

APPLICATIONS

The major application for ethyl chloride is in the manufacture of tetraethyl lead which is an anti-knock additive to gasoline. The increasing movement towards the use of lead-free gasoline has resulted in a rapid decline in this outlet. Other minor applications include the manufacture of dyes, cellulose plastics and pharmaceuticals. It can be used as a solvent and as a refrigerant.

Unless new outlets are found for ethyl chloride, demand is expected to continue to decline.

HEALTH AND HANDLING

Ethyl chloride vapour is irritating to the eyes, nose and throat, leading to dizziness, stomach cramps and narcosis. In contact with the skin, the liquid will cause frostbite and irritation. Contact lenses absorb and concentrate the vapour and must be avoided.

Store in a cool, well-ventilated, dry area away from oxidizing agents, powdered aluminium or zinc and potassium, sodium or calcium. All containers must be earthed to prevent static build-up and all handling and ventilation equipment must be explosion free.

If leaks occur, the area should be evacuated and attempts made to stop the flow. The liquid can be absorbed with paper towels which are placed in a closed container for disposal. Ethyl chloride must not be allowed to enter sewers because of the danger of explosion.

Ethyl chloride is a dangerous fire and explosion hazard. Carbon dioxide or dry chemical can be used to extinguish the fire but water must be avoided as it can scatter the flames. As the vapour is heavier than air, flashback is a hazard. All firefighting staff must wear protective clothing and self-contained breathing apparatus.

MAJOR PLANTS

Plants with capacities greater than 25 000 tonnes per year:

Elf Atochem	Le Point de Claix	France
Vestolit	Marl	Germany
EniChem	Novara	Italy
	Porto Maghera	Italy
Associated Octyl	Ellesmere Port	UK
PPG Industries	Lake Charles	US

MAJOR LICENSORS

Halcon-Scientific Design
Huels
Shell Development

Ethylene

$CH_2=CH_2$

SYNONYMS

ETHYLENE ethene, acetene

Ethylene is the largest volume hydrocarbon used in the petrochemical industry. Employed exclusively as a chemical intermediate, its impressive growth since World War II has been due to its ready availability at competitive prices from an abundance of economically priced feedstocks.

Ethylene is produced by steam cracking from a wide range of hydrocarbons including ethane, propane, butane, naphtha, liquefied petroleum gas (LPG) and gas oils. Refinery off-gases and light hydrocarbons recovered from natural gas are sources of ethane, propane and butane.

Feedstock patterns vary in different countries. In the US, ethane is the prime feedstock, but in West Europe and Japan most production arises from naphtha obtained by the distillation of crude oil. This situation is due to the demand for naphtha for catalytic reforming to gasoline and the ready supply of natural gas in the US. On the other hand, until the discovery of North Sea gas and oil, European producers lacked access to natural gas feedstocks and Middle East crude oil prices made naphtha attractive.

Crackers were constructed as part of a complex whereby the range of by-products, especially C_4s and aromatics, could be utilized. Although naphtha will still predominate as the prime feedstock, Europe has seen the construction of ethane or ethane–naphtha-based crackers due to the availability of North Sea gas. Ethylene plants based on ethane are cheaper to construct, less complicated to operate, and give high yields with minimal by-products and are attractive to companies not fully integrated to market propylene and aromatics associated with naphtha cracking. Decreasing availability of natural gas has led to an increase in the volume of ethane and naphtha being used.

Naphtha composition varies with the type of crude and refinery cracking conditions. Ethylene/olefin production from naphtha can be maximized by using:

- feedstocks with high *n*-paraffins content;
- high-severity cracking;
- quick quench of cracked gases.

Recent developments have concentrated on process optimization, computer control, reactor design, increasing ethylene yields by high temperature cracking at low reaction times and faster quenching systems. Attention has been paid to energy conservation by decreased refluxing and improved heat exchange. A notable introduction has been the Kellogg millisecond furnace.

Increasing feedstock costs and uncertainty of supply have led to research into alternative processes. The technology for crude or residual oil cracking has been developed by a number of companies and is in commercial operation in a few plants. Phillips Petroleum has a 'Triolefin' process for the disproportionation of propylene to ethylene and butylene.

Other routes to ethylene include: the catalytic hydrogenation of ethyl alcohol which is used in India, Brazil, Sweden and China; cracking feedstock derived from the gasification of coal utilized in SASOL's syngas plants in South Africa, and UOP/Mobil's methyl alcohol to olefins where methane is cheap.

Naphtha is the principal ethylene raw material in West Europe and Japan but in the US ethane is the prime feedstock. In the US, 52% of ethylene is produced from ethane, 22% from gas oil and the remainder from other feedstocks. Only 5% is obtained from naphtha unlike in Western Europe where the comparable figure is 71%. Gas oil and LPG each account for another 11%, the remainder coming from ethane cracking.

Future feedstock usage will be influenced by availability and price. Large sources of natural gas and natural gas liquids are to be found worldwide but are not fully exploited by petrochemical manufacturers due to their location and recovery.

Largest increases in new capacity are concentrated in regions which have access to large sources of oil and natural gas such as China, Mexico, and the Middle and Far East excluding Japan. Although all feedstock sources will grow, natural gas-based feedstocks will gain greater dominance in the market.

Capacities range from 50 000 to 1 870 000 tonnes per year.

PROCESSES

1. From hydrocarbons by steam cracking

The hydrocarbon stream is heated and then diluted by mixing with steam, before entering a tubular reactor. Depending on the feedstock used, cracking takes place at a temperature of 750–870°C under partial pressure with a residence time of up to one second. The presence of steam helps to reduce the amount of coking in the reactor tubes. The reaction is endothermic requiring considerable heat input. (*See Figure 58*)

The exit gases are rapidly quenched to 550–600°C to prevent secondary reactions. The heat is used to generate high-pressure steam. Raw gases from the quench tower are compressed in a multi-stage compressor system to around 32–38 bar. After each stage, liquid is removed and the remaining gases are treated with an aqueous caustic alcohol–amine mixture to remove any sulphurous gases and carbon dioxide. Water is condensed by further cooling and the gases dried by molecular sieves to prevent the formation of hydrates and ice. Hydrogen and methane are removed in the demethanizer. These gases are either burnt as fuel or purified prior to sale.

Bottoms from the demethanizer, containing C_2 and heavier products, are fed to the deethanizer, where acetylene, ethylene and ethane are separated overhead. Acetylene is hydrogenated and removed. In the C_2 splitter, ethylene is recovered by fractionation overhead while ethane in the bottom stream is recycled to the cracking furnaces.

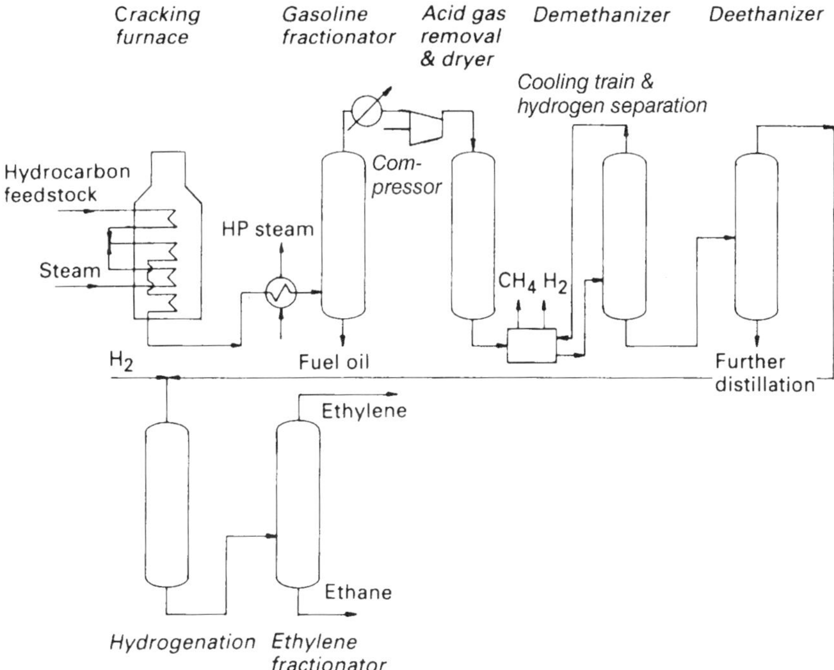

FIGURE 58 Ethylene from hydrocarbons by steam cracking

Effluent from the base of the deethanizer passes to the depropanizer and C_3 fractions are separated overhead from C_4 and higher carbon fractions. Propylene is separated from propane by fractionation. The propane stream is recycled for further cracking.

If naphtha feedstock is used, a wider range of products is produced. Bottoms from the depropanizer are fed to the debutanizer, where C_4s are extracted and sent for butadiene and butylenes recovery (see Butadiene). The remaining raw gasoline stream can be processed further to recover aromatics or used as gasoline feedstock (see Benzene).

Reaction

$C_2H_6 \rightarrow C_2H_4 + H_2$ ethylene

$2C_3H_8 \rightarrow C_3H_6 + H_2 + C_2H_4 + CH_4$ propylene

Product range and yield

The choice of feedstock and cracking conditions used determines the ratio of products obtained. Typical yields in wt% for various feedstocks are given below:

Product	*Ethane*	*Propane*	*Butane*	*Naphtha*	*Gas oil*
Ethylene	79–84	42–45	30–40	28–38	23–26
Propylene	1–3	14–18	16–20	13–18	13–14
Butadiene	2	2	2.5–3	4–5	4.8–5
Butanes/butenes	1	1	6.5–6.8	4–5	4.5–5.3
Aromatics	0.4	3.5	3.4	7–14	10–13
Yield	30–35%				

2. From propylene by disproportionation

A demethanized propylene–propane stream is fed into a reactor operating at 7 bar and a temperature of 360–450°C and containing a catalyst. A variety of catalysts can be used based on the oxides of molybdenum, cobalt, tungsten or rhenium on an alumina or silica support. (*See Figure 59*)

The effluent gases are fractionated and ethylene is recovered overhead. Any unconverted propane is recycled to the reactor. Bottoms from the fractionator pass into a butene purifier where high-purity butenes are collected overhead. Propylene conversion per pass is around 40%.

Reaction

$2C_3H_6 \rightarrow C_2H_4 + C_4H_8$

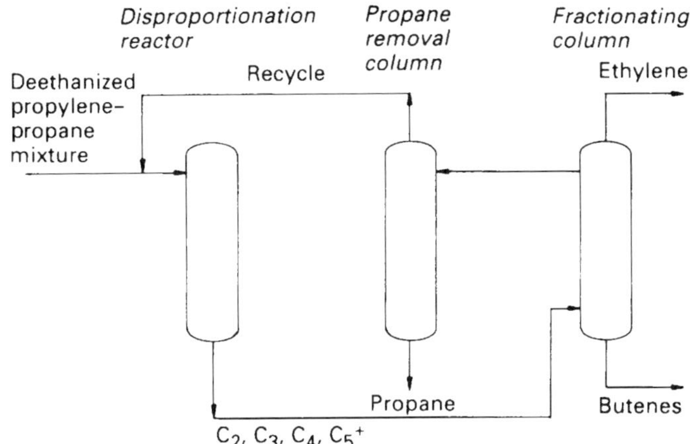

FIGURE 59 Ethylene from propylene by disproportionation

Raw material requirements and yield

Raw materials required per tonne of ethylene (theoretical):

Propylene	3000kg

Yield 95%

3. From ethyl alcohol by dehydrogenation

Ethyl alcohol is vaporized by preheating with high-pressure steam before passing over a fixed bed of activated alumina and phosphoric acid or alumina and zinc oxide contained in a reactor. The reactors can be either isothermal or adiabatic. The temperature is maintained at 296–315°C. Heat for the endothermic reaction is supplied by condensing vapour to give up its latent heat in the reactor shell. Accurate temperature control is required to minimize the formation of by-product acetaldehyde or ether. The catalyst is regenerated every few weeks by passing air and steam over it to remove carbon deposits. (*See Figure 60*)

Gases from the top of the reactor are quenched and water washed to remove any unreacted ethyl alcohol and acetaldehydes. This is followed by scrubbing with dilute caustic soda which absorbs any carbon dioxide. The gas is dried, compressed and purified by passing through an activated carbon bed which removes any C_4s.

A fluid-bed process has been developed which provides a more efficient means of temperature control and conversion rates of up to 99% are claimed.

Reaction

$$2C_2H_5OH \rightarrow (C_2H_5)_2O + H_2O$$

$$(C_2H_5)_2O \rightarrow 2C_2H_4 + H_2O$$

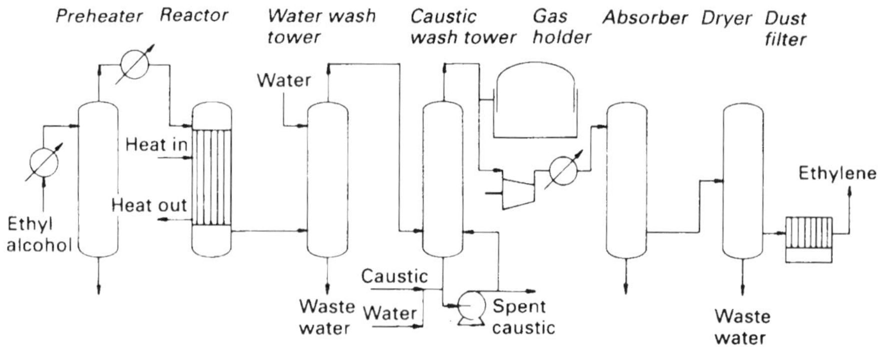

FIGURE 60 Ethylene from ethyl alcohol by dehydrogenation

Raw material requirements and yield

Raw materials required per tonne of ethylene:

Ethyl alcohol	1748kg
Yield 94%	

OTHER PROCESSES

From crude or residual oil

Several companies have developed processes which crack crude or residual oil. A few are in commercial operation, but by-product utilization and high operating costs have caused problems.

Crude oil or heavy oil distillate and super-heated steam are cracked at 920–960°C with a residence time of 0.01–0.02 seconds. The burner in the cracking furnace is fed with oil and oxygen to generate the heat required. Operating conditions can be varied to give a propylene/ethylene ratio from 0.2 to 0.6.

Lurgi developed a process where crude oil preheated to 340°C is injected together with steam into a bed of hot sand in a fluidized-bed reactor. The hydrocarbons reach a temperature of 1300–1550°C in 0.3–0.5 seconds. The sand is continuously removed and any coke is burnt off before the sand is recycled to the reactor.

From coal

In South Africa, ethylene is produced by the Fischer–Tropsch process from gases obtained via the Lurgi coal–gasification process. It uses a promoted iron catalyst in a recycled fluid bed operating at 150–450 bar pressure. Significant amounts of methane are formed in addition to light olefins and gasoline. The process is only economic if cheap coal is available.

PROPERTIES

Colourless gas with faintly pleasant odour. Highly flammable and forms explosive mixtures with air. Slightly soluble in water. Soluble in liquid hydrocarbons.

Molecular Weight	28.05
Density at 0°C	0.98
Melting Point	–169.2°C
Boiling Point	–103.7°C
Autoignition Temperature	450°C
Explosive Limits in air at 1 bar and 20°C	
lower	2.7vol%
upper	36.0vol%
Flash Point Closed Cup	–136°C
Vapour Density (air = 1)	1.8
Exposure Limit COSHH	Asphyxiant
Exposure Limit ACGIH	Asphyxiant

GRADES

Technical 99.9%.

INTERNATIONAL CLASSIFICATIONS

UN No.	1962
CAS Reg No.	74–85–1
EINECS No.	200–815–3
EC No.	601–010–00–3
Description	Flammable gas
Emergency Action Code	2PE
HI (Kemler Code)	23

APPLICATIONS

Ethylene is the prime hydrocarbon raw material for the production of petrochemicals due its ready availability at low cost. In the US, polyethylenes consume 53% of total ethylene demand. Polyethylene can be broken down to high-density, linear-low-density, and low-density polyethylene, which account for 24%, 14% and 15% respectively of ethylene consumption. Ethylene derivatives are also used in the preparation of two other high-volume polymers via their intermediates ethylene chloride and ethyl benzene. These chemical derivatives utilize 16% and 7% respectively of total ethylene consumption.

Ethylene oxide, consuming 15% of ethylene, is the highest tonnage chemical derivative. Ethylene oligomers take a further 5%. Ethyl alcohol and vinyl acetate employ another 1% each.

Most ethylene consumption is either captive or consumed locally. In order to provide continuity of supply, major pipelines have been constructed in Europe, North America and Russia.

World ethylene demand is expected to grow at 4% per year up to 2001.

HEALTH AND HANDLING

Although ethylene gas causes no hazard to the skin or eyes, it is a simple asphyxiant at high concentrations. The liquid can cause frostbite if in contact with skin or eyes. The use of protective clothing, goggles and gloves are advised when handling the liquid. When entering an ethylene-rich area, a safety line should be worn attached to a stand-by colleague who has self-contained breathing apparatus.

Ethylene is normally stored as a liquid in refrigerated tanks at a pressure up to 7 bar or at room temperature at a pressure of 56–100 bar in underground cavities. Stable in closed containers at room temperature, ethylene should be stored in a well-ventilated area to ensure dispersal of any gas leakage. All sources of ignition must be avoided and electrical equipment earthed to prevent static build-up. Contact with strong oxidizing agents must be avoided, and ethylene is spontaneously explosive with chlorine in the presence of sunlight.

Ethylene presents a dangerous fire and explosion hazard. Although small fires can be extinguished with carbon dioxide the safest course, if possible, is to close off the source of ethylene, and allow the fire to burn itself out. This procedure prevents the danger of reignition. Special care must be taken with liquid spillage which can lead to a build-up of vapour especially in traps or sumps.

MAJOR PLANTS

Plants with capacities greater than 800 000 tonnes per year:

Fina Olefins	Antwerp	Belgium
ROW	Wesseling	Germany
Vestolen	Gelsenkirchen	Germany
BP Chemicals/ICI	Wilton	UK
Dow Chemical	Terneuzen	Netherlands
DSM	Geleen	Netherlands
Amoco Chemicals	Chocolate Bayou	US
Chevron Chemicals	Port Arthur	US
Dow Chemical	Freeport	US
	Plaquemine	US
Exxon Chemical	Baton Rouge	US
	Baytown	US
Huntsman Corp.	Port Arthur	US
Millenium Petrochemicals	Channelview	US
	Deer Park	US

Phillips Petroleum	Sweeny	US
Shell Chemical	Deer Park	US
	Norco	US
Union Carbide	Taft	US
Nova Chemicals	Joffre	Canada
Copene	Camacari	Brazil
SABIC/Mobil	Yanbu	Saudi Arabia
SABIC/Shell	Al Jubail	Saudi Arabia
Mitsubishi Chemical	Kashima	Japan
Chinese Petroleum	Linyuan	Taiwan
Petrochemical Corp.	Palau Ayer Merbau	Singapore

MAJOR LICENSORS

ABB Lummus Global
Braun
Brown & Root
Halcon-Scientific Design
Kellogg
Linde
Lurgi
Phillips Petroleum
SASOL
Stone & Webster
Technip
UOP/Mobil

Ethylene Dichloride (EDC)

$ClCH_2CH_2Cl$

SYNONYMS

ETHYLENE DICHLORIDE 1,2-dichloroethane, ethane dichloride, glycol dichloride, Dutch Oil

Ethylene dichloride (EDC) was first synthesized in the late 18th century, but did not achieve its present importance until the arrival of polyvinyl chloride (PVC), for which it is the starting material.

It is produced industrially by the chlorination of ethylene, either directly with chlorine or by oxychlorination using hydrogen chloride. In practice, both routes are carried out together as part of an integrated ethylene–EDC–vinyl chloride process. Hydrogen chloride, generated as a by-product from the cracking of ethylene dichloride to vinyl chloride, is used in the oxychlorination process. Excess hydrogen chloride generated by other processes such as the production of perchloroethylene or carbon tetrachloride, can be utilized.

Direct chlorination can be carried out in the liquid phase at either high or low temperatures. Low-temperature chlorination reduces the quantity of by-product formation, but energy consumption is high due to steam requirements for ethylene dichloride recovery. In the high-temperature process, the heat of reaction is utilized for ethylene dichloride distillation leading to considerable energy savings.

Most plants employ the chlorination–oxychlorination route because of its economic advantages, but a few sites where acetylene is available combine the direct chlorination of ethylene with the hydrochlorination of acetylene (see Vinyl Chloride).

Although considerable research has been carried out on the replacement of ethylene by ethane, problems with catalyst selectivity and performance have yet to be resolved. Ethylene dichloride is produced as a by-product of the oxychlorination route to ethylene oxide, but this route is no longer of any importance.

Capacities range from 50 000 to 1 800 000 tonnes per year.

PROCESSES

1. From ethylene and chlorine by direct chlorination

Liquid chlorine (free from bromine to avoid the production of brominated by-products) and pure ethylene in a stoichiometric chlorine–ethylene ratio are reacted in the presence of ferric chloride. The catalyst concentration varies from 100 ppm to 0.5wt%. Other catalysts, such as antimony, copper, tellurium, bismuth and tin chlorides have been suggested as alternatives. Oxygen or air can be added to check substitution chlorination. (*See Figure 61*)

The chlorination reaction can be carried out at low or high temperature. In the low-temperature process, ethylene chlorination proceeds at 20–70°C in ethylene dichloride as the solvent for temperature control. Internal or external heat exchangers remove excess heat from the exothermic reaction. One of the major advantages of this process is the low amount of by-product formation.

In the high-temperature process, a gaseous mixture of ethylene and chlorine, together with recycled ethylene dichloride, are reacted at a temperature of 100–150°C, and the heat generated is used to distil the ethylene dichloride formed. Developments in reactor design have led to higher conversion rates comparable with those achieved at low temperatures. This route has been used successfully as part of an integrated direct chlorination–oxychlorination–vinyl chloride process. Heat generated by the reaction is sufficient to distil all the ethylene chloride produced by the various reactions, thus leading to considerable energy savings.

Catalysed direct chlorination of ethylene in the gaseous phase has been proposed, but control of the process is difficult because of the high amounts of heat generated by the reaction.

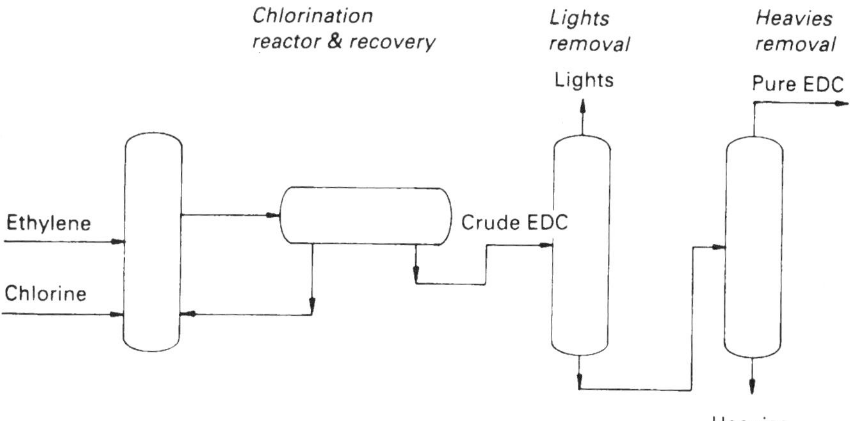

FIGURE 61 Ethylene dichloride from ethylene and chlorine by direct chlorination

Reaction

$$C_2H_4 + Cl_2 \rightarrow CH_2Cl-CH_2Cl \qquad \text{(direct chlorination)}$$

Raw material requirements and yield

Raw materials required per tonne of ethylene dichloride:

Ethylene	295kg
Chlorine	735kg

Yield

on ethylene	96–98%
on chlorine	98%

2. From ethylene by chlorination and oxychlorination

Pure ethylene (to minimize the formation of by-products) and hydrogen chloride are heated, mixed with oxygen and fed into a reactor. The reaction takes place at 200–300°C and 4–6 bar in the presence of a catalyst, usually 3–12wt% of cupric chloride, with a contact time of 1–40 seconds.

The reactor can be of either a fixed-bed or fluidized-bed type, but the former can lead to temperature control problems due to the high quantity of heat generated by the reaction. One method used to overcome this is to dilute the cupric chloride catalyst with alumina, silicon carbide or graphite; another is to vary the cupric salt concentration in the catalyst at different zones in the reactor. Efficient heat transfer is required to reduce the formation of hot spots which can lead to embrittlement of the reactor tubes.

Fluidized-bed reactors do not suffer from this problem and, as they operate at slightly lower temperatures, feed control is less crucial because the reaction takes place within the explosive limit. They are more flexible and catalyst life is longer, averaging over 3 years.

The reacted gases are quenched with water; the aqueous hydrogen chloride solution is sent to the recovery section where water is removed and the acid recovered. The remaining gases are cooled. The organic layer is washed with sodium hydroxide solution to remove chloral, and dried by azeotropic distillation. After scrubbing, the off-gases are either vented or compressed and recycled. Liquids from the base of the azeotropic distillation column are fed to the direct chlorination process.

If air is used instead of oxygen the reaction is easier to control because of the presence of inert nitrogen. However, oxygen-based processes operate at lower temperatures, considerably reducing the volume of vent gas. Ethyl chloride, 1,1,2-trichloroethane and chloral are produced as by-products.

Kellogg has developed an aqueous liquid-phase catalysed oxychlorination process. The reaction takes place at 170–185°C and a pressure of 17–19 bar in an aqueous solution of cupric salts. Because of corrosion problems caused by the aqueous catalyst solution at high temperatures, the process is not used commercially.

Reaction

$C_2H_4 + Cl_2 \rightarrow C_2H_4Cl_2$

$C_2H_4 + 2HCl + ½O_2 \rightarrow C_2H_4Cl_2 + H_2O$

Raw material requirements and yield

Raw materials required per tonne of ethylene dichloride:

Ethylene	302kg
Hydrogen chlorine	760kg

Yield

on ethylene	93–97%
on hydrogen chloride	96–99%

PROPERTIES

Colourless, highly volatile liquid with a pleasant odour. Toxic and flammable. Miscible with chlorinated hydrocarbons and most organic solvents.

Molecular Weight	98.96
Density at 20°C	1.257
Melting Point	–35.4°C
Boiling Point	83.7°C
Autoignition Temperature	413°C
Explosive Limits in air	
lower	6.2vol%
upper	15.9vol%
Flash Point Closed Cup	17°C
Vapour Density (air = 1)	3.42

Exposure Limit COSHH (Schedule 1) 5ppm 8 hour TWA (Maximum Exposure Limit)
Defined as a carcinogen under COSHH regulations.
Exposure Limit ACGIH 10ppm TLV-TWA

(OSHA have set PEL at 2ppm TLV-STEL and 1ppm TLV-TWA.)

GRADES

Inhibited >99%.

INTERNATIONAL CLASSIFICATIONS

UN No.	1184
CAS Reg No.	107–06–2
EINECS No.	203–458–1
EC No.	602–012–00–7

Description	Flammable liquid, toxic substance
Packing Group	II
Emergency Action Code	2YE
HI (Kemler Code)	336

APPLICATIONS

The major outlet for ethylene dichloride, accounting for around 98% of total consumption, is for the production of vinyl chloride.

The consumption of ethylene dichloride for the manufacture of chlorinated solvents of which 1,1,1-trichloroethane, tri- and tetra-chloroethylene are the most important, has declined rapidly due to legislation restricting their use in many countries. They are used as solvents for fats and other organics and as degreasants. Another minor use is as an intermediate in the production of ethylenediamines.

Global demand is expected to grow at 6% per year to 2000, with the greatest increase in the Asia–Pacific basin. Future growth will depend on the demand for PVC, which since the 1980s has been very variable due its outlets in the cyclical automotive and construction industries.

HEALTH AND HANDLING

Ethylene dichloride vapour is irritating to the eyes, nose and skin leading to dizziness, depression and vomiting. It is injurious to the liver, lungs, blood and the central nervous system. The material is absorbed by the skin. Liquid splashes in the eyes can lead to cornea opacity. Contact lenses must not be worn as they concentrate the vapour leading to eye damage. Protective clothing to prevent skin contact, goggles and shoes must be worn. Launder contaminated clothing before reuse and discard all footwear that comes into contact with the liquid.

Store in closed steel containers in a well-ventilated area free from sources of ignition. As hazardous explosions occur in contact with strong oxidizing agents, ammonia, and aluminium or magnesium powder, ethylene dichloride must not be stored in proximity to these materials. Protective clothing must be worn when handling ethylene dichloride and all equipment must be explosion-proof.

If spills occur, all sources of ignition must be extinguished, personnel evacuated and the liquid absorbed with sand or vermiculite. Any remaining traces can be flushed with water but care must be taken to ensure that no wastes reach sewers or watercourses. Clean-up staff must wear protective clothing, goggles and self-contained breathing apparatus and use non-sparking tools. Wastes should be disposed of in accordance with local regulations.

Ethylene dichloride is a dangerous fire and explosion hazard. Carbon dioxide, alcohol foam, sand or water spray can be used to fight fires. As ethylene dichloride vapours are heavier than air, flashbacks can occur. Toxic fumes are given off and all firefighters must wear self-contained breathing apparatus.

MAJOR PLANTS

Plants with capacities greater than 600 000 tonnes per year:

LVM	Tessenderlo	Belgium
Solvay	Jemeppe sur Sambre	Belgium
Elf Atochem	Lavera	France
Rovin	Botlek	Netherlands
EVC	Porto Marghera	Italy
Norsk Hydro	Rafnes	Norway
CONDEA Chemical	Lake Charles	US
Dow Chemical	Freeport	US
	Plaquemine	US
Formosa Plastics	Baton Rouge	US
	Point Comfort	US
Geon	La Porte	US
Georgia Gulf	Plaquemine	US
Occidental Chemical	Deer Park	US
Oxymar	Corpus Christi	US
PPG Industries	Lake Charles	US
Dow Chemical	Fort Saskatchewan	Canada
Petrobras-Fafen	Araucaria	Brazil
SADAF	Al Jubail	Saudi Arabia
Kashima VCM	Kashima	Japan
Tosoh Corp.	Shin-Nanyo	Japan
Formosa Plastics	Jenwu	Taiwan

MAJOR LICENSORS

ABB Lummus Global
Dow Chemical
EVC
Geon
Mitsui Petrochemical
PPG Industries
Stauffer Chemical
Tokuyama Soda
UOP
Vulcan Materials

Ethylene Glycol

$HOCH_2CH_2OH$

SYNONYMS

ETHYLENE GLYCOL 1,2-dihydroxyethane, 2-hydroxyethanol, glycol, 1,2 ethanediol, ethane-1,2-diol, ethylene dihydrate, ethylene alcohol

Ethylene glycol was first produced commercially by the hydrolysis of ethylene oxide obtained by the chlorination of ethylene via the chlorohydrin route. This was followed by the reaction of carbon monoxide with formaldehyde via glycolic acid and methyl glycolate, but both processes are now obsolete.

Direct oxidation of ethylene to ethylene glycol using a silver catalyst was utilized for a period but operating problems led to its fall from favour. Considerable research has been carried out to improve the process and overcome these problems. Acids and bases accelerate the reaction and allow lower reaction temperatures to be used. Other catalysts proposed are molybdates, vanadates and organic antimony compounds but none has been commercialized.

Amongst the newer processes, the reaction of ethylene with carbon dioxide to ethylene carbonate followed by hydrolysis looks promising. Catalysts suggested for the reaction include quaternary ammonium and phosphonium salts. Methods for product separation and catalyst recovery have yet to be resolved.

The results from new catalyst systems (based on mixtures of palladium chloride, lithium chloride and sodium nitrate–acetic acid/anhydride) for the direct oxidation of ethylene to glycol acetate anhydride are encouraging. Halcon has proposed tellurium oxide–bromine salts in a solution of acetic acid as the oxidizing agent to catalyse the reaction. In the presence of water, glycol acetate hydrolyses to ethylene glycol and acetic acid. Work has been carried out on alternative palladium chloride–copper chloride–copper acetate catalyst systems with yields of up to 90% claimed.

Research on the synthesis of ethylene glycol from carbon monoxide and hydrogen has been undertaken but yields have been low. High temperatures and pressures are required and catalyst systems based on rhodium–rhenium are expensive.

None of the newer processes is economically advantageous over the hydrolysis of ethylene oxide which is the only route in commercial use. By using water to oxide molar ratios in excess of 20:1, the production of higher glycols can be minimized.

Capacities range from 15 000 to 850 000 tonnes per year. Union Carbide is the world's largest producer of monoethylene glycol.

PROCESSES

1. From ethylene by air oxidation in the vapour phase

Ethylene is directly oxidized in the vapour phase with air or oxygen to ethylene oxide in the presence of a silver oxide catalyst (see Ethylene Oxide).

The ethylene oxide–water mixture is preheated to a temperature of 190–200°C before being transferred to a reactor. The reaction takes place in the aqueous phase, with ethylene glycol produced under a pressure of 14–22 bar. Di-, tri-, tetra- and poly-ethylene glycols are formed as by-products, but the proportion of higher homologues can be controlled by using an excess of water to minimize the reaction between ethylene oxide and glycols. In practice, a molar ratio of 1:20 ethylene oxide to water is employed. Pressure is controlled so that the vaporization of ethylene oxide from the liquid is minimized. (*See Figure 62*)

The water–glycol mixture from the reactor is fed to multiple evaporators where the pressure is progressively lowered; water is recovered as an overhead condensate and recycled. The water-free glycol mixture is separated by fractional distillation under vacuum to yield, successively, ethylene glycol which is removed as a side

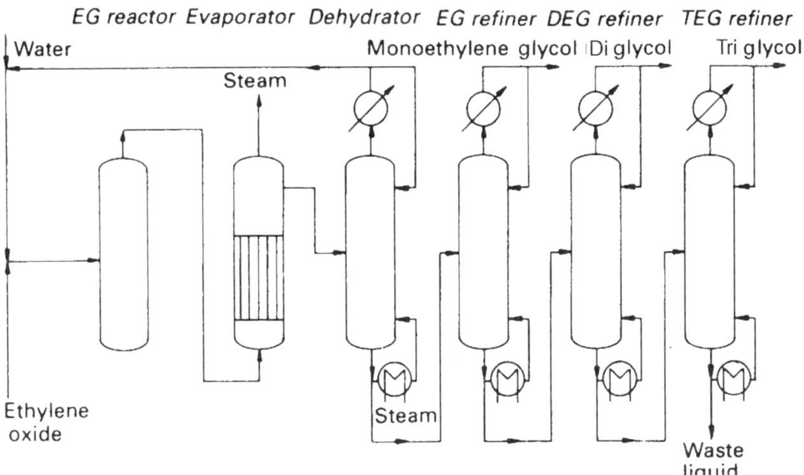

FIGURE 62 Ethylene glycol from ethylene by air oxidation

stream, and by-product di- and tri-ethylene glycols from the column bottoms. These are purified by further distillation. The water–ethylene oxide ratio is critical in determining the volumes of higher glycols produced.

Where ethylene oxide–glycol processes are carried out on the same site, energy is conserved by utilizing the heat generated by the exothermic oxidation reaction to generate steam for the evaporator unit.

Acids catalyse the hydration reaction and allow lower operating temperatures to be used. An ethylene oxide to water ratio of 1:10 is employed in the presence of 1% sulphuric acid. Typical reaction conditions are a temperature of 95°C and pressure of 14–21 bar with a contact time of 30 minutes. Conversions of 100%, including 10% of glycol ethers, are claimed. A disadvantage is that the solution becomes corrosive and excess acid has to be removed to prevent build-up.

Reaction

$$2H_2C=CH_2 + O_2 \rightarrow 2H_2C\underset{\diagdown\ \diagup \atop O}{-}CH_2$$

$$H_2C\underset{\diagdown\ \diagup \atop O}{-}CH_2 + H_2O \rightarrow HOCH_2CH_2OH$$

Raw material requirements and yield

Raw materials required per tonne of ethylene glycol:

Ethylene	940kg
Air	9800kg

Yield 67%

PROPERTIES

Clear, colourless, hygroscopic liquid with almost no odour. Very soluble in water, ethyl alcohol and ethers. Reacts violently with sulphuric acid.

Molecular Weight	62.07
Density at 20°C	1.113
Freezing Point	–12.6°C
Boiling Point	197.3°C
Autoignition Temperature	400°C
Flammability Limits in air	
lower	3.5vol%
upper	Not listed
Flash Point Closed Cup	116°C
Vapour Density (air = 1)	2.14
Exposure Limit COSHH	125ppm 15 minutes (vapour) 25ppm 8 hour TWA (vapour)
Exposure Limit ACGIH	60mg/m³ 8 hour TLV-TWA (vapour) 10mg/m³ 8 hour TLV-TWA (particulate)

GRADES

Technical, antifreeze with 3% triethanolamine phosphate, polyester >99%.

INTERNATIONAL CLASSIFICATIONS

UN No.	Not listed
CAS Reg No.	107–21–1
EINECS No.	203–473–3
EC No.	603–027–00–1

APPLICATIONS

The most important outlet for ethylene glycol is in the manufacture of polyester resins used for fibres, films, laminates and polyethylene terephthate (PET) soft drink bottles. Polyester fibres consume around 54% of total ethylene glycol production. PET packaging accounts for a further 20% of ethylene glycol demand.

Antifreeze, which consumes 14% of ethylene glycol production, is the second largest outlet. When combined with corrosion inhibitors, antifreeze (which lowers the freezing point of water) is used in motor vehicles, pumps, industrial heating and cooling units and solar energy systems. Other outlets include defrosting aircraft wings, de-icing runways, and in asphalt–emulsion paints to protect against freezing.

Films and resins consume a further 6% and 4% respectively of ethylene glycol production. Ethylene glycol's hygroscopic properties make it useful as a humectant for textile fibres, paper, leather and adhesives, as a plasticizer, and as a softening agent for ethyl ketal preparations.

Other outlets include its use in:

- hydraulic brake and shock absorber fluids;
- alkyd resins for surface coatings;
- as a stabilizer for water dispersions of urea – formaldehyde and melamine–formaldehyde.

World demand for ethylene glycol is expected to grow at 7% per year in the late 1990s. The increase in demand is due to the use of PET in packaging, the fastest growing outlet, and the rising consumption of polyester fibres, especially in the Asia–Pacific region.

Di- and tri-ethylene glycols are used as solvents for cellulose acetate derivatives and dyestuffs, and as drying agents for refinery gases.

HEALTH AND HANDLING

Liquid ethylene glycol causes slight irritation to the eyes but does not have any major effect on the skin. Repeated skin contact with the liquid can lead to irritation. Glycols are a chronic poison if ingested. Excessive exposure to the vapour causes irritation to the eyes, nose and throat leading to nausea, headaches and dizziness.

Ethylene glycol is a stable, non-corrosive liquid. Resin-coated steel, aluminium or stainless steel are used for storage containers to prevent colour change, which is important for product to be used in polyester manufacture. The liquid should be stored under nitrogen or alternatively containers must be sealed to prevent the ingress of moisture as the product is very hygroscopic. Keep in a cool, dry, ventilated area, and avoid contact with strong oxidizing agents.

Spills should be contained and transferred to a suitable container for disposal by mixing with a flammable solvent and incineration. Small spills can be absorbed with sand or earth followed by washing the area with water. Care should be taken to prevent the liquids from entering waterways or sewers. Clean-up staff should wear protective clothing to prevent skin contact. Ethylene glycol is combustible but not flammable. Dry chemical, carbon dioxide or water spray can be used to extinguish fires.

Ethylene glycol is classified as harmful by the EC Dangerous Substances Directive and all containers must be labelled accordingly.

MAJOR PLANTS

Plants with capacities greater than 300 000 tonnes per year:

Formosa Plastics	Pointe Comfort	US
Hoechst Celanese	Clear Lake	US
OxyChem	Bayport	US
PD Glycol	Beaumont	US
Shell Chemical	Geismar	US
Union Carbide	Seadrift	US
	Taft	US
Union Carbide	Prentiss	Canada
Alberta & Orient Glycol	Prentiss	Canada
SABIC	Al Jubail	Saudi Arabia
Saudi Yanpet Petrochemical	Yanbu	Saudi Arabia
Equate/Union Carbide	Shuaiba	Kuwait
Honam Petrochemical	Yeochon	South Korea

MAJOR LICENSORS

Elf Atochem
Halcon-Scientific Design
Japan Catalytic
Shell Development
Snamprogetti
Union Carbide

Ethylene Oxide

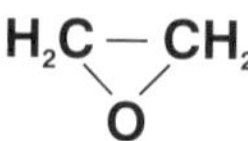

SYNONYMS

ETHYLENE OXIDE dihydro-oxirene, dimethylene oxide, oxirane, 1,2-epoxyethane

The early manufacture of ethylene oxide was via ethylene chlorohydrin as an intermediate, but this route has been superseded by the direct oxidation of ethylene in the presence of air or oxygen over a silver oxide catalyst.

Although the yield from the chlorohydrin route is higher than the direct oxidation process, most of the chlorine is lost as calcium chloride, which is an unwanted by-product.

In many of the larger plants, oxygen is favoured over air because it is more economic with higher yields and lower down time. Additionally, pollution problems caused by vent gases are markedly reduced. Currently, about 98% of world ethylene oxide capacity is based on the direct oxidation route. The three major direct oxidation processes employed worldwide are those of Shell, Union Carbide and Scientific Design. The remaining capacity is based on chlorohydrin.

In an attempt to improve the economics of ethylene oxide manufacture, considerable effort is being expended in trying to find ways of reducing energy consumption and prolonging catalyst life.

Capacities range from 20 000 to 640 000 tonnes per year. Because of its hazardous nature, the production of ethylene oxide is frequently integrated with that of its major derivative, ethylene glycol which is easily transported.

PROCESSES

From ethylene and oxygen

Ethylene (of a purity of 95% or greater), compressed oxygen and recycle gas are mixed and preheated by product gases from the reactor. An ethylene concentration of 15–40vol% is usually employed. The gas stream enters a tubular reactor containing a silver oxide catalyst supported on a porous carrier (such as aluminium oxide) which has been fired to a high temperature. The fixed-bed catalyst contains 7–20% of silver and promoters–compounds of alkalis or alkali earth metals, especially caesium and barium. (*See Figure 63*)

Selectivity is improved by the addition of chlorine compounds (such as chloroethane or vinyl chloride) which reduce the production of by-product carbon dioxide and ensure an even silver surface coating. As the activity of the catalyst declines, the reaction temperature is gradually increased to maintain ethylene oxide production. Ethylene and any other gases used must be free of sulphur, sulphur compounds or acetylene which are catalyst poisons. Catalyst life is 2–4 years.

The reaction conditions are a temperature of 200–300°C and pressure of 10–30 bar, with a residence time of 1 second. Small amounts of methane are added as a diluent which helps to increase the flammable limit in the reactor. The gas-phase reaction is highly exothermic; the oxidation temperature is controlled by boiling water or a high-boiling hydrocarbon in an outer reactor jacket to remove the heat produced. Any excess heat is used to generate steam which is circulated to the ethylene oxide recovery section.

The reactor gases are cooled and compressed before passing to a scrubber where chilled water absorbs the ethylene oxide. The dilute aqueous ethylene oxide solution is steam stripped, and the resultant ethylene oxide is purified by fractionation. Ethylene glycols, produced as by-products when ethylene oxide comes into contact with water in the scrubber, can be recovered from the waste streams.

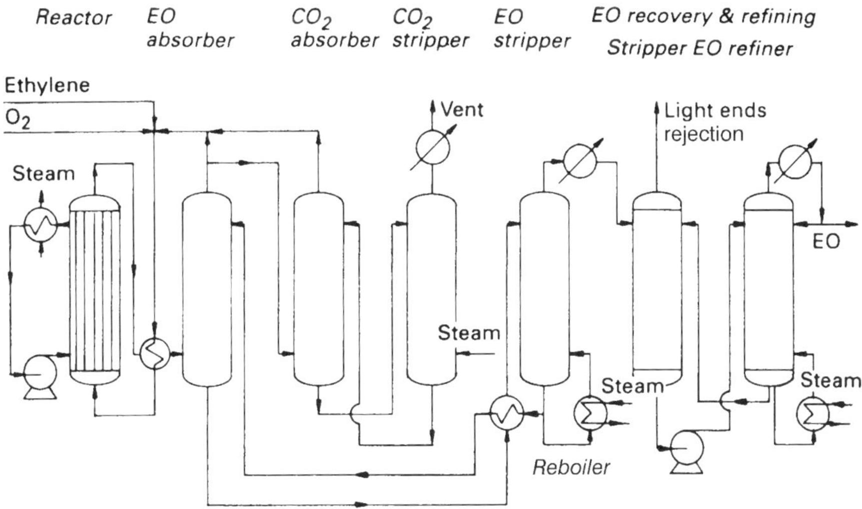

FIGURE 63 Ethylene oxide from ethylene and oxygen

By-product carbon dioxide in the gases from the top of the scrubber are removed by absorption with hot aqueous potassium carbonate. Steam stripping at atmospheric pressure releases the carbon dioxide which is vented, and the potassium carbonate solution is pumped back to the absorber.

A portion of the recycle gas stream from the absorber is removed and flared in order to avoid the build-up of inert gases present in the feed gases.

The air-based process is similar to the oxygen process but results in large amounts of nitrogen in the recycle gases. Ethylene concentrations of 2–10vol% are normally employed. The unabsorbed gases from the scrubber, which contain unreacted ethylene and inert gases, are diverted to a purge reactor to improve the ethylene yield before being vented to atmosphere. Although fluid-bed reactors have been tried, problems caused by abrasion have led to their abandonment in favour of fixed-bed reactors.

Reaction

$$CH_2 = CH_2 + \frac{1}{2}O_2 \rightarrow CH_2OCH_2$$

Raw material requirements and yield

Raw material requirement required per tonne of ethylene oxide:

Ethylene	1100kg
Air	13 100kg

Yield 60%

GRADES

Technical 99.5%.

PROPERTIES

Colourless, toxic and flammable gas or liquid. Miscible with water, ethyl alcohol and many organic solvents. Very reactive. Starts to decompose at a temperature of around 400°C.

Molecular Weight	44.05
Density at 20°C	0.869
Freezing Point	–112.5°C
Boiling Point	10.4°C
Autoignition Temperature	429°C
Flammability Limits in air	
lower	3vol%
upper	100vol%
Flash Point Open Cup	–18°C
Vapour Density (air = 1)	1.49

Exposure Limit COSHH (Schedule 1) 5ppm 8 hour TWA (Maximum Exposure Limit)
Defined as a carcinogen under COSHH regulations.
Exposure Limit ACGIH 1ppm 8 hour TLV-TWA
Classified A2, suspected carcinogen potential for humans.

INTERNATIONAL CLASSIFICATIONS

UN No.	1040
blanketed with nitrogen up to a total pressure of 10 bar at 50°C	
CAS Reg No.	75–21–8
EINECS No.	200–849–9
EC No.	603–023–00-X
Description	Toxic gas, flammable gas
Packing Group	Not allocated
Emergency Action Code	2PE
HI (Kemler Code)	263

UN No.	1041
carbon dioxide mixtures with >9% ethylene oxide but <87% ethylene oxide	
Description	Flammable gas
Packing Group	Not allocated
Emergency Action Code	2PE
HI (Kemler Code)	239

UN No.	1952
carbon oxide mixtures with <9% ethylene oxide	
Description	Non-flammable, non-toxic gas
Packing Group	Not allocated
Emergency Action Code	2PE
HI (Kemler Code)	20

UN No.	2983
propylene oxide mixtures with <30% ethylene oxide	
Description	Flammable liquid, toxic substance
Emergency Action Code	2PE
HI (Kemler Code)	336

UN No.	3070
dichlorofluoromethane mixture with <12.5% ethylene oxide	
Description	Non-flammable, non-toxic gas
Packing Group	Not allocated
Emergency Action Code	2RE
HI (Kemler Code)	20

UN No.	3297
chlorotetrafluoroethane mixture with <8.8% ethylene oxide	
Description	Non-flammable, non-toxic gas
Packing Group	Not allocated
Emergency Action Code	2RE
HI (Kemler Code)	20

UN No. pentafluoroethane mixture with <7.9% ethylene oxide	3298
Description	Non-flammable, non-toxic gas
Packing Group	Not allocated
Emergency Action Code	2RE
HI (Kemler Code)	20

UN No. tetrafluoroethane mixture with <5.6% ethylene oxide	3299
Description	Non-flammable, non-toxic gas
Packing Group	Not allocated
Emergency Action Code	2RE
HI (Kemler Code)	20

UN No. carbon oxide mixtures with >87% ethylene oxide	3300
Description	Toxic gas, flammable gas
Packing Group	Not allocated

APPLICATIONS

Because of its reactivity, ethylene oxide is an important raw material for the production of a wide range of intermediates and consumer products. Around 64% is converted to ethylene glycol, of which 10% is used in antifreeze and the remainder in the manufacture of polyesters. The next most important outlet, accounting for 12% of ethylene oxide consumption, is for the production of ethoxylates, which are surface-active agents used in detergent formulations, as a solvent for paints and lacquers, in the textile industry and for extracting sulphur compounds from refinery gases.

Ethanolamines, used in chemical synthesis and surfactants, take another 6%. Other derivatives each consuming around 5% of ethylene oxide, are glycol ethers, a solvent for paints and lacquers, and di- and tri-ethylene glycols. Polyglycols, polyols and choline chloride, an animal food supplement, are also synthesized from ethylene oxide and account for a further 4%. Ethylene oxide is an excellent disinfectant, fumigant, and sterilizing agent.

World demand is growing at around 5% per year.

HEALTH AND HANDLING

Ethylene oxide is toxic and a potential human carcinogen. Its vapour is extremely irritating to the eyes causing permanent damage, and to the respiratory tract.

Exposure can cause headaches, nausea, damage to the lungs and finally coma and death. In contact with skin, blisters are formed and absorption takes place. Sensitization can occur. Non-permeable protective clothing, polychloroprene rubber gloves and boots, chemical goggles and respirators must be worn at all times, and good working practices are essential. Polyvinyl chloride and nitrile rubbers have limited resistance to ethylene oxide and should not be used; neither should leather boots which will absorb the liquid.

The product is very hazardous as ethylene oxide gas or ethylene oxide mixed with air or an inert gas can decompose explosively – pressure, temperature and concentration being of importance. Staff therefore require training in the handling of explosive materials because of the dangers involved.

Store under an inert gas, such as nitrogen, in a welded stainless steel container of 4.5 bar minimum working pressure. Storage should be kept well away from the production facilities. Liquid ethylene oxide polymerizes easily in the presence of alkalis, mineral acids, metal chlorides, metal oxides, iron, aluminium or tin. Care is required when transferring ethylene oxide, and all equipment must be free from rust and other contaminants. Regular checks must be made for vapour leakage and build-up prevented by efficient extraction systems.

In the event of spills, extinguish all sources of ignition and evacuate personnel. Contain with dry sand or earth and disperse vapours with a water spray. Ethylene oxide must not be allowed to flow into waterways, drains or sewers.

Ethylene oxide is highly flammable and its vapour will decompose violently at high temperatures. Fires should be extinguished with water spray or alcohol foam. Firefighters and clean-up staff must wear full protective clothing and self-contained breathing apparatus.

The storage, use and transportation of ethylene oxide is governed by national regulations in most countries.

MAJOR PLANTS

Plants with capacities greater than 240 000 tonnes per year:

BASF	Antwerp	Belgium
Dow Chemical	Plaqemine	US
Formosa Plastics	Point Comfort	US
Hoechst Celanese	Clear Lake	US
Huntsman Corp.	Port Neches	US
OxyChem	Bayport	US
PD Glycol	Beaumont	US
Shell Chemical	Geismar	US
Union Carbide	Seadrift	US
	Taft	US
Union Carbide	Prentiss	Canada

SABIC	Al Jubail	Saudi Arabia
Saudi Yanpet Petrochemical	Yanbu	Saudi Arabia
Equate/Union Carbide	Shuaiba	Kuwait
Nippon Shokubai	Kawasaki	Japan
Honam Petrochemical	Yeochon	South Korea

MAJOR LICENSORS

Dow Chemical
Elf Atochem
Halcon-Scientific Design
Japan Catalytic
Shell Development
Snamprogetti
Toyo Engineering/ABB Lummus Global
Union Carbide

Ethyl Ether

$C_2H_5OC_2H_5$

SYNONYMS

ETHYL ETHER ether, ethoxyethane, diethyl ether, spirit ether, diethyl oxide

Ethyl ether is produced as a by-product in the direct hydration process for the production of ethyl alcohol. Approximately 5–10% by weight is obtained as the by-product (see Ethyl Alcohol).

Ethyl ether (formed by the interaction of ethylene and sulphuric acid) is hydrolysed to give ethyl alcohol which may react with further ethyl sulphate to yield diethyl ether. A considerable amount of ether is recovered as a by-product from ethyl alcohol synthesis, and in many countries this is sufficient to meet local demand.

It can also be produced by the dehydrogenation of ethyl alcohol with either sulphuric acid or an alum catalyst.

Capacities range from 5000 to 90 000 tonnes per year.

PROCESSES

From ethyl alcohol by dehyrogenation

Ethyl alcohol and 96% sulphuric acid in the ratio 1:3 are heated in a lead-lined steel reactor containing steam coils. The temperature is maintained between 125 and 130°C by controlling the rate of flow of ethyl alcohol vapour. (*See Figure 64*)

Overhead gases from the reaction containing ether and unreacted ethyl alcohol are washed with dilute alkali to neutralize any sulphuric acid and sulphur dioxide. The alkaline solution is fed into the bottom of the distillation column where any ether and ethyl alcohol are removed. Gases from the top of the scrubber also pass to the fractionation column from where the ethyl ether is condensed and collected. The crude ethyl ether is purified to remove remaining impurities consisting of aldehyde, alcohols, sulphur dioxide and water.

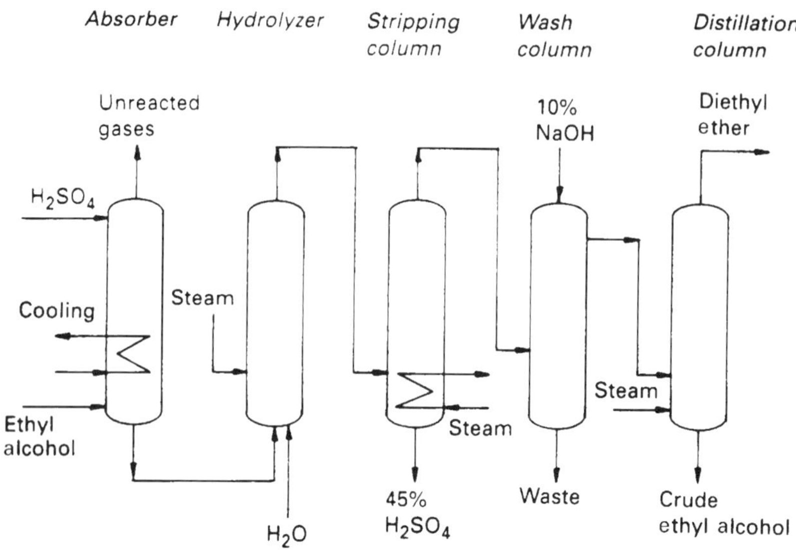

FIGURE 64 Ethyl ether from ethyl alcohol by dehydrogenation

Any unreacted ethyl alcohol is recycled, and water from the base of the fractionator is discharged to waste. When the acid contains considerable quantities of tarry products it is recharged.

Reaction

$$C_2H_5OH + H_2SO_4 \rightarrow C_2H_5HSO_4 + H_2O$$

$$C_2H_5OH + C_2H_5HSO_4 \rightarrow C_2H_5OC_2H_5 + H_2SO_4$$

Raw material requirements and yield

Raw materials required per tonne of ethyl ether:

Ethyl alcohol	1320kg
Sulphuric acid (96%)	14–23kg

Yield 95%

OTHER PROCESSES

From ethyl alcohol by the alum process

In the alum process, ethyl alcohol vapour at a temperature of 180–230°C reacts under pressure in a steel tube containing a special alum catalyst to yield ethyl ether. The catalyst consists of ground-fused potassium aluminium sulphate which is heated further until the water content is less than 20%. Ethyl ether is recovered by fractionation and purified as in the sulphuric acid hydrogenation process.

Reaction and yield

$$2C_2H_5OH \rightarrow C_2H_5OC_2H_5 + H_2O$$

Yield 75–80%

PROPERTIES

Colourless, mobile liquid with characteristic ethereal odour and burning taste. Very highly flammable with vapour which, being denser than air, tends to roll along the ground. Can form peroxides in storage.

Soluble in most organic liquids but only slightly soluble in water. Forms binary mixtures with methyl formate, isoprene and pentanes.

Molecular Weight	74.12
Density at 20°C	0.713
Melting Point	–116.3°C
Boiling Point	34.5°C
Flash point Closed Cup	–45°C
Autoignition Temperature	180°C
Explosive Limits in air	
lower	1.7vol%
upper	48.0vol%
Vapour Density (air = 1)	2.55
Exposure Limits COSHH	500ppm 15 minutes 400ppm 8 hour TWA
Exposure Limit ACGIH	500ppm TLV-STEL 400ppm TLV-TWA

GRADES

Technical 95% minimum, solvent, analytical and anaesthesic.

INTERNATIONAL CLASSIFICATIONS

UN No.	1155
CAS Reg No.	60–29–7
EINECS No.	200–467–2
EC No.	603–022–00–4
Description	Flammable liquid
Packing Group	I
Emergency Action Code	3YE
HI (Kemler Code)	33

APPLICATIONS

Ethyl ether is used as a solvent for oils, fats, dyes, gums and resins, and as an extractant of colorants and natural perfumery products. It can be mixed with ethyl alcohol and used to gel nitrocellulose.

In the chemical industry, ethyl ether is a good low-boiling solvent, often used as a reaction medium. It is being replaced as an anaesthetic. Demand is stagnant.

HEALTH AND HANDLING

Little ethyl ether is absorbed through the skin, but it can cause irritation. However, it is rapidly absorbed when inhaled causing headache, nausea and vomiting, and prolonged exposure can lead to unconsciousness and death. Contact lenses must not be worn as they concentrate the vapour leading to eye irritation. Good ventilation is essential when handling the product.

The handling of ethyl ether is hazardous because of its high flammability. Its movement is subject to special regulations. As flowing ether can generate static charges, all tanks and pipelines must be earthed to avoid the build-up of static which could result in a vapour explosion. As ether vapour rolls along the ground, good ventilation is important to avoid explosive mixtures with air. Ether is liable to explode at temperatures above 100°C.

Peroxides can be formed on exposure to air and sunlight. Ethyl ether should be stored in closed air-tight mild steel tanks under a nitrogen atmosphere. Product which has been stored should be tested for purity before use.

Strong acids oxidize ethyl ether, risking explosion. The risk can be reduced by the addition of 1–30mg/kg of phenols, such as 2.6 di-*tert*-butyl-*p* cresol or diethylthiocarbamate sodium salt.

In the event of spills, eliminate all sources of ignition and evacuate personnel. The liquid should be absorbed with sand or vermiculite, collected with non-sparking tools and disposed of promptly by controlled burning in an approved incinerator. Water can be used to wash away any remaining liquid, but care must be taken to ensure that it does not pollute waterways or enter sewers. Clean-up staff should wear protective clothing, eye protection and a respirator. Any contaminated clothing must be washed before reuse.

Use dry chemical, carbon dioxide or alcohol foam to extinguish fires. Water is ineffective and can only be used to cool fire-exposed containers. Surroundings must be cooled to below 180°C before fires can be extinguished. As ether vapour is heavier than air it can roll long distances and flashback is a dangerous hazard. Toxic gases are given off during burning and all firefighting staff must wear self-contained breathing apparatus.

MAJOR PLANTS

Plants with capacities greater than 10 000 tonnes per year:

SODES	Lillebonne	France
Huels	Herne	Germany
Den Norske Eterfabrikk	Oslo	Norway
Synthesia	Pardubue	Czech Republic
Millenium Petrochemicals	Tuscola	US
Showa Ether	Aiko-gun	Japan
Lucky	Yosu	South Korea
Chi Ming Chemical	Taipei	Taiwan

Ether is often extracted as a by-product from ethyl alcohol production.

MAJOR LICENSORS

Huels
Millenium
Shell Development
Union Carbide

2-Ethyl Hexyl Alcohol $CH_3(CH_2)_3CH(C_2H_5)CH_2OH$

SYNONYMS

2-ETHYL HEXYL ALCOHOL 2-ethyl-1-hexanol, iso-octyl alcohol, ethylhexyl alcohol, 1-ethyl-*n*-amylcarbinol, iso-octanol

2-Ethyl hexyl alcohol has been produced commercially since the 1930s and is the best known and most widely used of the higher aliphatic alcohols. It is produced by the aldol condensation of butyraldehyde followed by dehydration and hydrogenation.

A number of routes to butyraldehyde have been employed over the years. In the oldest, butyraldehyde was obtained from acetaldehyde via ethylene but this was superseded by the oxo process from propylene, which is now the preferred route. The development of rhodium catalysts instead of conventional cobalt catalysts has resulted in lower reaction temperatures and pressures and higher yields of *n*-butyraldehyde. Around 95% of 2-ethyl hexyl alcohol is produced from propylene, the remainder coming from acetaldehyde.

Capacities range from 10 000 to 200 000 tonnes per year. Capacities of most plants are flexible and capable of producing a range of products.

PROCESSES

From propylene via butyraldehyde

Butyraldehyde is produced by the exothermic, liquid-phase reaction between 95% propylene and synthesis gas at 130–150°C and a pressure of 100–300 bar. The ratio of hydrogen to carbon monoxide in the synthesis gas is regulated to give a molar ratio of 1:1. The catalyst consists of a mixture of cobalt hydrocarbonyl and dicobalt octacarbonyl, or cobalt tetracarbonyl hydride. If rhodium-based catalysts or complexes based on rhodium carbonyls and triphenyl phosphine are used, the reaction conditions are 100°C with a pressure of 7–24 bar. The phases are mixed to ensure rapid reaction and the heat produced is used for steam generation. (*See Figure 65*)

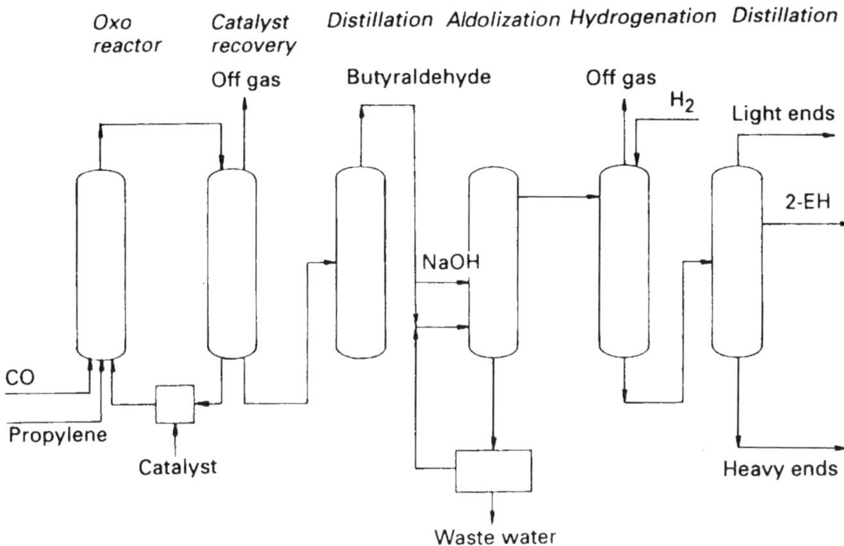

FIGURE 65 2-Ethyl hexyl alcohol from propylene via butyraldehyde

The reaction products are cooled and any unreacted gases are flashed off and recycled. The catalyst is removed by treatment with an alkaline solution prior to isolation of the aldehydes.

Pure *n*-butyraldehyde (free from isobutyraldehyde) is subjected to an aldol condensation at a temperature of 80–130°C in the presence of a dilute solution of sodium hydroxide. The ratio of aldehyde to alkali is maintained in the ratio of 1:10 to 1:20. The reaction mixture is passed to a separator where the upper organic phase is separated off and hydrogenated in the presence of a fixed nickel catalyst in the liquid or gas–liquid phase. Reaction conditions are a temperature of 80–100°C and a pressure of 3–10 bar. Excess water is removed from the aqueous phase. Additional sodium hydroxide is added to bring the concentration up to 2–4% before recycle to the aldolization section.

The 2-ethyl hexyl alcohol formed is purified by distillation. Isobutyl alcohol is produced as a by-product.

Reaction

$$2CH_3CH_2CH_2CHO \rightarrow CH_3(CH_2)_2CH(OH)CH(C_2H_5)CHO$$

$$CH_3(CH_2)_2CH(OH)CH(C_2H_5)CHO \rightarrow CH_3(CH_2)_2CH=C(C_2H_5)CHO + H_2O$$

$$CH_3(CH_2)_2CH=C(C_2H_5)CHO + 2H_2 \rightarrow CH_3(CH_2)_3CH(C_2H_5)CH_2OH$$

Raw material requirements and yield

Raw materials required per tonne of 2-ethyl hexyl alcohol:

Butyraldehyde 100%	1145kg
Hydrogen	360m^3
Nickel catalyst	1kg

Yield 93%

PROPERTIES

Colourless, slightly viscous liquid. Miscible with ethyl alcohol and ether.

Molecular Weight	130.23
Density at 20°C	0.833
Melting Point	–76°C
Boiling Point	184°C
Autoignition Temperature	250°C
Flammability Limits in air	
lower	1.1vol%
upper	7.4vol%
Flash Point Closed Cup	75°C
Vapour Density (air = 1)	4.5
Exposure Limit COSHH	Not established
Exposure Limit ACGIH	Not established

GRADES

Technical >98%.

INTERNATIONAL CLASSIFICATIONS

UN No.	Not allocated
CAS Reg No.	140–76–7
EINECS No.	203–234–7
EC No.	Not allocated
Description	Flammable liquid

APPLICATIONS

The major application of 2-ethyl hexyl alcohol is for the manufacture of plasticizers, especially di-2-ethylhexyl phthalate (DOP) which is used in vinyl resins. The next largest outlet is for the production of the acrylate esters which are used in adhesives and surface-coating materials such as acrylic paints, in printing inks and as impregnating agents.

2-Ethyl hexyl nitrate finds use as an octane number improver and the phosphate derivatives are employed as lubricating oil additives.

Other products manufactured from 2-ethyl hexyl alcohol include surfactants for use in antifoamants, dispersants, and flotation agents, for antioxidants, in herbicides, and as a solvent for nitrocellulose, gums and resins, and in mixtures with other solvents for the paint and lacquer industry.

World growth rate to 2000 is forecast at 2.8% per year.

HEALTH AND HANDLING

2-Ethyl hexyl alcohol is considered as having low toxicity but when heated the vapour can cause irritation to the eyes, nose and throat. Prolonged exposure causes headaches, nausea and giddiness. It can be absorbed through the skin causing irritation.

2-Ethyl hexyl alcohol is not corrosive to metals and can be stored in iron, mild steel, copper or aluminium containers, under an inert gas to prevent ingress of water and oxygen. Store in a well-ventilated area away from strong oxidizing agents. Gloves and eye protection should be worn to prevent skin contact when handling.

Absorb spills with sawdust or paper and dispose of promptly by incineration. Care must be taken to keep the liquid away from sewers and waterways. Clean-up staff must wear protective clothing to prevent skin contact.

2 Ethyl hexyl alcohol is considered to be a moderate fire hazard. Blanket fires with carbon dioxide, dry chemical or foam. Staff should wear self-contained breathing apparatus because of the fumes given off during burning.

MAJOR PLANTS

Plants with capacities greater than 100 000 tonnes per year:

Oxochimie	Lavera	France
BASF	Ludwigshafen	Germany
Hoechst	Oberhausen	Germany
Huels	Marl	Germany
Zaklady Azotowe	Kedzierzyn	Poland
Eastman Chemical	Longview	US
Aristech Chemical	Pasadena	US
SAMAD	Al Jubail	Saudi Arabia
Mitsubishi Chemical	Mitushima	Japan
Kyowa Yuka	Yokkaichi	Japan
Han Yang Chemical	Yeochon	South Korea

Plants produce a range of products and capacities are therefore flexible.

MAJOR LICENSORS

BASF
Hoechst
Huels
Rhone-Poulenc
Union Carbide

Formaldehyde

$$\begin{array}{c} \phantom{H-{}}H \\ \phantom{H-{}}| \\ H-C=O \end{array}$$

SYNONYMS

FORMALDEHYDE methanal, formic aldehyde, methyl aldehyde, methylene oxide, oxomethane, oxomethylene

Commercial production of formaldehyde began in Germany in the 1880s, but it was the development of a process for the synthesis of methyl alcohol in the 1920s which provided the spur for its large-scale manufacture.

The two major commercial processes for the production of formaldehyde from methyl alcohol are:

- oxidation–dehydrogenation with air in the presence of a silver catalyst;
- complete oxidation with excess air in the presence of a metal oxide catalyst (Formox process).

Silver catalyst processes can be either the complete conversion of methyl alcohol (BASF) or the incomplete conversion of methyl alcohol as developed by ICI, Degussa and Borden. Competition between the two technologies has led to many improvements in catalyst performance, especially activity, life, and overall yield.

Yields on both processes are around 92%, but as the oxidation reaction takes place at a lower temperature and the metal catalyst used is cheaper than silver, this process has economic advantages over the oxidation–dehydrogenation route. However, the partial oxidation–dehydrogenation process is still the most important. All formaldehyde produced from methyl alcohol is usually marketed as an aqueous solution.

Although propane, butane, ethylene, propylene, butylene or ethers have been used as starting materials for formaldehyde manufacture, they are not of major commercial importance for economic reasons. The non-catalytic oxidation of propane–butane mixtures to formaldehyde accounts for a minor amount of total production.

Formaldehyde can be produced from natural gas or methane, but the mixture of products obtained present problems in subsequent separation stages. It is also a by-product of the oxidation of naphtha to acetic acid.

Capacities range from 5000 to 440 000 tonnes per year.

PROCESSES

1. From methyl alcohol by oxidation–dehydrogenation

There are two variants of this process. In the complete conversion of methyl alcohol to formaldehyde, a methyl alcohol-water mixture in a 3:2 ratio, enters an evaporator column into which air is sent by a blower under flow rate control. The balance between the two reactions is maintained by employing a slight deficiency of air. This also ensures the reaction mixture is kept outside the upper explosion limits. (*See Figure 66*)

A heat exchanger evaporates the methyl alcohol–water mixture and the vapours are superheated with steam before entering the reactor. The reaction is carried out in contact with a fixed-bed catalyst of silver crystals at a temperature between 600–720°C depending on the methyl alcohol concentration. Superheated steam entering from below cools the reaction gases to 150°C. The gases are cooled and the

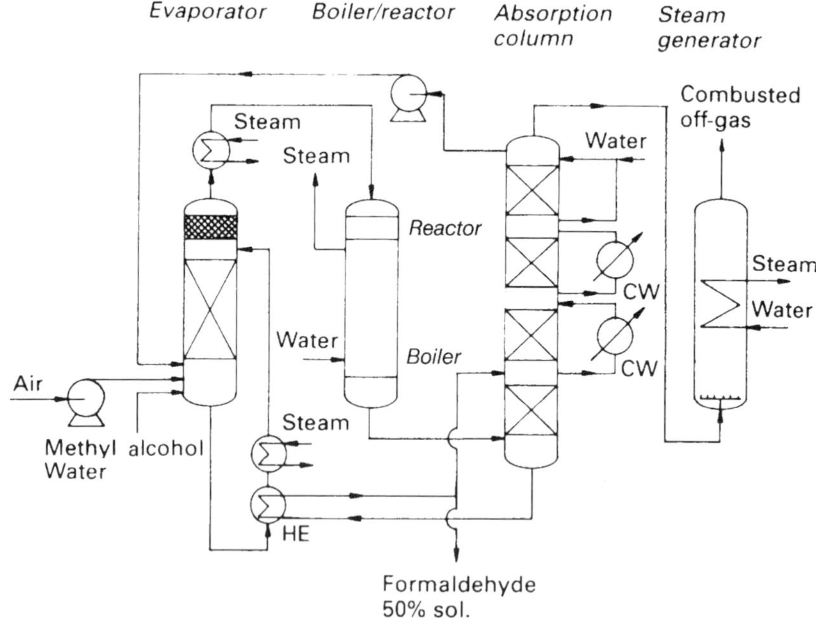

FIGURE 66 Formaldehyde from methyl alcohol by oxidation–dehydrogenation

heat used to generate steam before entering a condenser. Controlled volumes of water are fed into the top and a 40–50% aqueous formaldehyde solution is obtained from the base of the condenser. The solution is concentrated and purified by distillation under reduced pressure. Part of the off-gas from the top of the condenser is recycled while the remainder is burnt as fuel.

The catalyst bed is regenerated electrolytically after 4 to 8 months of operation depending on the level of impurities present in the feed gases.

In the incomplete conversion process, pure methyl alcohol and air are fed into an evaporator and the vapour mixture is combined with superheated steam. The gaseous mixture, containing an excess of methyl alcohol, is introduced into the reactor at a temperature of 590–620°C where it passes through a catalyst bed consisting of silver crystals or layers of silver gauze. The formation of by-products is suppressed by lower temperatures.

The exit gases are cooled and methyl alcohol, water and formaldehyde condense. Any remaining gases are washed with water to remove residual formaldehyde before being used to generate steam. The aqueous formaldehyde–methyl alcohol solution is distilled; methyl alcohol is recovered from the top and recycled to the evaporator. Formaldehyde solution from the bottom of the distillation column is passed through an ion exchange bed to remove any acid before further concentration by distillation as in the complete oxidation process.

Reaction

$$CH_3OH \rightarrow CH_2O + H_2$$

$$H_2 + \frac{1}{2}O_2 \rightarrow H_2O$$

$$CH_3OH + \frac{1}{2}O_2 \rightarrow CH_2O + H_2O$$

Raw material requirements and yield

Raw materials required per tonne of formaldehyde:

Methyl alcohol	1176–1215kg

Yield 89–92%

The yield of formaldehyde obtained depends largely on the methyl alcohol feed concentration and the catalyst temperature.

2. From methyl alcohol by oxidation with air (Formox process)

Gaseous methyl alcohol is mixed with air and preheated to 340°C before entering the reactor containing tubes filled with a catalyst based on iron, molybdenum, or vanadium oxide. In the reactor, methyl alcohol is oxidized to formaldehyde at a temperature of 470°C. An oil-transfer medium is used to remove excess heat generated by the reaction. (*See Figure 67*)

The gas exiting at approximately 280°C, is cooled to 110–130°C in a heat exchanger before entering the absorber. The product gases are absorbed in water, and an aqueous formaldehyde solution is withdrawn from the bottom. Overhead gases are scrubbed with water in the top section, before being catalytically incinerated.

The aqueous condensate and liquor from the scrubber is passed through an anion exchange column to absorb any formic acid, prior to fractionation. A 37% formaldehyde solution is recovered, and any remaining methyl alcohol is removed from the top and recycled.

An excess of air is used in the process to ensure that the air–methyl alcohol ratio is below the lower explosion limits. As the reaction is highly exothermic, the heat recovered is used in the gas preheat stage.

Reaction

$$CH_3OH + \tfrac{1}{2}O_2 \rightarrow CH_2O + H_2O$$

$$CH_3OH \rightarrow CH_2O + H_2$$

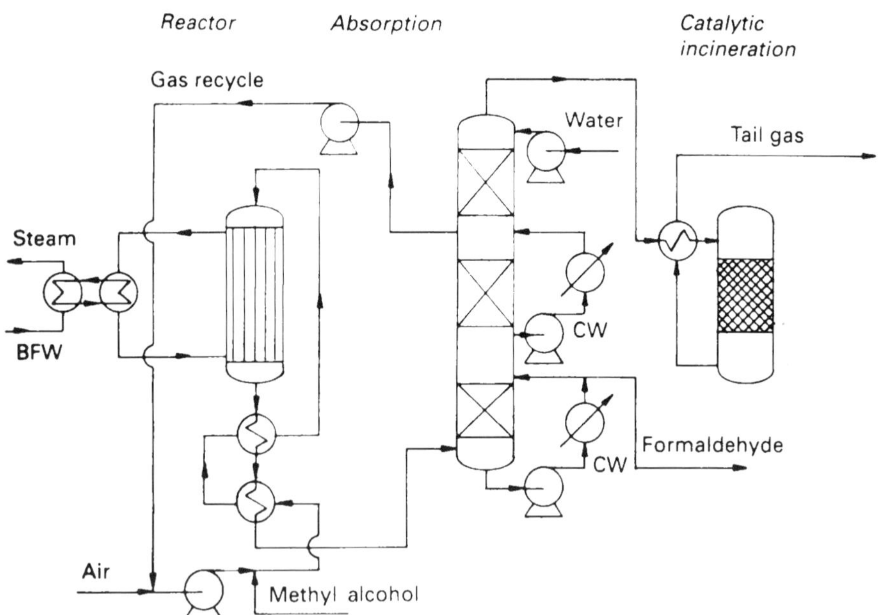

FIGURE 67 Formaldehyde from methyl alcohol by oxidation with air (Formox process)

Raw material requirements and yield

Raw materials required per tonne of formaldehyde:

Methyl alcohol	1162kg

Yield 90–92%

PROPERTIES

Colourless, pungent gas which can be liquefied by cooling. Polymerizes slowly at temperatures below 100°C, but rapidly in the presence of traces of polar impurities, alkalis, acids and water. Soluble in water and ethyl alcohol but partially soluble in benzene, acetone and ether.

Molecular Weight	30.03
Density at 20°C	0.8153
Melting Point	–118°C
Boiling Point	–19.2°C
Autoignition Temperature	430°C
Explosive Limits in air	
lower	7vol%
upper	72vol%
Flash Point Closed Cup	50°C
Vapour Density (air = 1)	1.04

Exposure Limit COSHH (Schedule 1)	2ppm 15 minutes (Maximum Exposure Limit)
	2ppm 8 hour TWA (Maximum Exposure Limit)
Exposure Limit ACGIH	0.3ppm TLV-TWA (Ceiling Exposure Limit)

Classified A2, suspected human carcinogen.

POLYMERS OF FORMALDEHYDE

Formaldehyde forms two principal polymers: a linear polymer of variable composition, paraformaldehyde, and a cyclic trimer, trioxane.

Paraformaldehyde

White powder with a formaldehyde odour. Soluble in dilute acids, alkalis and hot water.

Trioxane

Colourless crystals. Soluble in water.

Molecular Weight	90.05
Density at 65°C	1.39
Melting Point	63°C
Boiling Point	115°C
Autoignition Temperature	410°C
Explosive Limits in air	
lower	3.6vol%
upper	28.7vol%
Flash Point Closed Cup	45°C

GRADES

Anhydrous 99.85%, aqueous 37–55wt% unstablized or stabilized with 0.5–12wt% methyl alcohol. Other stabilizers which can be used include guanamine, melamine, urea, and methyl cellulose.

INTERNATIONAL CLASSIFICATIONS

Formaldehyde solution, flammable

UN No.	1198
CAS Reg No.	50–00–0
EINECS No.	200–001–8
EC No.	605–001–00–5
Description	Flammable liquid, corrosive
Packing Group	III
Emergency Action Code	2YE
HI (Kemler Code)	38

Formaldehyde solution with >25% formaldehyde

UN No.	2209
CAS Reg No.	50–00–0
EINECS No.	200–001–8
EC No.	605–001–00–5
Description	Corrosive
Emergency Action Code	2Z
HI (Kemler Code)	80

Formaldehyde solution with 5–25% formaldehyde

EC No.	605–001–01–2 conc. 5–25%
EC No.	605–001–02-X conc. 1–5%

APPLICATIONS

Formaldehyde has many and varied uses but its main outlet, amounting to over 50% of total demand, is in the preparation of resins and adhesives. These adhesives are used in the production of plywood and particle board, while the resins find an outlet as binders for foundry sand, in brake liners, phenol formaldehyde moulding materials, slow-release nitrogen fertilizers and auxiliaries in the textile, leather, rubber and cement industries.

Around 40% of formaldehyde production is used as an intermediate for chemical synthesis. It is employed in the manufacture of polyacetal and polyurethane resins, surface coatings, plasticizers, cross-linking agents and slow-release fertilizers

Other smaller outlets include the manufacture of dyes, tanning agents, dispersants, vitamins, flavourings and pharmaceuticals. Components found in new detergent formulations such as nitrilotriacetic acid (NTA) and ethylenediaminetetraacetate, are derived from formaldehyde. About 2% of formaldehyde is used directly as a corrosion inhibitor in the metals industry.

Formaldehyde is an excellent preservative and disinfectant used in cosmetics, soap, and sugar syrup recovery to prevent bacterial growth, and in tanning liquors and wood preservatives.

Growth in consumption worldwide is around 2–3% per year, due to the increasing demand for building materials and resins.

HEALTH AND HANDLING

Formaldehyde is irritating to the eyes, nose, throat and skin, and continued exposure can lead to dermatitis. Concern has been expressed about its carcinogen potential but although this has been examined it has not been proven. Suitable protective clothing should be worn when handling formaldehyde and exposure to the vapour minimized by adequate ventilation and extraction systems.

Because formaldehyde polymerizes readily, it is commercially available in the aqueous form. Storage containers should be made of stainless steel, aluminium or polyester resin. Copper or iron must be avoided because of traces of formic acid in formaldehyde which could cause corrosion. Higher concentrations of aqueous formaldehyde are inhibited to prevent the formation of paraformaldehyde. Care must be taken to avoid contamination with acids or alkalis which can cause rapid polymerization.

Spills should be contained and absorbed with sand or earth and the waste incinerated. Fires can be extinguished with carbon dioxide, water spray, dry chemical or foam. In both cases, protective clothing and breathing apparatus must be worn by firefighters and clean-up staff.

MAJOR PLANTS

Plants with capacities greater than 230 000 tonnes per year (based on 37% content):

Elf Atochem	Toulouse	France
BASF	Ludwigshafen	Germany
Degussa	Arnsberg	Germany
Bayer	Krefeld	Germany
Elf Atochem	Leuna	Germany
Neste Resins	Delfzijl	Netherlands
	Europort	Netherlands
Sadepan Chimica	Viadana	Italy
Formol y Derivados	Almusafes	Spain
PERSTOP	Perstorp	Sweden

Krems Chemie	Krems	Austria
Borden	Fayetheville	US
	Geismar	US
Du Pont	Belle	US
Hoechst-Celanese	Bishop	US
	Rock Hill	US
Georgia-Pacific	Houston	US
	Vienna	US
Celanese Canada	Edmonton	Canada
Mitsui Toatsu Chemical	Omuta	Japan
Mitsubishi Gas Chemical	Tokyo	Japan

MAJOR LICENSORS

BASF
Borden Inc.
Degussa
Du Pont
EniChem
Haldor Topsoe/Nippon Kasei Chemical
Hiag/Lurgi
ICI
Mitsubishi Chemical/ABB Lummus Global
Mitsui Petrochemical
Reichold Chemicals

Formic Acid

HCOOH

SYNONYMS

FORMIC ACID methanoic acid, hydrogen carboxylic acid, aminic acid, formylic acid

Formic acid, the first in the series of aliphatic acids, is a useful chemical intermediate. It occurs naturally in the secretions of many insects such as ants, beetles and bees, as well as in plant leaves and roots.

Formic acid is produced commercially by:

- hydrolysis of formamide;
- hydrolysis of methyl formate;
- acidolysis of formate salts;
- oxidation of *n*-butane or naphtha where it is a by-product.

Historically, large amounts of formic acid were made from formamide but the formation of by-product ammonium sulphate has made this route unattractive. Large quantities were recovered as a by-product from naphtha or *n*-butane oxidation, but the advent of the carbonylation of methyl alcohol to acetic acid process, which does not produce formic acid, has resulted in a decline in this source.

A new process to manufacture pure formic acid from impure carbon monoxide, commercialized by several companies during the 1980s, has ensured that, from the raw material cost point of view, methyl formate will be the route of choice in the future.

Methyl alcohol can be dehydrogenated in the vapour phase at atmospheric pressure in the presence of a catalyst, to methyl formate. The direct synthesis of formamide from carbon monoxide and ammonia (with sodium methoxide in a methyl alcohol solvent as catalyst) has been developed. Neither process has been commercialized.

Formic acid is also obtained from a by-product in the reaction between formaldehyde and acetaldehyde to give pentaerythritol.

Over half of formic acid production comes from methyl formate, some from formates and a little from formamide. The remainder is obtained as a by-product from the oxidation of naphtha or butane to acetic acid. The volume being produced as a by-product of acetic acid manufacture will continue to decline.

Capacities range from 6000 to 150 000 tonnes per year.

PROCESSES

1. From methyl formate by hydrolysis

Dilute or impure anhydrous carbon monoxide is reacted with methyl alcohol at 80°C and 45 bar pressure in the presence of catalyst in the liquid phase. Sodium methoxide is the catalyst normally used in a 2.5% concentration. (*See Figure 68*)

The methyl formate formed is degassed and any unreacted carbon monoxide is recycled, before being hydrolysed with water. In order to overcome the unfavourable equilibrium constant for the methyl formate–formic acid reaction a large excess of water is used. The reaction is carried out at 80°C and under increased pressure.

The products formed are flashed off and separated by distillation. Methyl alcohol and methyl formate are recovered overhead and fed back to the reactor. The formic acid–water mixture remaining is taken off from the base of the column into an extraction tower. A secondary amide is used to extract the formic acid and some

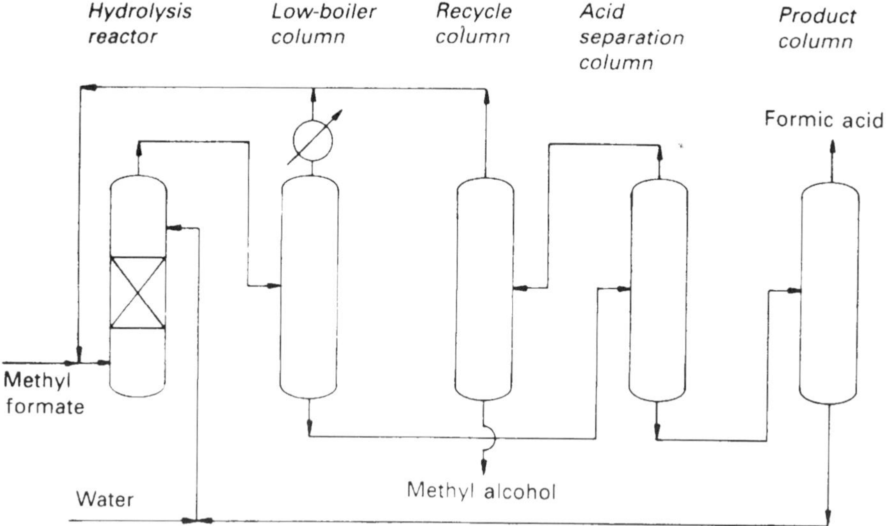

FIGURE 68 Formic acid from methyl formate by hydrolysis

water. The mixture is distilled and excess water removed, leaving a 90% solution of formic acid.

Various processes aimed at reducing the energy requirements of the dehydration step have been developed. Formic acid has been concentrated by extractive distillation using N-formylmorpholine, or by salt formation with a tertiary amine such as l-*n*-pentylimidazole. The salt-like compound formed is distilled from the base.

In the Leonard process, hydrolysis is carried out in two stages at a temperature of approximately 120°C and a pressure of 9 bar. Equimolar ratios of formate and water are used in the first stage; the formic acid produced is added to excess methyl formate in the second stage. After removal of methyl alcohol and methyl formate, dehydration of the formic acid formed is carried out by distillation under a pressure of 3 bar.

Reaction

$CH_3OH + CO \rightarrow HCOOCH_3$

$HCOOCH_3 + H_2O \rightarrow HCOOH + CH_3OH$

Raw material requirements and yield

Raw materials required per tonne of formic acid:

Methyl alcohol	40kg

64% conversion per pass.

2. From sodium formate by acidolysis

Aqueous sodium hydroxide and carbon monoxide are mixed in a countercurrent tower. Operating conditions are 180°C and a pressure of 15–20 bar. The sodium formate produced is pulverized before being fed into a vessel containing sulphuric acid. Formic acid and sodium sulphate are produced. (*See Figure 69*)

Sodium formate is also produced as a by-product of pentaerythritol manufacture and, following solvent extraction, can be used for formic acid recovery. In practice, very little formic acid is obtained in this way.

Reaction

$CO + H_2O \rightarrow HCOOH$

$HCOOH + NaOH \rightarrow HCOONa + H_2O$

$2HCOONa + H_2SO_4 \rightarrow 2HCOOH + Na_2SO_4$

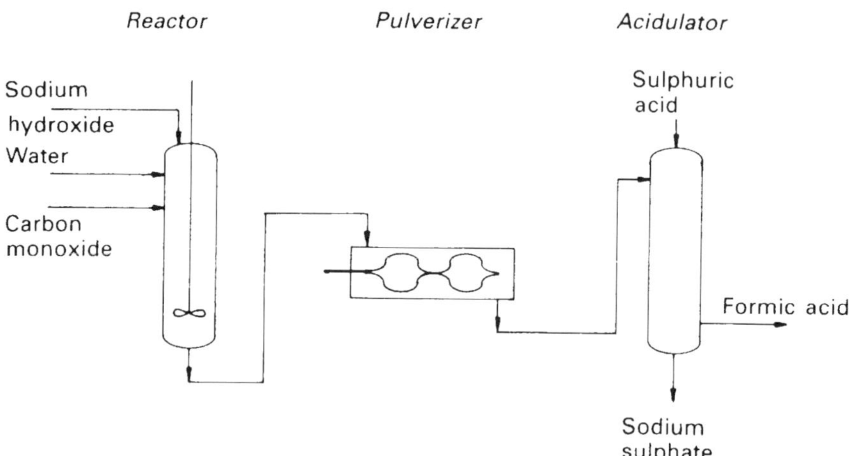

FIGURE 69 Formic acid from sodium formate by acidolysis

Raw material requirements and yield

Raw materials required per tonne of formic acid:

Carbon monoxide	610kg
Sodium hydroxide	950kg
Sulphuric acid	1175kg

Yield 90–95%

3. From *n*-butane or naphtha by liquid-phase oxidation

The feedstock, which can be *n*-butane or naphtha, is oxidized in the presence of a catalyst consisting of manganese or cobalt acetate in acetic acid. Reaction conditions for the liquid-phase oxidation are 150–200°C and 56 bar pressure. (*See Figure 70*)

The resultant range of products are separated from the reaction mixture, where acetic acid is the main product, by azeotropic distillation (see Acetic Acid).

Reaction water produced is removed as an azeotrope (using diisopropyl ether) prior to formic acid recovery by azeotropic distillation using benzene. Pure acid can be recovered by further distillation. With butane feedstock, approximately 1kg of formic acid is formed for every 20kg of acetic acid. The figures for naphtha feedstock are up to 1kg of formic acid for every 4kg of acetic acid although this ratio can vary depending on operating conditions.

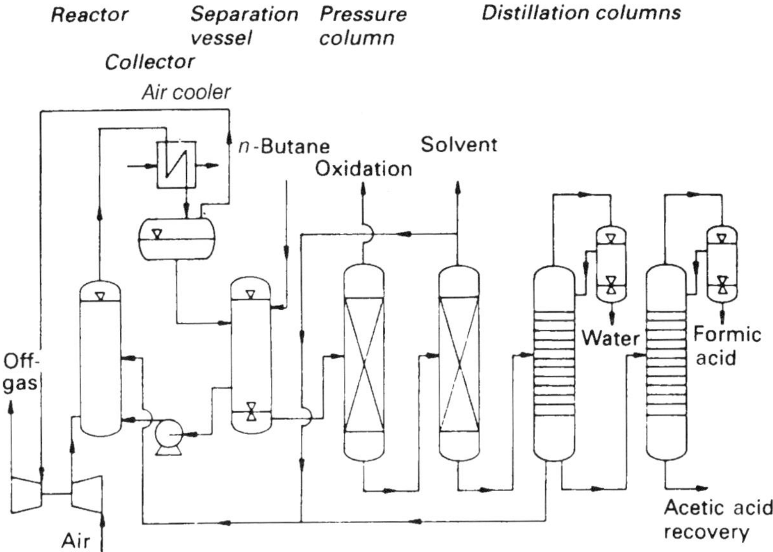

FIGURE 70 Formic acid from *n*-butane by liquid-phase oxidation

4. From formamide

The production of formic acid takes place in three stages:

- carbonylation of methyl alcohol to methyl formate;
- conversion of methyl formate to formamide;
- hydrolysis of formamide to formic acid.

Anhydrous methyl alcohol and carbon monoxide are introduced into a reactor together with the catalyst, sodium methoxide in methyl alcohol. Any water in the feed would cause the catalyst to be hydrolysed to insoluble sodium formate leading to possible clogging. The reaction takes place at 80°C and 45 bar pressure, the concentration of the catalyst being maintained around 2wt% by regulating the feed rate. (*See Figure 71*)

Overhead vapours from the reactor are cooled, methyl formate and methyl alcohol condensed and fed into a separator. Here they join the reaction mixture drawn from the base of the reactor. A small amount of carbon monoxide is purged to remove impurities. Off-gas from the separator is cooled and scrubbed with cold methyl alcohol to remove any residual methyl alcohol and methyl formate vapours.

Cold methyl alcohol from the scrubber and liquids from the separator are combined and fed into a distillation column where methyl formate is recovered overhead. The liquid from the bottom of the column, which contains methyl alcohol and catalyst, has any insoluble inactive catalyst removed before being recycled.

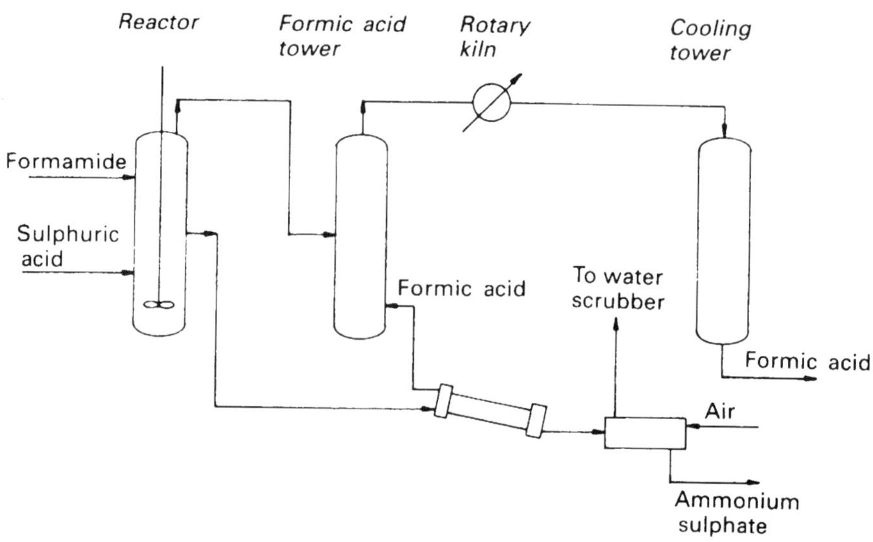

FIGURE 71 Formic acid from formamide

In the second stage, methyl formate and ammonia are passed into a second reactor operating at a temperature of 65°C and a pressure of 13 bar. Heat from the exothermic reaction is removed by heat exchangers. Overhead gases from the reactor are condensed and methyl alcohol is removed before any unreacted ammonia is recycled.

Formamide is separated from the reaction mixture (which contains additionally ammonia, methyl formate and methyl alcohol) by distillation.

Ammonia, methyl formate and methyl alcohol are collected overhead and separated by further fractionation before the ammonia and methyl formate are recycled. Bottoms from the distillation tower pass to a finishing column and formamide is recovered as a residue stream.

In the final stage, formamide and 68–74% sulphuric acid in equal quantities are introduced into a stirred reactor. The formic acid produced is vaporized by the heat from the exothermic reaction. The reaction products, consisting of ammonium sulphate and the remaining formic acid, are fed into an externally heated rotary tubular kiln. Formic acid vapours are produced and combined with those from the reactor before passing into a tower where they are condensed.

Dry ammonium sulphate, recovered from the end of the kiln, is freed from any residual formic acid by blown air. The air is scrubbed with water to remove any acid before being vented to the atmosphere.

The production of formamide directly from carbon monoxide and ammonia (in the presence of sodium methoxide in an alcohol solvent) has been achieved at high pressure. This process has yet to be commercialized.

Reaction

$CO + CH_3OH \rightarrow HCOOCH_3$

$HCOOCH_3 + NH_3 \rightarrow HCONH_2 + CH_3OH$

$2HCONH_2 + H_2SO_4 + 2H_2O \rightarrow 2HCOOH + (NH_4)_2SO_4$

Raw material requirements and yield

Raw materials required per tonne of formic acid:

Methyl alcohol	31kg
Carbon monoxide	702kg
Ammonia	314kg
Sulphuric acid	1010kg

Yield	93%
methyl formate from methyl alcohol	98%
formamide from methyl formate	98%
formic acid from formamide	93%

OTHER PROCESSES

Methyl alcohol can be dehydrogenated over a copper catalyst in the vapour phase to methyl formate. Catalyst compositions suggested include copper–zirconium–zinc, copper with either chromium–manganese or magnesium, and copper and zinc on silica. The endothermic reaction is carried out at a temperature of 280–330°C at atmospheric pressure. A conversion per pass of 50%, based on methyl alcohol, is claimed.

The direct synthesis of formamide from carbon monoxide and ammonia in the presence of an alkaline catalyst has been developed. The liquid-phase reaction takes place in methyl alcohol using sodium methoxide as the catalyst, at a temperature of 70°C and 300 bar.

A process to produce formic acid directly from carbon monoxide and hydrogen, using anhydrous rhodium chloride as the catalyst and methyl iodide as the promoter, has been described.

None of these processes has been commercialized.

PROPERTIES

Colourless, mobile liquid with a pungent odour. Miscible with water, acetone, ether, methyl alcohol and ethyl acetate. Partially soluble in benzene, carbon tetrachloride and toluene. In the vapour phase, it exists with its dimer. Glacial acid slowly decomposes to carbon monoxide and water.

Glacial formic acid 98+%

Molecular Weight	46.03
Density at 20°C	1.221
Melting Point	8.4°C
Boiling Point	100.6°C
Autoignition Temperature	1114°C
Flammability Limits in air	
lower	12vol%
upper	38vol%
Flash Point Closed Cup	69°C
Vapour Density (air = 1)	1.6
Exposure Limit COSHH	5ppm 8 hour TWA
Exposure Limit ACGIH	5ppm TLV-TWA

GRADES

Commercial 90%, 95%, 98% (diluted with water), glacial 99%.

INTERNATIONAL CLASSIFICATIONS

UN No.	1779
CAS Reg No.	64–18–6
EINECS No.	200–580–7
EC No.	607–001–00–0
Description	Corrosive substance
Packing Group	II
Emergency Action Code	2X
HI (Kemler Code)	80

APPLICATIONS

Around 25% of formic acid is used in the tanning and treatment of leather to prevent mould formation.

Formic acid has a number of industrial uses including:

- the manufacture of aspartamine, a synthetic sweetener, and in pharmaceuticals;
- as a coagulant of rubber latex;
- in textile dying and finishing.

Pharmaceuticals are a fast-growing application. In Europe, the most important end use is in silage preservation.

HEALTH AND HANDLING

Formic acid vapour is irritating to the skin, eyes and lungs, and splashes of the liquid can cause skin burns. When handling the acid, rubber gloves, goggles and protective clothing should be worn. Contaminated clothing must be laundered before reuse and shoes discarded.

Containers made of stainless steel, ceramic, glass or lined with rubber or resin are suitable. Formic acid of 98% concentration slowly decomposes on storage to carbon monoxide and water leading to a pressure build-up in unvented containers. The storage area should be well ventilated and containers must be kept away from oxidizing agents.

Spills should be diluted with water until the liquid is non-flammable; residues can be neutralized with sodium bicarbonate. Formic acid is a moderate fire hazard and fires should be extinguished with dry foam, water spray or carbon dioxide. Operators require protective clothing, self-contained breathing equipment and eye protection.

MAJOR PLANTS

Plants with capacities greater than 10 000 tonnes per year:

BASF	Ludwigshafen	Germany
Polioli	Vercelli	Italy
BP Chemicals	Hull	UK
Kemira Oy	Oulu	Finland
Norsk Hydro	Porsgrunn	Norway
PERSTORP	Perstorp	Sweden
Hoechst Celanese	Pampa	US
	Wilmington	US
GNFC	Ranipet	India
Daicel Chemical	Otaka	Japan
Mitsubishi Gas	Toyko	Japan
Lee Chang Yung Chemical	Kaohsiung	Taiwan
Asidken	Kuala Lumpur	Malaysia

MAJOR LICENSORS

Acid-Amine Technologies
BASF
BP Chemicals
Celanese
Glitsch
Halcon-Scientific Design
Huels
Mitsubishi Gas Chemical

Glycerol

$$\underset{\text{OH}}{\underset{|}{CH_2}} - \underset{\text{OH}}{\underset{|}{CH}} - \underset{\text{OH}}{\underset{|}{CH_2}}$$

SYNONYMS

GLYCEROL glycerine, 1,2,3-trihydroxypropane, 1,2,3-propanetriol, glycyl alcohol

Glycerol is a simple triol occurring in natural fats and oils. Until the early 1940s it was produced as a by-product of soap manufacture.

There are three synthetic processes for the manufacture of glycerol from propylene via the following intermediates:

- allyl chloride–epichlorohydrin;
- acrolein–allyl alcohol–glycidol;
- propylene oxide–allyl alcohol–glycidol.

Although the acrolein and propylene oxide processes do not involve the use of chlorine, the epichlorohydrin route is the only one which is important industrially.

Glycerol can be produced from sugar by the fermentation of alcohol, or by the hydrogenation of carbohydrates such as starch, cellulose or sugar. Neither of these routes is of economic importance.

Around 90% of glycerol is produced from natural sources and under 10% is synthetic. By-product natural glycerol production is growing, forcing some synthetic plants to close.

Synthetic glycerol capacities range from 8000 to 63 000 tonnes per year. Dow Chemical is the leading producer of synthetic glycerine in the world.

PROCESSES

1. From propylene via allyl chloride

Propylene gas is dried and preheated to 400°C before being mixed with chlorine vapour in a ratio of 4:1. The mixed gases enter a tubular steel reactor where the exothermic reaction takes place at a temperature of 400–500°C. With pressure kept

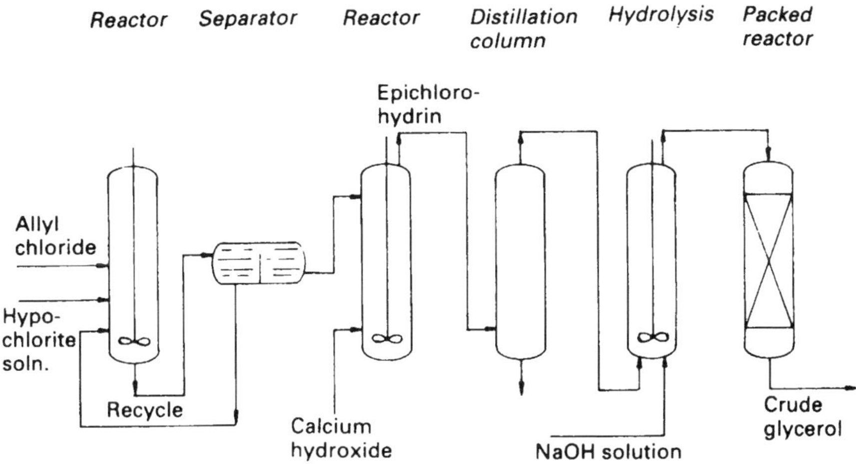

FIGURE 72 Glycerol from propylene via allyl chloride

at 1 bar, and a residence time of 2–3 seconds, around 99% of the chlorine present is consumed. (*See Figure* 72)

The reaction mixture is cooled before going to the first fractionating column where excess propylene and hydrogen chloride are separated by refluxing with liquid propylene. The overhead gases are fed into an absorber; hydrogen chloride is recovered as the acid and propylene is scrubbed with dilute sodium hydroxide to remove any residual acid before being recycled.

The bottoms from the fractionator are distilled to separate off the light ends before a second distillation to remove allyl chloride overhead. Allyl chloride is oxidized with hypochlorite solution at 30–40°C in a stirred tank to yield a mixture of dichlorohydrins. The mixture passes to a separator, where the aqueous layer is recycled; the organic layer is reacted with calcium hydroxide at 50°C in a second stirred tank.

The epichlorohydrin formed is separated by steam distillation as the azeotrope and purified by further distillation. Hydrolysis of the epichlorohydrin is carried out with a 10% aqueous solution of sodium hydroxide at 80–150°C in a stirred reactor at atmospheric pressure. The resultant aqueous glycerol solution is concentrated and purified by evaporation and distillation under vacuum.

Reaction

$$CH_2 = CHCH_3 + Cl_2 \rightarrow CH_2 = CHCH_2Cl + HCl$$
(allyl chloride)

$$CH_2 = CHCH_2CL + HOCl \rightarrow CH_2ClCHClCH_2OH$$
(dichlorohydrin)

$$2CH_2ClCHClCH_2OH + Ca(OH)_2 \rightarrow 2CH_2OCHCH_2Cl + CaCl_2 + 2H_2O$$
(epichlorohydrin)

$$CH_2OCHCH_2Cl + NaOH + H_2O \rightarrow HOCH_2CHOHCH_2OH + NaCl$$
(glycerol)

Raw material requirements and yield

Raw materials required per tonne of glycerol:

Propylene	620kg
Chlorine	2000kg
Sodium chloride	450kg
Calcium hydroxide	450kg

Yield 80%

2. From propylene via acrolein

Propylene, steam and air are reacted in the presence of multi-component metal oxides to yield acrolein (see Acrolein). (*See Figure 73*)

Purified acrolein and isopropyl alcohol are reacted in the liquid phase at a temperature of 400°C using the catalyst *sec*-butoxide. The mixture of allyl alcohol and methyl ethyl ketone formed is separated by distillation. The allyl alcohol passes to a reactor where it is converted to glycidol by hydrogen peroxide containing 0.2% of tungstic acid, before being hydrolysed to glycerol. The aqueous glycerol solution is distilled under vacuum to yield the pure product. The catalyst solution is recovered and recycled.

Reaction

$$H_2C = CHCH_3 + O_2 \rightarrow H_2C = CHCHO + H_2O$$
(acrolein)

$$H_2C = CHCHO + (CH_3)_2CHOH \rightarrow H_2C = CHCH_2OH + (CH_3)_2CO$$
(allyl alcohol)

$$H_2C = CHCH_2OH + H_2O_2 \rightarrow H_2COCHCH_2OH + H_2O$$
(glycidol)

$$H_2COCHCH_2OH + H_2O \rightarrow HOCH_2CHOHCH_2OH$$
(glycerol)

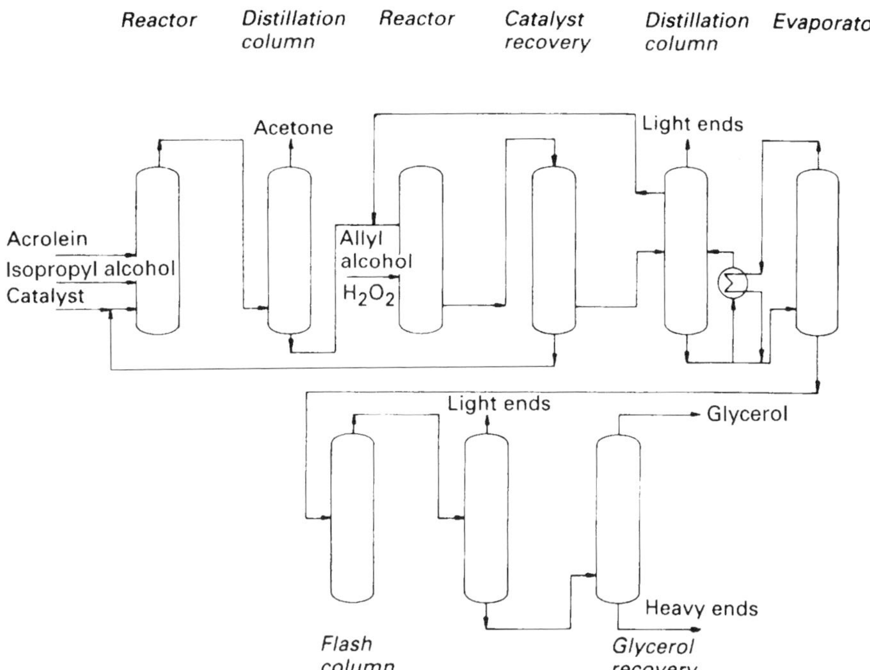

FIGURE 73 Glycerol from propylene via acrolein

Raw material requirements and yield

Raw materials required per tonne of glycerol:

Propylene	920kg
Oxygen	230kg
Hydrogen peroxide (100%)	480kg

Yield 51%

3. From propylene via allyl alcohol

Propylene is oxidized to propylene oxide which is then isomerized in the vapour phase to allyl alcohol. The reaction takes place at 280°C in the presence of a lithium phosphate catalyst (see Propylene Oxide). (*See Figure 74*)

Allyl alcohol is fed into the base of a reactor and a solution of peracetic acid (in a solution of ethyl acetate containing a stabilizer) is introduced about halfway up. Water in a 10–50% mole excess to the acid enters at the top. The reaction takes place at 50–70°C under reduced pressure and the heat generated is removed by azeotropic distillation of the ethyl acetate. The ethyl acetate vapour is collected overhead and recycled to the peracetic acid feed. The glycidol formed is immediately hydro-

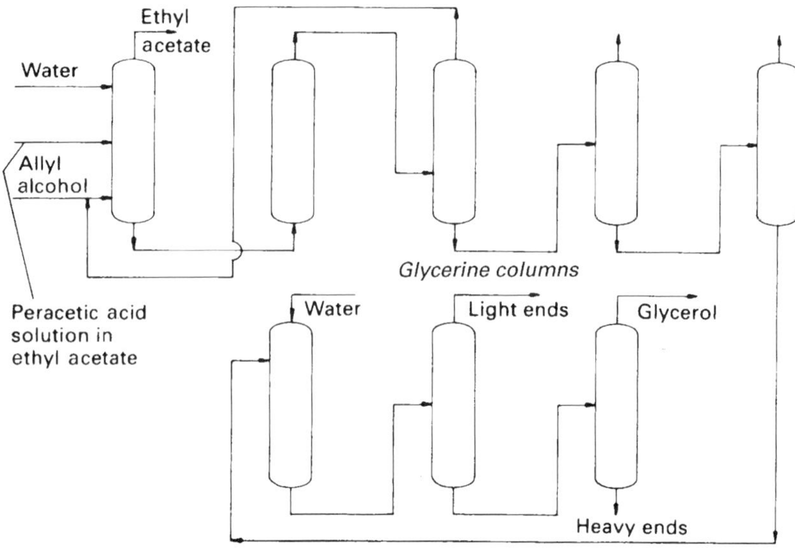

FIGURE 74 Glycerol from propylene via allyl alcohol

lysed to glycerine by the excess water present. Bottoms pass to a distillation column where unreacted allyl alcohol is removed overhead; acetic acid is extracted with a solvent, purified and recovered. The aqueous solution of glycerol is dehydrated before being purified by distillation. Yields are stated to be around 50%.

Reaction

$$2H_2C=CHCH_3 + O_2 \rightarrow 2H_2COCHCH_3$$
(propylene oxide)

$$H_2COCHCH_3 \rightarrow H_2C=CHCH_2OH$$
(allyl alcohol)

$$H_2C=CHCH_2OH + CH_3COOOH \rightarrow H_2COCHCH_2OH + CH_3COOH$$
(glycidol)

$$H_2COCHCH_2OH + H_2O \rightarrow HOCH_2CHOHCH_2OH$$
(glycerol)

PROPERTIES

Colourless, odourless, oily liquid with a sweet taste. Hygroscopic. Soluble in water, ethyl alcohol, phenol and glycols. Almost insoluble in chloroform.

Molecular Weight	92.09
Density at 20°C	1.26

Melting Point	17.97°C
Boiling Point	290°C (decomposes)
Autoignition Temperature	429°C
Flash Point Closed Cup	180°C
Vapour Density (air = 1)	3.17
Exposure Limit COSHH	10mg/m³ 8 hour TWA (mist)
Exposure Limit ACGIH	10mg/m³ TLV-TWA (mist)

GRADES

Industrial 99%, food 99.5%, refined 99.5%.

INTERNATIONAL CLASSIFICATIONS

UN No.	Not allocated
CAS Reg No.	56–81–5
EINECS No.	200–289–5
EC No.	Not allocated

APPLICATIONS

Glycerol is used in a large number of applications especially in the pharmaceuticals, toiletries and cosmetics industries. Because of its low toxicity, it finds outlets in toothpaste, hair colorants, as a skin moisturizer, and to hinder crystallization in food products. Glycerol fatty esters are used as emulsifiers for ice cream, dairy products and in pharmaceuticals. Its other important applications are in the manufacture of alkyd resins and cross-linked polyesters.

Other outlets include its use as a humeticant in tobacco and food, as a lubricant especially for equipment used in the textile and food industries, and explosives. Small amounts are still consumed for the manufacture of nitroglycerine.

Synthetic production is steadily declining and some small plants have closed due to the growth of by-product natural material.

HEALTH AND HANDLING

Glycerol mist can cause irritation to the eyes, skin and respiratory tract. Protective clothing should be worn when handling the product to minimize skin contact and also goggles to prevent splashes reaching the eyes. Contact lenses must be avoided as they can absorb glycerol vapour. Any contaminated clothing should be laundered before reuse.

Store glycerol in closed containers in a dry, well-ventilated area away from strong oxidizing agents. Glycerol reacts violently with acetic anhydride, chromium oxides, calcium oxychloride and alkali metal hydrides.

Spills should be contained and absorbed with vermiculite or sand. Dry chemical, alcohol foam or carbon dioxide should be used to extinguish fires, as water may ineffective. Hazardous gases are formed in the event of fire and firefighters must wear protective clothing and self-contained breathing apparatus.

MAJOR PLANTS

Plants with capacities greater than 12 000 tonnes per year:

Solvay	Tavaux	France
Dow Chemical	Stade	Germany
Dow Chemical	Freeport	US
Kashima Chemical	Kashima	Japan

MAJOR LICENSORS

Daicel Chemical
FMC
Shell Development

Hexamethylenediamine (HMDA)

$H_2N(CH_2)_6NH_2$

SYNONYMS

HEXAMETHYLENEDIAMINE 1,6-diaminohexane, 1,6-hexanediamine

Hexamethylenediamine (HMDA) became important following the discovery and growth in demand for nylon. It reacts with adipic acid to form nylon salt, the raw material used for the manufacture of nylon 66.

At one time hexamethylenediamine was produced from furfural, but this was abandoned after the discovery of the adipic acid route. For several years Celanese in the US utilized a direct route to hexamethylenediamine based on the amination of 1,6-hexanediol, but this plant was closed in 1984.

All commercial processes in current use are based on the catalytic hydrogenation of adiponitrile which can be obtained from:

- adipic acid by amination;
- butadiene by catalytic addition of hydrogen cyanide;
- acrylonitrile by electrolytic dimerization.

With a few exceptions, hexamethylenediamine plants form part of an integrated complex and the product is used captively by a small number of companies.

Capacities range from 75 000 to 250 000 tonnes per year.

PROCESSES

From adiponitrile by hydrogenation

Adiponitrile, ammonia and catalyst are injected into a reactor which can be of the fixed-bed or powder-suspension type. A range of catalysts such as Raney nickel, promoted cobalt, ruthenium or iron oxide have been proposed. Fresh and recycle hydrogen are introduced into the base of the reactor which is maintained at a temperature 80–160°C and a pressure of 200–400 bar. (*See Figure 75*)

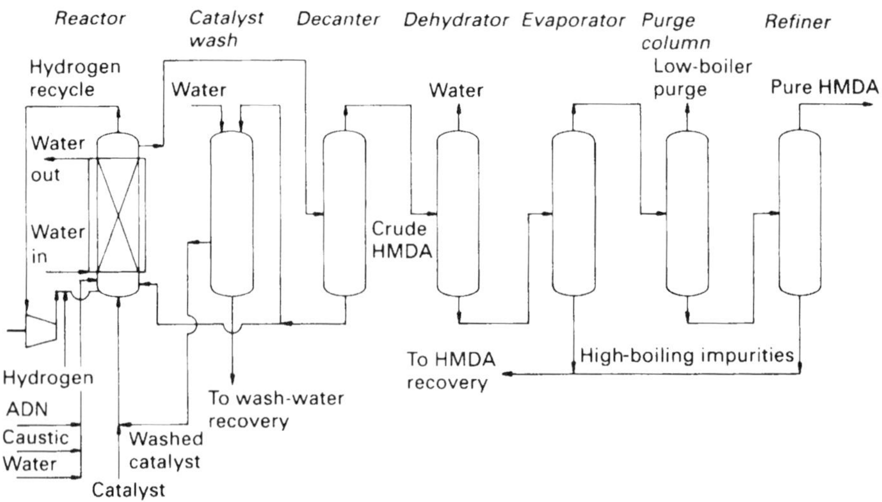

FIGURE 75 HMDA from adiponitrile by hydrogenation

The reaction proceeds rapidly, the heat produced being removed by using ammonia as the heat-transfer medium and by water-cooling jackets. Product from the reactor is passed into a decanter where crude hexamethylenediamine is separated from the solid catalyst which is returned to the reactor. Periodically, some of the catalyst is purged for washing to remove accumulated impurities.

Crude hexamethylenediamine is dehydrated before being evaporated to separate it from high-boiling impurities. Hexamethylenediamine is purified by further distillation during which low-boiling impurities are removed.

By-products 1,2-diaminocyclohexane, hexamethyleneimine, and bishexamethylenetriamine can be recovered and sold.

Reaction

$$CN(CH_2)_4CN + 4H_2 \rightarrow (CH_2)_6(NH_2)_2$$

Raw material requirements and yield

Raw materials required per tonne of hexamethylenediamine:

Adipic acid	930kg
Ammonia	40kg
Hydrogen	70kg

Yield 95%

PROPERTIES

Colourless powder. Soluble in water, benzene and ethyl alcohol.

Molecular Weight	116.14
Density at 20°C	0.88
Melting Point	40.9°C
Boiling Point	200°C
Vapour Density (air = 1)	4.01
Exposure Limit COSHH	Not listed
Exposure Limit ACGIH	Not listed

GRADES

Fibre.

INTERNATIONAL CLASSIFICATIONS

Solid

UN No.	2280
CAS Reg No.	124–09–4
EINECS No.	204–679–6
EC No.	612–104–00–9
Description	Corrosive substance
Packing Group	III
Emergency Action Code	2X
HI (Kemler Code)	80

Solution

UN No.	1783
CAS Reg No.	124–09–4
EINECS No.	204–679–6
EC No.	Not allocated
Description	Corrosive substance
Packing Group (Allocated in accordance with Approved Requirements)	II or III
Emergency Action Code	2X
HI (Kemler Code)	80

APPLICATIONS

About 95% of hexamethylenediamine production is used as the chemical intermediate for the manufacture of nylon 66 fibres and resins. Future demand is tied to the market for this product.

HEALTH AND HANDLING

Contact with hexamethylenediamine can cause irritation to the eyes, skin and respiratory tract. High levels of exposure may lead to sensitization, conjunctivitis, and skin burns. Repeated exposure can lead to anaemia and kidney and liver damage. Personnel should avoid contact with the product by wearing rubber protective clothing and goggles.

Store in closed containers in a cool, well-ventilated area. Hexamethylenediamine will attack zinc, copper and bronze. The dust can form explosive mixtures in air and all containers must be earthed to prevent static build-up.

Spills should be contained and the discharge stopped if possible. Hexamethylenediamine floats and mixes with water and must not be allowed to enter drains or watercourses. Care must be taken not to generate dust during the clean-up operation. Waste material must be disposed of in accordance with local regulations.

Extinguish fires with carbon dioxide, dry chemical, foam or water spray. Firefighters should wear protective clothing and self-contained breathing apparatus.

MAJOR PLANTS

Plants with capacities greater than 90 000 tonnes per year:

Rhone-Poulenc	St. Fons	France
SNC Butachimie	Chalampe	France
BASF	Middlesbrough	UK
Du Pont	Wilton	UK
Du Pont	Orange	US
	Victoria	US
Monsanto Chemicals	Decatur	US
	Pensacola	US

MAJOR LICENSORS

BASF
Du Pont
EniChem
ICI
Monsanto
Nippon Shokubai
Rhone-Poulenc
Zimmer

Isopropyl Alcohol (IPA)

$(CH_3)_2CHOH$

SYNONYMS

ISOPROPYL ALCOHOL isopropanol, *sec*-propyl alcohol, 2-propanol, propane-2-ol, dimethylcarbinol

Isopropyl alcohol (IPA) was one of the first chemicals manufactured from petroleum and formed the beginning of the petrochemical industry.

There are two commercial routes to isopropyl alcohol, both of which are based on propylene. The older sulphuric acid process proceeds via the indirect hydration of propylene, while in the newer process, propylene is hydrated directly. The sulphuric acid process can use a feedstock with a concentration as low as 65% and although the isopropyl alcohol produced tends to be freer of by-products, the direct catalytic process is the preferred route for new plants.

Direct hydration can be carried out in three different ways:

- in the vapour phase over a fixed-bed catalyst incorporating phosphoric acid on an inert support;
- in the liquid phase in the presence of a soluble tungsten catalyst;
- in the mixed vapour–liquid phase using a strongly sulphonated polystyrene cationic exchange resin catalyst.

Operating conditions vary according to the process chosen. Because low temperature and high pressure favour the formation of isopropyl alcohol, considerable effort has been spent on finding catalyst systems with increased activity at low temperature and high pressure.

At the present time, most isopropyl alcohol is produced from propylene, with a very small amount from acetone.

Capacities range from 30 000 to 300 000 tonnes per year.

PROCESSES

1. From propylene by indirect hydration

In the first stage of a two-step process, a C_2 stream containing 65–90% propylene, but free from impurities, is fed into the bottom of a reactor. Sulphuric acid of 70–85% concentration is introduced into the top of the reactor. Pressure is maintained at 20–30 bar to keep the reactants in the liquid state at the reaction temperature of 45–60°C. Heat produced by the exothermic reaction is removed by a cooling jacket containing brine around the reactor. (*See Figure 76*)

In order to increase the rate of reaction between the counterflows of acid and hydrocarbon, the liquids are stirred vigorously. The reaction products, consisting of isopropyl hydrogen sulphate and diisopropyl sulphate, are withdrawn to a settler tank where water is added until the acid content is reduced to around 60% before depressurization.

In the second stage, the sulphates are hydrolysed with steam under vacuum to sulphuric acid and isopropyl alcohol. Crude isopropyl alcohol is recovered from the mixture by steam stripping. The acid is concentrated and recycled. After scrubbing with alkali to remove any residual acid, the crude isopropyl alcohol is condensed prior to being sent to a distillation column. The water–isopropyl alcohol azeotrope, containing 91% isopropyl alcohol, is collected overhead and any higher alcohols are absorbed by passing the distillate through mineral oil. A major by-product, diisopropyl ether, is recovered to prevent build-up and recycled to the hydrolyser. Vent gases are recycled.

Isopropyl alcohol is dehydrated by further distillation with diisopropyl ether, which acts as an entrainer. Isopropyl alcohol is collected from the bottom of the

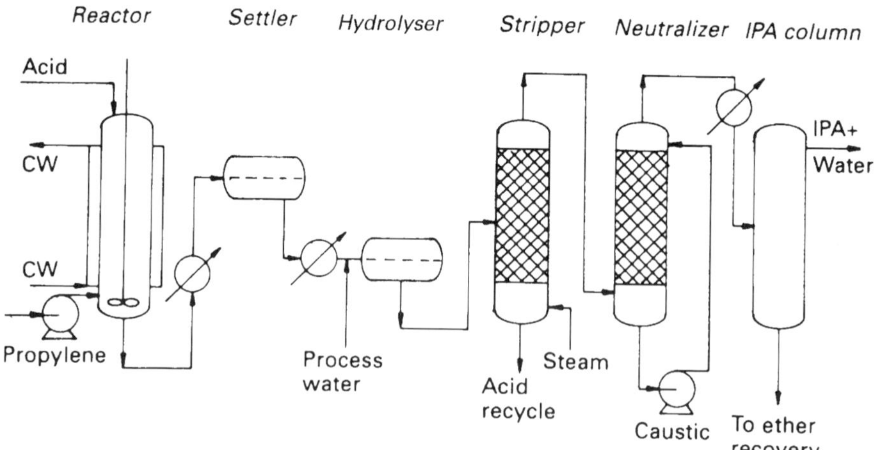

FIGURE 76 Isopropyl alcohol from propylene by indirect hydration

distillation column. For some uses, the isopropyl alcohol is passed through ion exchange resins or activated carbon which remove any undesirable odours.

Reaction

$$CH_3CH{=}CH_2 + H_2SO_4 \rightarrow (CH_3)_2CH(OSO_3H)$$

$$(CH_3)_2CH(OSO_3H) + H_2O \rightarrow (CH_3)_2CHOH + H_2SO_4$$

Raw material requirements and yield

Raw materials required per tonne of 91% isopropyl alcohol:

Propylene	950kg
Sulphuric acid (85%)	13kg

Yield 93–95%

2. From propylene by direct hydration

Propylene (with a purity greater than 90%) and water in a mole ratio of 0.3–0.5:1 are superheated to 170–190°C in two stages, firstly by passing through heat exchangers and secondly by high-pressure steam. The partially liquid reactants, under a pressure of between 30 and 40 bar, pass into a trickle-flow reactor containing a packed bed consisting of sulphonated polystyrene cation exchange resin. Alternatively, the reaction can be carried out in the vapour phase over a catalyst incorporating phosphoric acid on an inert support. Catalyst life is about 8 months. *(See Figure 77)*

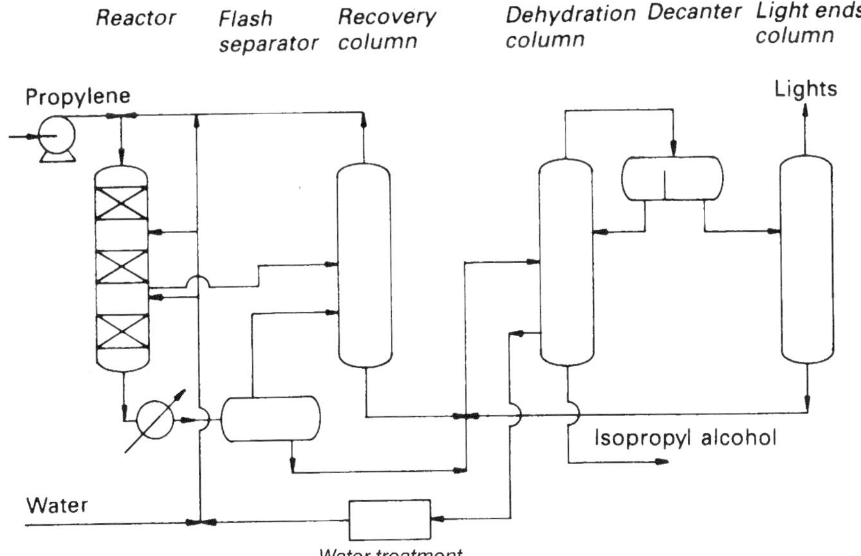

FIGURE 77 Isopropyl alcohol from propylene by direct hydration

The reaction mixture from the base of the reactor (containing unreacted propylene, water, isopropyl alcohol and diisopropyl ether) is cooled and depressurized. After scrubbing with water, unreacted propylene and by-product isopropyl ether are flashed off. Propylene is separated overhead in the propylene recovery column, compressed and recycled. If a lower concentration propylene feed is used, then part of the recycle gases are purified from inerts by passage through a make-up column prior to mixing with fresh propylene.

Isopropyl alcohol is obtained from the aqueous solution from the bottom of the flash column by distillation. Water removal and purification are carried out in a similar manner as for the indirect sulphonation process.

Around 5% of by-product isopropyl ether is formed during the reaction and this can be recovered from the bottoms from the propylene recovery column by further distillation.

The mixed vapour–liquid process has overcome many of the disadvantages of the gas-phase route – high temperature and pressure and low conversion rate. The catalyst used in the liquid-phase process is recycled resulting in considerable cost savings on the other two routes.

Reaction

$$CH_3CH=CH_2 + H_2O \rightarrow (CH_3)_2CHOH$$

Raw material requirements and yield

Raw materials required per tonne of isopropyl alcohol:

Propylene (100% purity) 715kg

Yield 94%

OTHER PROCESSES

Acetone can be hydrogenated in the liquid phase over a fixed-bed Raney-liquid catalyst. Conversion rates of 99% are claimed. An alternative catalyst system, consisting of a mixture of copper–chromium oxides, can be used but the conversion rate is around 95% in this case.

These routes are only of value where excess acetone is available, for example as a by-product of the phenol from the cumene process.

PROPERTIES

Colourless, mobile liquid with an alcoholic odour. Miscible with water and ethyl alcohol.

Molecular Weight	60.10
Density at 20°C	0.786
Melting Point	–88.5°C
Boiling Point	82.5°C
Autoignition Temperature	399°C
Explosive Limits in air	
lower	2.5vol%
upper	12.1vol%
Flash Point Closed Cup	11.8°C
Vapour Density (air = 1)	2.07
Exposure Limit COSHH	500ppm 15 minutes 400ppm 8 hour TWA
Exposure Limit ACGIH	500ppm TLV-STEL 400ppm TLV-TWA

GRADES

Technical 99%, 95%, and azeotrope 87.9%.

INTERNATIONAL CLASSIFICATIONS

UN No.	1219
CAS Reg No.	67–63–0
EINECS No.	200–661–7
EC No.	03–003–00–0
Description	Flammable liquid
Packing Group	II
Emergency Action Code	2SE
HI (Kemler Code)	33

APPLICATIONS

Isopropyl alcohol is an excellent low-cost solvent used in many consumer and industrial products and as an extractant. In the US, around 24% is used in cosmetics, personnel care and household products. A further 15% goes into printing inks and surface-coating applications and another 10% is used in pharmaceuticals.

Isopropyl alcohol is much in demand in industry as a solvent especially for oils and gums. A further 20% is consumed as a raw material for the production of acetone.

Several chemical compounds are synthesized from isopropyl alcohol, in particular methyl isobutyl ketone and a range of esters. Isopropyl alcohol is used as a de-icer, and in the manufacture of fishmeal concentrates. Low-grade isopropyl alcohol is used in motor fuels.

Annual global growth rate to the year 2000 is expected to be around 2–3%.

HEALTH AND HANDLING

Isopropyl alcohol vapour is a mild irritant to the eyes, nose and throat. Prolonged exposure can lead to nausea, headaches and mild narcosis. It can be absorbed through the skin leading to irritation and dermatitis. Contact lenses must not be worn as they tend to concentrate the vapour.

Dry isopropyl alcohol is not corrosive to metals so that storage containers can be made of iron, steel, or copper. Because of its flammability, containers should be stored in a well-ventilated, cool, explosion-proof area away from sources of ignition and strong oxidizing agents. Handling equipment must be spark-proof and earthed to avoid static build-up.

In the event of a leak, remove all sources of ignition and heat. Contain and absorb with paper towels, vermiculite or dry sand. Use non-sparking tools to clean up the waste and place in closed bins for disposal by burning. The contaminated area can be washed with water but care must be taken to keep the washings away from sewers or water outlets. Clean-up staff should wear impervious clothing and eye protection to prevent skin contact. Any contaminated clothing must be laundered before reuse.

Isopropyl alcohol is a dangerous fire and moderate explosion hazard. Carbon dioxide, water spray, dry chemical or foam can be used to extinguish fires. Flashback is a real hazard as vapours can roll for considerable distances. Firefighters must wear full protective clothing, eye protection and self-contained breathing apparatus.

MAJOR PLANTS

Plants with capacities greater than 100 000 tonnes per year:

Shell Chimie	Berre	France
CONDEA Chemical	Moers	Germany
Shell Nederland Chemie	Pernis	Netherlands
Shell Chemicals	Ellesmere Port	UK
Exxon Chemical	Baton Rouge	US
Shell Chemical	Deer Park	US
Union Carbide	Texas City	US
Shell Canada	Corunna	Canada
Nippon Petrochemicals	Kawasaki	Japan

MAJOR LICENSORS

Edeleanu
Exxon
Halcon-Scientific Design
Huels
ICI
Shell Development
Texaco Development
Tokuyama Soda
Union Carbide
UOP
Veba Chemie/BP Chemicals

Maleic Anhydride

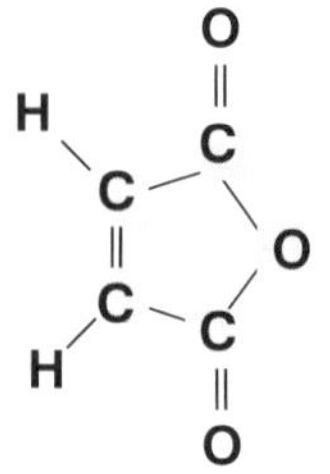

SYNONYMS

MALEIC ANHYDRIDE cis butenedioic anhydride, 2,5-furandione, toxilic anhydride, dihydro-2,5-dioxofuran

Production of maleic anhydride has grown rapidly since its commercialization in the US in the early 1930s.

Until the 1970s, almost all maleic anhydride was derived from the partial oxidation of benzene over a promoted vanadium catalyst, or obtained as a by-product from phthalic anhydride manufacture. Following the increase in benzene prices and environmental pressures leading to emission control, routes using alternative feedstocks were developed. Improved catalysts and lower raw material costs made *n*-butane attractive. Mitsubishi Chemical in Japan and Bayer and BASF in Germany developed processes based on *n*-butylene extracted from C_4 streams obtained from naphtha crackers but these are no longer commercial.

Oxidation of *n*-butane or *n*-butane–butene mixtures, in the presence of a catalyst, can be carried out in a fixed or fluidized bed. Considerable research has been carried out on the use of fluidized-bed processes because of the advantages they offer over fixed-bed routes. These advantages are:

- lower air-to-hydrocarbon concentration in the feedstock;
- no premixing required thus avoiding the risk of explosion;
- good temperature profile with no hot spots;
- greater capacity at lower cost.

However, their disadvantages are the remixing of products, abrasion of the catalyst resulting in problems with selectivity, catalyst life, conversion rates and by-product formation. Although fluidized-bed processes are growing, it is unlikely that they will surpass fixed-bed processes. Catalyst and reactor improvements by Huntsman, which allows higher productivity, could lead to changes in the future position.

Monsanto and Du Pont have developed a transport-bed process for the production of maleic anhydride also using butane feedstock. The maleic anhydride formed is recovered as maleic acid for further hydrogenation to tetrahydrofuran and a plant using this route came on-stream in Spain in 1997.

Cyclopentene has been suggested as a starting material for maleic anhydride but this route has not been commercialized. Some maleic anhydride is recovered commercially from phthalic anhydride manufacture.

The oxidation of *n*-butane is the preferred route and several plants have been converted from benzene to accept the new feedstock. In the US, all maleic anhydride production is based on butane and Europe is converting from benzene feedstock to butane. Future growth of maleic anhydride production will be based on butane as the raw material.

Currently, over 75% of maleic anhydride is produced from *n*-butane and 20% from benzene. Around 3% is recovered from phthalic anhydride manufacture.

Capacities range from 3000 to 100 000 tonnes per year.

PROCESSES

1. From benzene by catalytic oxidation

The oxidation of benzene is carried out in a fixed-bed reactor in the presence of air. (*See Figure 78*)

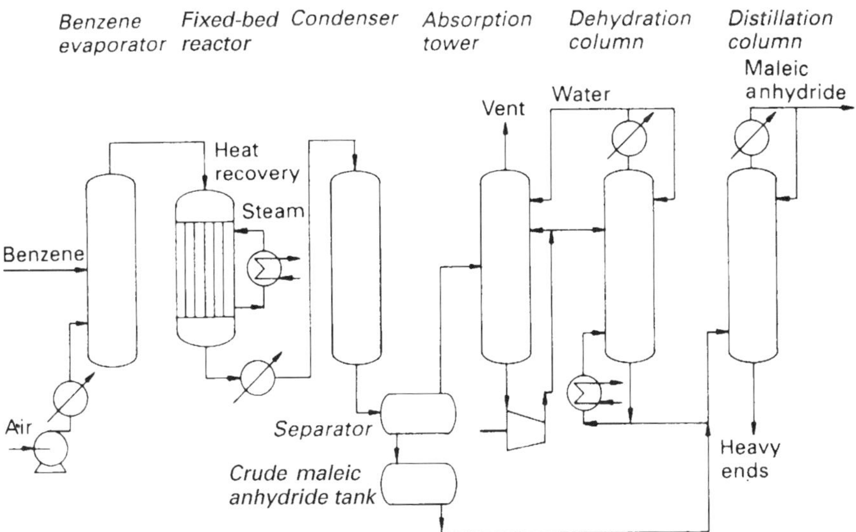

FIGURE 78 Maleic anhydride from benzene by catalytic oxidation

Filtered air is preheated, compressed from a blower and mixed with benzene vapour to give a concentration of 3wt% benzene. It is then fed into a multi-tubular reactor containing a vanadium oxide–molybdenum oxide catalyst on an inert substrate such as alumina or silica. An excess of air is required to keep the vapours outside explosive limits. Silver, cobalt, nickel, titanium and sodium oxide and salts have been used as promoters to increase yields. Heat from the strongly exothermic reaction is removed by molten salts as in the butane process, circulated by a pump, which pass through a heat exchanger to generate high-pressure steam. The reaction temperature is kept at 350–450°C.

The exit gases (containing around 1% maleic anhydride, carbon monoxide, carbon dioxide and traces of phenols and carboxylic acids) are initially cooled to 150–160°C so that neither the anhydride nor the steam condenses. The reaction gas temperature is then reduced to 55–65°C when about 60% of the maleic anhydride condenses as a liquid which is removed from the vapour stream in order to reduce maleic acid formation. It is separated from the gases and sent to the crude maleic anhydride tank. The remaining gases are scrubbed countercurrently with water in an absorption tower. The resultant maleic acid is dehydrated by azeotropic distillation with *o*-xylene, and water is removed overhead. Organic solvents have also been used to replace water for maleic acid recovery in order to reduce the volume of by-product disposal.

The dehydrated product is combined with the condensed anhydride before being recovered by distillation. Purification of the crude maleic anhydride is carried out by vacuum distillation or sublimation. Maleic anhydride is either sold in the molten state or formed into pellets and flakes, and bagged.

Reaction

$$C_6H_6 + 4\tfrac{1}{2}O_2 \rightarrow C_4H_2O_3 + 2H_2O + 2CO_2$$

Raw material requirements and yield

Raw material requirements per tonne of maleic anhydride:

Benzene	1136kg
Air	16 000m^3

Yield 88%

2. From *n*-butanes by direct oxidation

Air (at a pressure slightly above atmospheric and in a proportion below the 1.8mol% explosive limit) is vaporized, mixed with superheated butane (or *n*–butane–*n*–butene mixtures with a high paraffin content) and fed into a tubular fixed-bed reactor. The catalyst consists of vanadium–phosphorus oxide supported on silica. The oxidation takes place at a temperature of 400–430°C. Heat from the exothermic reaction is removed by molten salts, usually a mixture consisting of 53% potassium nitrate, 40% sodium nitrite and 7% sodium nitrate, circulated through the reactor shell. Excess heat is used to generate steam. (*See Figure 79*)

The reaction gases are cooled to 120°C; 30% of the maleic anhydride formed is condensed and removed from the vapour stream. This stage is carried out as quickly as possible in order to reduce the amount of maleic acid formed from contact with the high water content of the reaction gas stream. The remainder of the maleic anhydride made is recovered as maleic acid from the water scrubber. Waste gases are burnt.

The crude acid is dehydrated using *o*-xylene as the entrainer under vacuum distillation. The crude anhydride is added to the condensed product and purified by distillation under reduced pressure.

Alusuisse Italia have developed a fluidized-bed catalytic process which uses an organic solvent for maleic anhydride recovery. The process is claimed by the company to be more economic than its fixed-bed route as it has a lower air-to-butane ratio, greater single reactor capacity, good temperature profile, lower steam demand and less waste water with reduced by-product disposal. However, it does suffer from the disadvantage of catalyst erosion leading to poor conversion performance and the higher costs associated with fluidized beds.

Monsanto and Du Pont's new transport-bed process uses two reactors. In the first, spent catalyst obtained from the oxidation of butane is regenerated with oxygen. The regenerated catalyst, isolated from the oxygen-containing gas and without any additional air, is fed together with new butane feed into the second reactor.

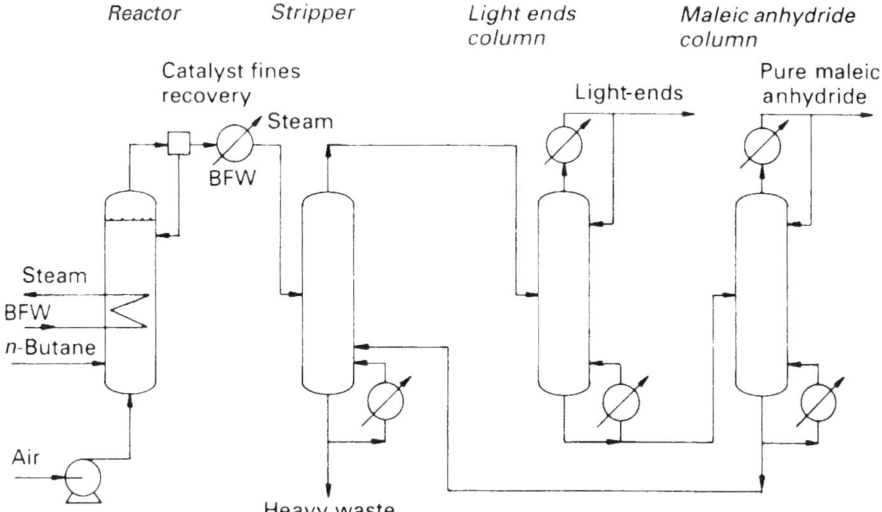

FIGURE 79 Maleic anhydride from *n*-butanes by direct oxidation

The catalytic reaction takes place with only the oxygen entrapped in the catalyst. Crude maleic anhydride is purified by batch or continuous distillation under reflux until any maleic acid produced is dehydrated.

Reaction

$$2CH_3 - CH_2 - CH_2 - CH_3 + 7O_2 \rightarrow 2C_4H_2O_3 + 8H_2O$$

Raw material requirements and yield

Raw material requirements per tonne of maleic anhydride:

n-Butane (100%)	1105kg
Yield 63%	

PROPERTIES

White crystalline flakes with strong acrid odour. Soluble in acetone, ether and petroleum fractions. Reacts with water to form maleic acid. Dimerizes to 1,2,3,4-cyclobutane tetracarboxylic dianhydride in the presence of UV light. A decomposition-polymerization explosion hazard can occur in the presence of alkalis at temperatures above 150°C.

Molecular Weight	98.06
Density at 20°C (solid)	1.43
Melting Point	52.85°C
Boiling Point	202°C sublimes
Autoignition Temperature	477°C
Explosive Limits in air	
lower	1.4vol%
upper	7.1vol%
Flash Point Closed Cup	103°C
Vapour Density (air = 1)	3.4
Exposure Limit COSHH	3mg/m^3 15 minutes (Maximum Exposure Limit) 1mg/m^3 8 hour TWA (Maximum Exposure Limit)
Respiratory sensitizer.	
Exposure Limit ACGIH	0.25ppm 8 hour TLV-TWA

GRADES

Commercial 99% as flakes, formed into different shapes, or molten.

INTERNATIONAL CLASSIFICATIONS

UN No.	2215
CAS Reg No.	108–31–6
EINECS No.	203–571–6

EC No.	607–096–00–9
Description	Corrosive substance
Packing Group	III
Emergency Action Code	2X
HI (Kemler Code)	80

APPLICATIONS

In the US, the most important outlet, accounting for over 64% of maleic anhydride consumption, is for the manufacture of unsaturated polyester resins. Formed by the reaction of maleic anhydride with glycols, these resins are reinforced with fibre glass to go into boat hulls, vehicle bodies, building panels and storage tanks, or are cast to form putty resins for automobile repairs, and used as sealants.

Lubricating oil additives (such as polyisobutylene succinate) which are used as viscosity index improvers and dispersants, consume around 19% of maleic anhydride production. Other outlets include fumaric acid, used as a food additive, in resins and paper size, in the manufacture of alkyd resins, maleic acid, water-soluble polycarboxylated polymers (which are used in detergents as a replacement for tripolyphosphates) and agricultural chemicals such as fungicides, insecticides and growth regulators.

Future growth for maleic anhydride will be closely tied to the demand for polyester resins.

HEALTH AND HANDLING

Maleic anhydride is toxic and an irritant to the skin and mucous membranes leading to burning and ulceration. Care must be taken to avoid inhalation of the product and protective clothing should be worn. Any contaminated clothing must be washed before reuse.

Store in stainless steel containers in a cool, dry area.

As the product has a limited shelf-life, good stock control must be practised. Earthing is required to prevent static build-up which can lead to dust explosions. Considerable quantities of maleic anhydride are transported in the molten state at temperatures of 60–70°C which reduces handling problems. Pellets are better than flakes which tend to produce dust when handled.

Spills should be allowed to solidify and then placed into bins, and recovered or disposed of according to local regulations. Carbon dioxide, alcohol foam or water spray should be used to fight fires and all firefighters must wear full protective clothing.

MAJOR PLANTS

Plants with capacities greater than 25 000 tonnes per year:

Pantochim	Feluy	Belgium
CONDEA Chemical	Bottrop	Germany
Lonza	Ravenna	Italy
Polioli	Vercelli	Italy
Complex	Novomoskovsk	Russia
Amoco	Joliet	US
Ashland Petrochemical	Neal	US
	Neville Island	US
Bayer	Baytown	US
	Houston	US
Huntsman Chemical	Pensacola	US
SVCG	Al Jubail	Saudi Arabia
Nippon Shokubai	Himeji	Japan
Tonen Chemical	Kawasaki	Japan
Shinwha Petrochemical	Ulsan	South Korea

MAJOR LICENSORS

Benzene	*Bayer/Lurgi*
	Halcon-Scientific Design
	Lonza
	Mitsubishi Chemical
	Mitsui Petrochemical/BP Chemicals
	Nippon Shokubai
n-Butane	*Alusuisse Italia/ABB Lummus Global*
	BP Chemicals
	Du Pont
	Huntsman Chemical
	Lonza
	Monsanto
	Scientific Design
	Sisas
Butylene	*BASF*
	Bayer
	Lurgi
	Mitsubishi Chemical
By-product recovery	*BASF*
	Lonza
	Nippon Shokubai
	UCB

Methyl Alcohol CH_3OH

SYNONYMS

METHYL ALCOHOL carbinol, methanol, methyl hydroxide

In recent years, methyl alcohol processes have undergone a number of modifications designed to improve energy balances and hence the cost of production. Simultaneously, the capacities of new plants have increased so that the higher investment costs which these improvements require can be distributed across the product price.

Although the distillation of wood was the original source of methyl alcohol, around 97% of production is now based on natural gas, naphtha or refinery light gas. The discovery of new gas deposits and the increasing use of naphtha as a petrochemical have made natural gas the dominating raw material source. As a result, there has been a shift in the location of plants from industrialized countries to those where large sources of low-cost natural gas are available.

Large-scale methyl alcohol processes based on hydrogen–carbon oxide mixtures were introduced in the 1920s. Natural gas replaced gases obtained from coking operations after World War II. The major breakthrough in the early 1970s was the development of low-pressure processes, to replace the high-pressure route, and these are the choice for all new plants.

More recently, the emphasis on energy efficiency and product cost reduction have resulted in the development of an autothermal reactor in combination with a steam reformer to produce the raw material, synthesis gas. This has reduced natural gas consumption per tonne of methyl alcohol by nearly 8%. The falling price of methyl alcohol and increasing cost of natural gas feedstock have accelerated these trends.

Liquid-phase processes using copper powder slurry catalysts have reached pilot-plant scale. Although they are highly selective, none has been adopted commercially.

ICI's adiabatic route with 60% and Lurgi's isothermal route with 30%, dominate the world's methyl alcohol production processes. Methyl alcohol can be used as an alternative fuel or as an octane booster for blending with gasoline for motor cars, the output of some plants being used solely for this purpose. Nearly 85% of methyl alcohol is consumed in the chemical industry, the remainder going into the fuel sector.

In the future, methyl alcohol production could be seen as an alternative to flaring natural gas, especially in new oil-field developments and where environment considerations are important. It can be transported with fewer problems than that required for many other petrochemicals.

Capacities range from 60 000 to 2 250 000 tonnes per year.

PROCESSES

1. From carbon monoxide and hydrogen

The hydrocarbon feedstock, desulphurized by passing over activated carbon or hot zinc oxide at 400°C, is mixed with process steam at 2 bar before entering a tubular reformer. The tubes, packed with a promoted nickel catalyst, are heated from the outside. (*See Figure 80*)

The resultant synthesis gas (a mixture of carbon monoxide, carbon dioxide and hydrogen, with hydrogen present in excess) leaves the reformer at a temperature of 800–850°C. The gas is cooled to ambient temperature and the heat, recovered by boiler and heat exchangers, is used for preheating boiler feed water to generate high-pressure steam and for distillation. The ratio of hydrogen and carbon is adjusted to give a stoichiometric ratio of 2:1 by purging excess hydrogen or adding

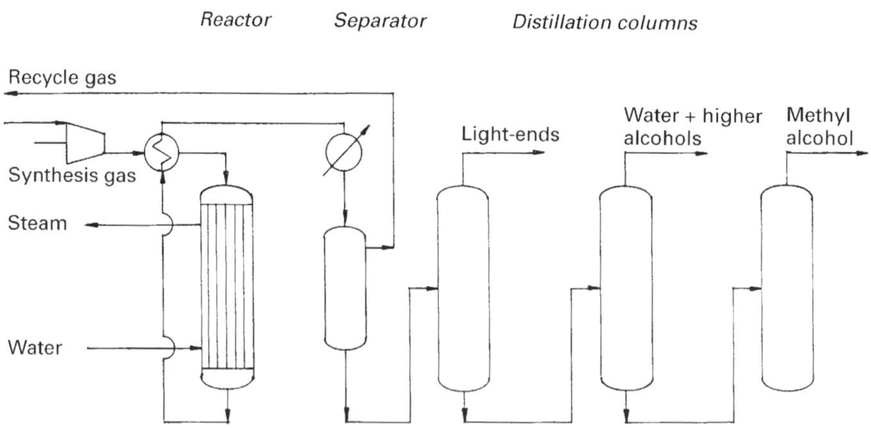

FIGURE 80 Methyl alcohol from carbon monoxide and hydrogen

carbon dioxide. The mixture then enters the methyl alcohol convertor where the gases are compressed to 50–100 bar, mixed with recycled gas and preheated.

The methyl alcohol synthesis takes place in the presence of copper-based catalysts at a controlled temperature of 250–260°C. A range of catalyst systems has been recommended, copper with zinc–boron or zinc–chromium being the most popular. The resultant gases are cooled in a condenser, pressure is reduced and dissolved gases are flashed off. Crude methyl alcohol is recovered in a separator and purified by distillation. A small proportion of the residual gas stream is purged for use as hydrogenation gas for desulphurization before the remainder is mixed with fresh synthesis gas and recycled.

If heavy fuel oil is used as a feedstock, it is partially oxidized by oxygen and steam at 1400°C and 60 bar pressure. The product gas is scrubbed with water, desulphurized and adjusted to the optimum ratio of hydrogen and carbon monoxide before being fed directly into the methyl alcohol synthesis loop.

Reaction

$$CO + 2H_2 \rightarrow CH_3OH$$

Raw material requirements and yield

Raw materials required per tonne of methyl alcohol:

Carbon monoxide	1160m^3
Hydrogen	2340m^3

Yield 61%

2. From methane by two-step reforming

The discovery of natural gas deposits has led to the modification of conventional methyl alcohol technology by the development of a two-step or combined reforming process. (*See Figure 81*)

Desulphurized natural gas and steam are passed to the primary reformer and the exit gases are led directly into an autothermal reformer. Preheated oxygen mixed with steam is fed into the oxygen-blown autothermal reformer. The amounts of oxygen and natural gas are adjusted to generate a synthesis gas with a stoichiometric ratio of just over two, with low inerts content.

The following reactions are involved.

$$CH_4 + H_2O \rightarrow 3H_2 + CO \qquad \text{Steam reforming}$$
$$CH_4 + \tfrac{1}{2}O_2 \rightarrow 2H_2 + CO \qquad \text{Catalytic autothermal reforming}$$

Lurgi, Haldor Topsoe and Mitsubishi Gas Chemical have modified their existing technology to incorporate these new developments. As well as a reduction in nat-

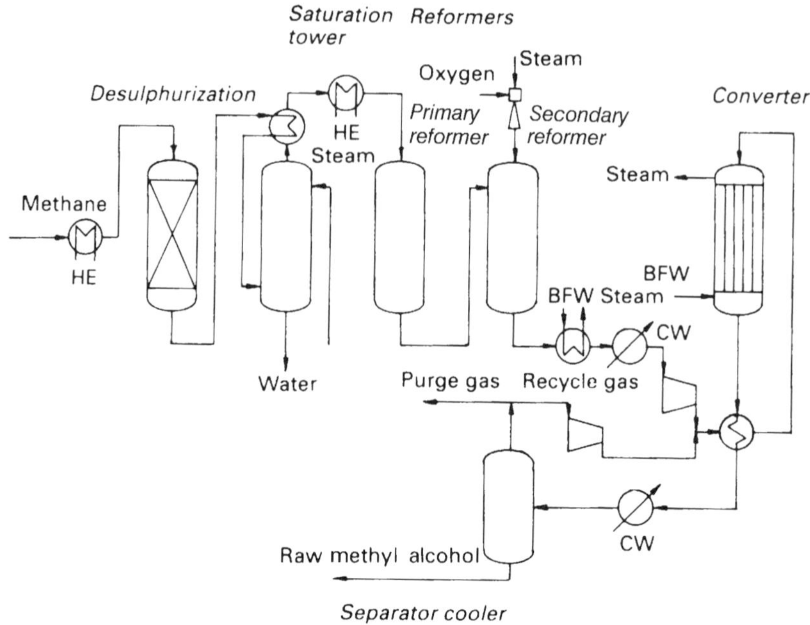

FIGURE 81 Methyl alcohol from methane by two-step reforming

ural gas consumption, the steam reformer for both processes needs to be around one-quarter of the size of a conventional unit with a resultant saving in costs.

Reaction

$CO + 2H_2 \rightarrow CH_3OH$ Yield 99%

$CO_2 + 3H_2 \rightarrow CH_3OH + H_2O$ Yield 70–90%

(depending on the composition of the synthesis gas)

OTHER PROCESSES

In a recent development, hydrogen and carbon monoxide have been converted to methyl alcohol and C_2–C_5 alcohols by reaction over a molybdenum sulphide catalyst supported by activated carbon containing small quantities of potassium, iron, nickel, or cobalt metals. In order to increase the production of by-product higher alcohols, the hydrogen:carbon monoxide ratio has to be below 2. These mixtures, which are cheaper to manufacture than blending separate materials, can be added directly to gasoline without the need for co-solvents.

PROPERTIES

Clear, colourless, volatile and extremely flammable liquid. Soluble in water, ethyl alcohol and ether.

Molecular Weight	32.04
Density at 20°C	0.792
Melting Point	–97.68°C
Boiling Point	64.7°C
Autoignition Temperature	470°C
Flammability Limits in air	
lower	5.5vol%
upper	44.0vol%
Flash Point Closed Cup	12.2°C
Vapour Density (air = 1)	1.11
Exposure Limit COSHH	250ppm 15 minutes 200ppm 8 hour TLV-TWA
Exposure Limit ACGIH	250ppm 15 minutes TLV-STEL 200ppm 8 hour TLV-TWA

GRADES

Pure 99.8%, British 95%, denatured 98%.

INTERNATIONAL CLASSIFICATIONS

UN No.	1230
CAS Reg No.	67–56–1
EINECS No.	200–659–6
EC No.	603–001–00-X
Description	Flammable liquid, toxic substance
Packing Group	II
Emergency Action Code	2WE
HI (Kemler Code)	336

APPLICATIONS

Methyl alcohol is one of the world's major chemicals and its most important outlet is in the manufacture of formaldehyde which accounts for 37% of total demand. Another 28% of methyl alcohol consumption is used for the production of methyl *tert*-butyl ether (MTBE), an octane enhancer, and 3% is employed directly in fuels. With the increasing importance of the carbonylation of methyl alcohol to acetic acid process, 8% is employed in this outlet. Methyl alcohol derivatives include the intermediates dimethyl terephthalate, methyl methacrylate, accounting for 2% and 3% respectively. A further 4% is used as a solvent. Other outlets include the production of methylamines and ethyl halides, and in antifreeze.

Future demand in the chemical sector will be heavily dependent on the oil price and the level of the world's economy. The mid-1990s has seen a consolidation of the market with the emergence of Methanex as the world leader controlling over 20% of global methyl alcohol capacity.

The use of methyl alcohol in fuel is dependent on the price of crude oil. However, if the price of crude oil rises, then methyl alcohol based on cheap natural gas becomes attractive. With the trend towards unleaded gasoline, the blending of methyl alcohol with solubilizers could escalate. It is possible to produce alcohol blends from the Lurgi and Haldor Topsoe processes, which do not require the addition of expensive solubilizers such as MTBE or isopropyl alcohol (IPA).

World consumption is forecast to grow at 3.7% per year up to 2000.

HEALTH AND HANDLING

Taken orally, methyl alcohol is poisonous and blindness and death can result. Absorbed by skin contact, the vapour can cause eye and nose irritation, dizziness, headache, nausea and narcosis. The toxic effects are compounded by repeated exposure. Gloves, aprons, boots, face shields and goggles must be worn, and staff should be given training on the correct handling of the product. Contact lenses, which can concentrate the effects of the vapour, must be avoided. All staff working with methyl alcohol should receive regular medical checks.

Methyl alcohol should be stored in closed containers in a dry, well-ventilated, cool area, away from strong oxidizing agents. As methyl alcohol is slightly corrosive to metals, steel, aluminium or lead-lined containers are required. Methyl alcohol will attack some polymers and rubber, so seals and jointing materials must be checked for their suitability. Containers must be bonded to prevent the build-up of static and all electrical equipment must be explosion-proof.

If spills occur, extinguish all forms of ignition and evacuate personnel. Absorb with paper, vermiculite or dry sand and place in a container using non-sparking tools for disposal by burning in an approved incinerator or for reclamation. Any residues can be flushed with water but care must be taken to prevent the waste liquids from entering sewers or water intakes.

Methyl alcohol is a dangerous fire and moderate explosion hazard. Carbon dioxide, dry chemical or alcohol foam is used to extinguish fires. Water should not be used as it tends to scatter the flames. Flashback is a hazard. Firefighters and clean-up staff must wear protective clothing, full face protection and self-contained respirators when dealing with methyl alcohol.

MAJOR PLANTS

Plants with capacities greater than 660 000 tonnes per year:

Methanor	Delfzijl	Netherlands
Atochem	Leuna	Germany
Statoil	Tjeldbergodden	Norway
Complex	Tomsk	Russia
Borden	Geismar	US
Lyondell	Channelview	US
Terra Industries	Beaumont	US
Celanese Canada	Edmonton	Canada
Methanex	Medicine Hat	Canada
Methanex	Cabo Negro	Chile
Trintoc	Pointe Lisas	Trinidad
Ecofuel	Jose	Venezuela
Pequiven/Mitsubishi	Jose	Venezuela
National Methanol	Al Jubail	Saudi Arabia
Saudi Methanol	Al Jubail	Saudi Arabia
Petrochemical Industries Corp.	Shuaiba	Kuwait
Methanex	New Plymouth	New Zealand

MAJOR LICENSORS

Acid-Amine Technologies
BASF
Cascale
Haldor Topsoe
ICI Katalco
Inventa
Kellogg
Linde
Lurgi
Mitsubishi Gas Chemical
Raytheon Engineering
Snamprogetti
Uhde

Methylamines

Mono-methylamine	**CH_3NH_2**
Di-methylamine	**$(CH_3)_2NH$**
Tri-methylamine	**$(CH_3)_3N$**

SYNONYMS

MONO-METHYLAMINE	MMA, aminomethane, methanamine
DI-METHYLAMINE	DMA
TRI-METHYLAMINE	TMA

Methylamines are produced commercially by the catalytic alkylation of anhydrous ammonia with methyl alcohol. All three amines are formed because mono-methylamine will react with more alcohol to yield the di- and tri- isomers. Although the product distribution can be varied by the choice of reaction conditions, by the catalyst used and by recycling the unwanted products, it is not economic to produce only one of the amines. Most commercial manufacturers market all three isomers. Recent research has concentrated on the use of shape-selective zeolite catalysts to improve the production of di-methylamine, the most desired isomer.

Methyl alcohol can be replaced by formaldehyde and the reaction carried out in two stages. Formaldehyde is reacted with ammonia and then hydrogenated to the amine. The choice of route depends mainly of the cost of the raw materials and varies from country to country.

The synthesis of methylamines from carbon monoxide, hydrogen and nitrogen over a zirconium-based catalyst or the hydrogenation of hydrogen cyanide, have attracted interest but both routes are still under investigation.

Capacities range from 10 000 to 100 000 tonnes per year.

PROCESSES

From methyl alcohol and ammonia

Vaporized methyl alcohol and ammonia in the molar ratio of 1:2 are mixed with recycle amine feed and preheated to 350°C under a pressure of 14 bar. The vapours pass to a reactor containing a catalyst which consists of amorphous silica–aluminium oxide. Thorium oxide, chromium oxide, tungsten oxide or a mixture of oxides can be used instead. (*See Figure 82*)

The reaction takes place at 390–450°C and the mixture formed (containing methylamines, unreacted methyl alcohol and ammonia) is cooled prior to entering a rectifier. Under a pressure of 14 bar, unreacted ammonia is removed overhead and recycled to the synthesis section. The methylamine mixture from the bottom of the rectifier is extractively distilled under pressure with water, and tri-methylamine is recovered overhead. Mono-methylamine is obtained overhead by distillation in a third column and di-methylamine can be separated from high-boiling impurities by further distillation in a fourth column.

The methylamine product distribution can be changed by varying the methyl alcohol:ammonia ratio from 1:4 for mono-methylamine to 1:1.5 for tri-methylamine. As the market demand is for mono- and di-methylamines rather than the tri-methylamine favoured by the reaction, excess tri- and frequently mono-methylamine are recycled over the catalyst. Dismutation and disproportionation of the recycled amines results in the production of the most desired di-methylamine. If mono-methylamine is the desired product, an alternative method is to add small

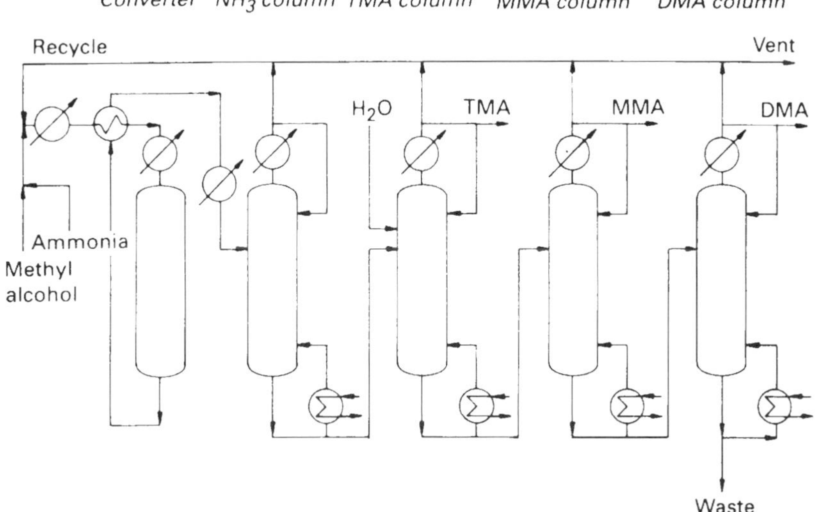

FIGURE 82 Methylamines from methyl alcohol and ammonia

amounts of tri-methylamine or water to the raw feed to suppress the formation of higher homologues.

Recently, work on shape-selective zeolite catalysts which only permit certain products to diffuse through them has opened up a new method to increase the percentage of di-methylamine. Although methyl alcohol conversion to methylamine is higher with zeolites than the conventional oxides with a lower reaction temperature, coke formation in the catalyst seriously limits its life.

Reaction

$$CH_3OH + NH_3 \rightarrow CH_3NH_2 + H_2O$$

$$CH_3OH + CH_3NH_2 \rightarrow (CH_3)_2NH + H_2O$$

$$CH_3OH + (CH_3)_2NH \rightarrow (CH_3)_3N + H_2O$$

Raw material requirements and yield

Raw materials required per tonne of methylamine:

	MMA	*DMA*	*TMA*
Methyl alcohol	1087kg	1497kg	1713kg
Ammonia anhydrous	577kg	328kg	303kg

Yield 95%

OTHER PROCESSES

Formaldehyde and ammonia are reacted together in the vapour phase at a temperature of 100–160°C and pressure slightly above atmospheric. The water produced is removed in a stripper before the reaction products are hydrogenated. The same catalysts are used as for the methyl alcohol reaction and the proportions of the three methylamines formed can be controlled by the volume of excess ammonia used.

Separation is carried out as in the methyl alcohol based process.

Reaction

$$HCHO + NH_3 \rightarrow HCNH_2 + H_2O$$

$$HCNH_2 + H_2 \rightarrow CH_3NH_2$$

PROPERTIES

All methylamines are gases at room temperature with an odour of ammonia. Highly flammable.

	MMA	*DMA*	*TMA*
Molecular Weight	31.06	45.08	59.11
Density at 20°C	0.66	0.66	0.63
Melting Point	–93.5°C	–92.2°C	–117°C
Boiling Point	–6.3°C	6.9°C	2.87°C
Autoignition Temperature	430°C	400°C	190°C
Explosive Limits in air			
lower (vol%)	4.95	2.8	2.0
upper (vol%)	20.8	14.4	11.6
Flash Point Closed Cup	0°C	–18°C	–7°C
Vapour Density (air = 1)	1.1	1.6	
Exposure Limit COSHH			
ppm 15 minutes			15
ppm 8 hour TWA	10	10	10

Di-methylamine is on the current Advisory Committee on Toxic Substances (ACTS) and the Working Group on the Assessment of Toxic Chemicals (WATCH) work programme.

Exposure Limit ACGIH			
ppm TLV-STEL	15	15	15
ppm TLV-TWA	10	10	10

GRADES

Anhydrous 99.5%, aqueous solution 25%, 40%, 50%, 60%.

INTERNATIONAL CLASSIFICATIONS

Mono-methylamine – anhydrous

UN No.	1061
CAS Reg No.	74–89–5
EINECS No.	200–820–0
EC No.	612–001–00–9
Description	Flammable gas
Packing Group	Not allocated
Emergency Action Code	2PE
HI (Kemler Code)	23

Mono-methylamine – aqueous solution

UN No.	1235
CAS Reg No.	74–89–5
EINECS No	200–820–0
EC No.	612–001–01–6
Description	Flammable liquid, corrosive
Packing Group	II
Emergency Action Code	2WE
HI (Kemler Code)	338

Di-methylamine – anhydrous

UN No.	1032
CAS Reg No.	124–40–3
EINECS No.	204–697–4
EC No.	612–001–00–9
Description	Flammable gas
Packing Group	Not allocated
Emergency Action Code	2PE
HI (Kemler Code)	23

Di-methylamine – aqueous solution

UN No.	1160
CAS Reg No.	124–40–3
EINECS No.	204–697–4
EC No.	612–001–01–6
Description	Flammable liquid, corrosive
Packing Group	II
Emergency Action Code	2PE
HI (Kemler Code)	338

Tri-methylamine – anhydrous

UN No.	1083
CAS Reg No.	75–50–3
EINECS No.	200–875–0
EC No.	612–001–00–9
Description	Flammable gas
Packing Group	Not allocated
Emergency Action Code	2PE
HI (Kemler Code)	23

Tri-methylamine – aqueous solution <50% by mass

UN No.	1297
CAS Reg No.	75–50–3
EINECS No.	200–875–0
EC No.	612–001–01–6
Description	Flammable liquid, corrosive
Packing Group (allocated in accordance with Approved Requirements)	I or II
Emergency Action Code	2PE
HI (Kemler Code)	338

Tri-methylamine – aqueous solution <50% by mass

UN No.	1297
CAS Reg No.	75–50–3
EINECS No.	200–875–0
EC No.	612–001–01–6
Description	Flammable liquid, corrosive
Packing Group (allocated in accordance with Approved Requirements)	III
Emergency Action Code	2PE
HI (Kemler Code)	38

APPLICATIONS

The prime outlet for mono-methylamine is for the preparation of insecticides especially 1-naphthyl-*n*-methylcarbamate. It is also used for hair removal in the tanning of hides, in surfactants, as a precursor of some photographic developers and for the manufacture of dyestuffs, explosives and analgesics.

Around half of total di-methylamine consumption is converted to di-methylacetamide and di-methylformamide which are used as spinning solvents for acrylic fibres. Other outlets include: the manufacture of lauryl di-methylamine, a surfactant; chemical accelerators used in the vulcanization of rubber and di-methylamino-ethanol; an intermediate for the manufacture of pharmaceuticals and fungicides; an emulsifier and corrosion inhibitor. Di-methylamines are utilized in catalysts for the production of urethanes.

Tri-methylamine is used for the manufacture of choline chloride (a supplement for animal feeds), cationic starches and disinfectants. Other outlets include the manufacture of flocculation agents, artificial sweetners and ion exchange resins.

HEALTH AND HANDLING

Methylamine vapour is a severe irritant to the eyes, nose, throat and skin. It can cause conjunctivitis, oedema of the lungs and dermatitis. All personnel working with methylamines must always wear protective clothing, goggles, gloves, gauntlets, aprons and boots. Contact lenses should not be worn in the vicinity of the vapour. Contaminated clothing must be laundered before reuse.

Store in an explosion-proof, isolated area away from mercury and strong oxidizing agents. Methylamine is stable in closed pressurized containers. If leaks occur, evacuate personnel and extinguish all forms of ignition. Local regulations must be followed in clean-up operations.

Methylamines are a dangerous fire and explosion hazard. Attempts should be made to stop the flow of gas; water spray, carbon dioxide, dry chemical or alcohol foam can be used to extinguish the flames. As toxic fumes are given off during burning, all firefighting staff must be fully protected and use self-contained breathing apparatus.

MAJOR PLANTS

Plants with capacities greater than 30 000 tonnes per year:

UCB	Gent	Belgium
BASF	Ludwigshafen	Germany
UCB	Leuna	Germany
Air Products & Chemicals	Billingham	UK
Air Products & Chemicals	Pace	US
Du Pont	Belle	US
Mitsubishi Gas	Niigata	Japan
Nitto Chemical	Yokohama	Japan
Korea Fertilizer	Ulsan	South Korea

MAJOR LICENSORS

Acid-Amine Technologies
Mitsubishi Gas Chemical
UCB

Methyl Chloride

CH_3Cl

SYNONYMS

METHYL CHLORIDE monochloromethane

There are two commercial routes in use for the production of methyl chloride: the hydrochlorination of methyl alcohol and the chlorination of methane.

The methyl alcohol process has become increasingly important due to the ready availability at economic cost of the feedstock. It also has the additional advantages of:

- utilization of hydrogen chloride, a by-product of a number of chemical processes;
- no expensive separation from other chlorinated hydrocarbons as only methyl chloride is produced;
- manufacture can take place at any location due to the transportability and easy storage of the methyl alcohol feedstock.

As a result, the use of methane-based processes has declined, accelerated by increasing environmental pressure for the safe disposal of the waste chlorine-containing residue. It is only attractive where sources of low-cost methane and chlorine are available. Since one route produces hydrochloric acid while the other consumes it, combination processes have been developed for the production of a range of chlorinated methanes with minimal formation of by-product acid.

Methyl chloride can be formed during a number of chemical reactions such as dimethyl sulphate and aluminium chloride, methane and phosgene, and methyl alcohol with ammonium chloride, but none of these are of any commercial importance. It can also be recovered as a by-product of silicone manufacture but is normally reused in the silane process.

Capacities range from 6000 to 170 000 tonnes per year. Capacities are flexible as all plants can produce a range of chlorinated products.

PROCESSES

1. From methyl alcohol by hydrochlorination

The methyl alcohol process can be carried out in the vapour or liquid phase. In the original liquid-phase route, a 70% solution of zinc chloride was used as the catalyst, with the reaction carried out at 150°C and moderate pressure. This was followed by the non-catalytic reaction route between methyl alcohol and hydrogen chloride at 130–160°C. However, yields in both cases are low and vapour-phase processes are preferred. (*See Figure 83*)

High-purity methyl alcohol (99.9%) and hydrochloric acid (free from impurities) are vaporized by passage through a heat exchanger prior to mixing. A slight excess of hydrogen chloride is used to decrease the volume of by-product dimethyl ether formed. The gases are subjected to further heating before they enter a multi-tubular reactor packed with activated aluminium oxide catalyst. The vapour-phase reaction takes place at 300–350°C and a pressure of 3–6 bar. Heat generated by the exothermic reaction is used to heat the incoming feed.

The reaction gases are quenched by contact with hydrochloric acid solution, and some by-product water separates out to form a dilute hydrochloric acid stream. The gases pass through a heat exchanger where most of the remaining water and acid are condensed. Small quantities of by-product dimethyl ether and any residual water are removed from the gaseous methyl chloride by passage through 96% sulphuric acid. The remaining gases are compressed and distilled under 20 bar pressure to yield pure methyl chloride. The crude hydrogen chloride solution can be purified to yield 20% acid for sale.

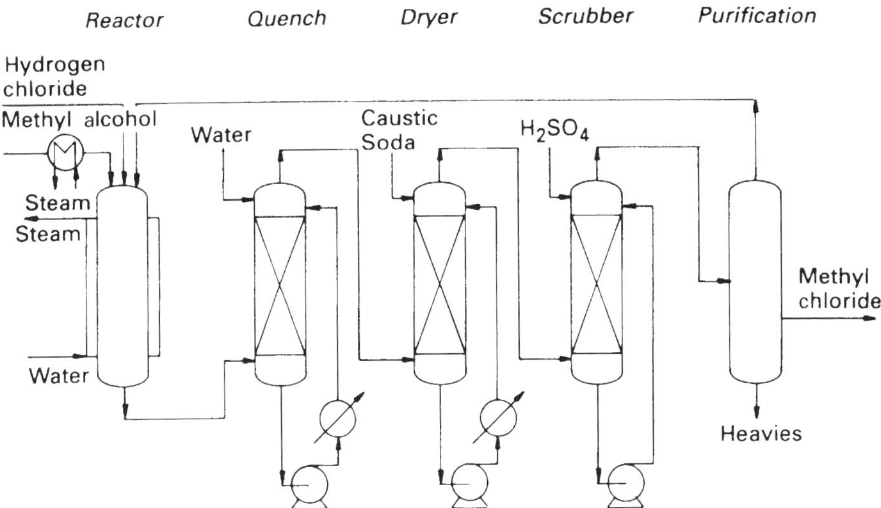

FIGURE 83 Methyl chloride from methyl alcohol by hydrochlorination

In order to prolong catalyst life, the feed gases must be free of impurities, the addition of small amounts of oxygen reducing the build-up of carbon deposits. Catalyst life is normally 1–2 years.

Reaction

$CH_3OH + HCl \rightarrow CH_3Cl + H_2O$

Raw material requirements and yield

Raw materials required per tonne of methyl chloride:

Methyl alcohol (99.9%)	660kg
Hydrogen chloride	800kg

Yield 95%

2. From methane by chlorination

Methane and chlorine will react to yield a range of chloromethanes; if methyl chloride is the desired product, then a large excess of methane must be used. The reaction can be carried out either thermally or in the presence of UV light, but the former is normally employed. (*See Figure 84*)

Methane and chlorine in a molar ratio of 1.7:1 are preheated and fed into a reactor containing a catalyst consisting of partially reduced cupric oxide on pumice. At 360°C and atmospheric pressure a mixture of chloromethanes are formed. The gases are cooled, scrubbed with methyl chloride and any unreacted gases are

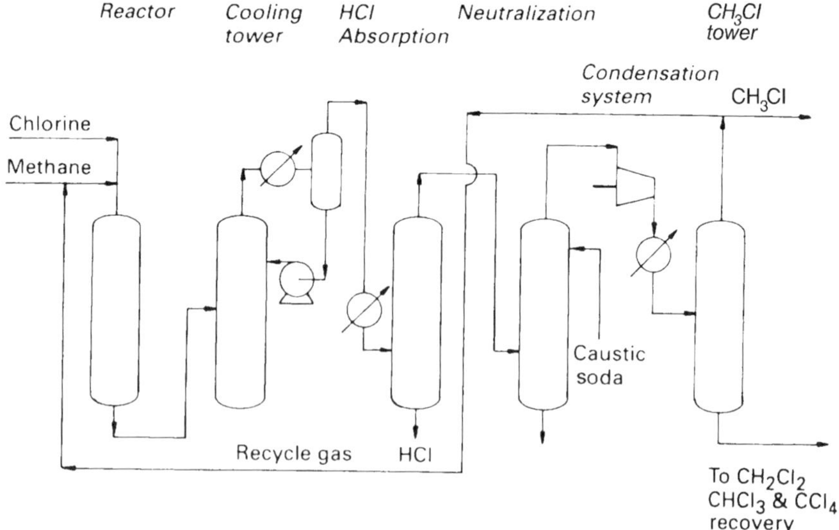

FIGURE 84 Methyl chloride from methane by chlorination

removed overhead. After the removal of by-product hydrogen chloride by passage through an acid–water mixture, the gases are dried, compressed, and methyl chloride is separated by distillation under pressure.

Reaction

$CH_4 + Cl_2 \rightarrow CH_3Cl + HCl$

Raw material requirements and yield

Raw materials required per tonne of methyl chloride (theoretical):

Methane	443m^3
Chlorine	1405kg

Yield

based on methane	90%
based on chlorine	99%

PROPERTIES

Colourless, almost odourless gas. Highly explosive and flammable.

Molecular Weight	50.49
Density at 20°C	0.916
Melting Point	–97.7°C
Boiling Point	–24.2°C
Autoignition Temperature	625°C
Flammability Limits in air	
lower	8.1vol%
upper	17.2vol%
Vapour Density (air = 1)	1.8
Exposure Limit COSHH	100ppm 15 minutes 50ppm 8 hour TWA
Exposure Limit ACGIH	100ppm TLV-STEL 50ppm TLV-TWA

GRADES

Technical 99%.

INTERNATIONAL CLASSIFICATIONS

UN No.	1063
CAS Reg No.	74–87–3
EINECS No.	200–817–4
EC No.	602–001–00–7
Description	Flammable gas
Emergency Action Code	2WE
HI (Kemler Code)	23

APPLICATIONS

The outlet which dominates methyl chloride consumption is the manufacture of silicones. Due to the phasing out of anti-knock lead additives in gasoline, its use in tetramethyl lead production has declined rapidly. Other applications include its usage in butyl rubber, herbicides, methyl cellulose, and quaternary amine manufacture.

HEALTH AND HANDLING

Methyl chloride is a very explosive and flammable gas. Inhalation of the gas causes dizziness and nausea, leading to damage to the central nervous system. As methyl chloride is almost odourless, there is no warning of its toxic effects. It is a known carcinogen. The liquid will cause burns in contact with the skin. As contact lenses absorb and concentrate the chemical, they must not be worn.

Store in steel or iron containers in a well-ventilated area preferably in a separate location and away from oxidizing agents, amines, amides, potassium, sodium, aluminium, magnesium and zinc. Neoprene or polyvinyl chloride (PVC) gloves, aprons, boots, full protective clothing, goggles and respirator must be worn at all times.

Methyl chloride is normally shipped in pressure vessels, earthed to prevent static build-up. Automatic vapour warning equipment is required where methyl chloride is stored or handled. Spills should be dealt with in accordance with local regulations.

Protective clothing and self-contained breathing apparatus must be worn when dealing with fires as toxic fumes, mainly hydrogen chloride and phosgene, are given off during burning. Carbon dioxide or dry chemical can be used to blanket fires. Water should be used to cool containers and firefighting personnel. As methyl chloride gas is denser than air, it can collect in sumps with the danger of flashback and explosion being great hazards.

Because of the toxicity of methyl chloride, emergency and handling procedures should be rehearsed and checked prior to the operation of the plant. Its storage, handling and transportation are governed by strict regulations relating to compressed gas.

MAJOR PLANTS

Plants with capacities greater than 45 000 tonnes per year:

Elf Atochem	Jarrie	France
Bayer	Leverkusen	Germany
Celanese	Frankfurt	Germany
Dow Chemical	Stade	Germany
Wacker Chemie	Burghausen	Germany
Dow Chemical	Barry	UK
ICI	Runcorn	UK
Complex	Sterlitansk	Russia
Complex	Ufa	Russia
Dow Chemical	Plaquemine	US
Dow Corning	Carrolton	US
GE Plastics	Waterford	US
Vulcan Chemicals	Geismar	US
Shin-Etsu Chemical	Naoetsu	Japan
Korea Fertilizer	Ulsan	South Korea

All capacities are flexible.

MAJOR LICENSORS

Asahi Glass
Elf Atochem
Halcon-Scientific Design
Hoechst
Huels
Solvay
Vulcan Materials

Methylene Dichloride CH_2Cl_2

SYNONYMS

METHYLENE DICHLORIDE methylene chloride, dichloromethane, Freon 30, methane dichloride, methylene bichloride

Methylene dichloride, the most important of the chloromethanes, is produced by the chlorination of methane–methyl chloride in the gaseous phase. This process produces a mixture of products: in order to obtain an optimal yield of methylene dichloride, a large excess of methane or methyl chloride is employed. By keeping the residual concentration of chlorine at a low level, the separation of the chloromethanes formed is facilitated.

Various kinds of reactor have been developed ranging from loop to tubular types. Although loop reactors (by their internal circulation of the gases) avoid the formation of explosive mixtures and offer good temperature control, better selectivity is obtained from tubular reactors. To overcome this problem, reactors are frequently operated in parallel or in series interposed with a condensation unit to remove high-boiling-point chloromethanes.

Chloromethanes can be produced by the oxychlorination of methane – the Transcat process. The composition of the chloromethanes formed is dependent on the level of conversion and the degree of recycling employed.

Capacities range from 10 000 to 100 000 tonnes per year but are flexible because most plants are capable of producing a range of chloromethanes.

PROCESSES

From methane by chlorination

A mixture of methane, methyl chloride and chlorine are introduced into a nickel-loop reactor. At a temperature of 350–400°C, the gaseous mixture reacts to form a range of chloromethanes (see Chloroform). The methane must be pure to prevent a range of chlorinated hydrocarbons being formed which would complicate the separation process. Equally the presence of inerts creates large volumes of off-gas which reduces the yields obtained. (*See Figure 85*)

The gases are cooled and washed with dilute acid. After compression, drying and cooling, the chloromethanes are separated by distillation. Unreacted methane and methyl chloride are recycled to the reactor. By careful choice of the methyl chloride to methane ratio, up to 70wt% of methylene dichloride can be obtained.

In a variant of the process, methyl chloride can be chlorinated selectively to methylene dichloride by irradiation with a UV lamp at –20°C.

Reaction

$CH_4 + Cl_2 \rightarrow CH_3Cl + HCl$

$CH_3Cl + Cl_2 \rightarrow CH_2Cl_2 + HCl$

Raw material requirements and yield

Raw materials required per tonne of methylene dichloride (theoretical):

Methane 190kg

Yield
- on methane 90%
- on chlorine 99%

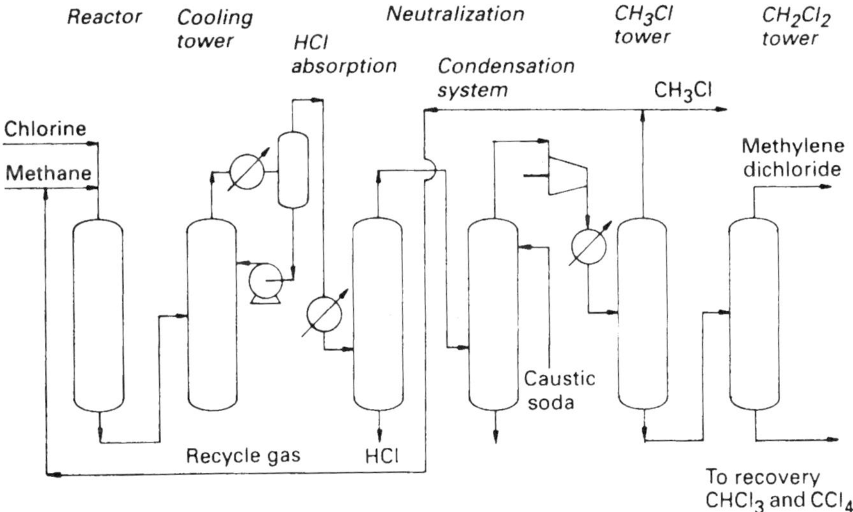

FIGURE 85 Methylene dichloride from methane by chlorination

OTHER PROCESSES

In order to overcome the problems of hydrochloric acid stream disposal, an integrated oxychlorination–chlorination process has been developed. Methyl chloride is reacted with chlorine at 350–450°C and 8–15 bar pressure. The chloromethanes formed are separated and methyl chloride is returned to the chlorinator. Hydrogen chloride by-product is condensed and reacted with methyl alcohol to yield methyl chloride which is recycled (see Chloroform).

PROPERTIES

Clear, colourless, volatile liquid with a distinctive penetrating ether-like odour.

Molecular Weight	84.94
Density at 20°C	1.326
Melting Point	–96.7°C
Boiling Point	39.75°C
Autoignition Temperature	556°C
Flammability Limits in air	
lower	12vol%
upper	19vol%
Flash Point Closed Cup	Flammable air mixtures formed at 100°C
Vapour Density (air = 1)	2.9

Exposure Limit COSHH (Schedule 1) 300ppm 15 minutes (Maximum Exposure Limit)
100ppm 8 hour TWA (Maximum Exposure Limit)
Substance on the current Advisory Committee on Toxic Substances (ACTS) and the Working Group on the Assessment of Toxic Chemicals (WATCH) work programme.

Exposure Limit ACGIH 50ppm 8 hour TLV-TWA
Classified A2, suspected human carcinogen.
ACGIH action level has been changed to 12.5ppm TWA-PEL

GRADES

Technical 99% stabilized with 0.1–0.2wt% ethyl alcohol or methyl alcohol.

INTERNATIONAL CLASSIFICATIONS

UN No.	1593
CAS Reg No.	75–09–2
EINECS No.	200–838–9
EC No.	602–004–00–3
Description	Toxic substance
Packing Group	III
Emergency Action Code	2Z
HI (Kemler Code)	60

APPLICATIONS

The major outlet for methylene chloride, accounting for almost half of total demand, is as a solvent degreaser and paint remover. It is used as an extractant–solvent for plastics and a blowing agent in the production of polyurethane foam. Its application in propellant mixtures for aerosols has declined with the phasing out of chlorinated hydrocarbons.

Methylene dichloride growth is stagnant due to concerns about its toxicity and effects on the environment.

HEALTH AND HANDLING

Methylene dichloride liquid is irritating to the skin and eyes. Contact lenses must not be worn as they tend to absorb and concentrate the vapour. Exposure to the vapour can lead to headaches and dizziness, but 24–48 hours can elapse before symptoms appear. Prolonged exposure can lead to high carboxyhaemoglobin levels in the blood. Methylene dichloride is a suspected carcinogen. Personnel working with methylene dichloride should receive annual medical checks.

Store in closed steel or galvanized iron containers to prevent the ingress of moisture. Aluminium or copper are not recommended. Methylene dichloride is stable at room temperature but care must be taken to avoid exposure to heat as its high vapour pressure can cause containers to rupture. Containers should be kept in a well-ventilated, explosion-proof area away from oxygen, sodium, potassium, lithium and finely powdered aluminium or magnesium.

If spills occur, evacuate all personnel and extinguish all sources of ignition. Contain and absorb the liquid with dry sand, earth or vermiculite and, using non-sparking tools, transfer the waste to containers for disposal according to local regulations. Methylene dichloride must be kept away from sewers and waterways because of its toxicity.

Methylene chloride is not flammable under normal storage conditions, but it can form flammable mixtures with air at 100°C. Fires can be extinguished with carbon dioxide, foam or dry chemicals. Strict rules apply to the filling, transfer and transportation of methylene chloride because of its toxicity.

MAJOR PLANTS

Plants with capacities greater than 30 000 tonnes per year:

Elf Atochem	Lavera	France
Solvay	Tavaux	France
Dow Chemical	Stade	Germany
Celanese	Frankfurt	Germany
ICI	Runcorn	UK
Complex	Sterlitansk	Russia
	Ufa	Russia
Dow Chemical	Freeport	US
	Plaquemine	US
Vulcan Chemicals	Geismar	US
	Wichita	US
Mitsui Toatsu Chemicals	Nagoya	Japan
Shin-Etsu Chemical	Niigata	Japan
Tokuyama Corp.	Tokuyama	Japan
Korea Fertilizer	Ulsan	South Korea

MAJOR LICENSORS

ABB Lummus Global
Asahi Glass
Halcon-Scientific Design
Hoechst
Huels
Solvay
Stauffer Chemical
Vulcan Materials

Methyl Ethyl Ketone (MEK)

$$C_3CH_2 - \underset{\underset{O}{\|}}{C} - CH_3$$

SYNONYMS

METHYL ETHYL KETONE   2-butanone, MEK, butan-2-one, methyl acetone

Interest in methyl ethyl ketone grew with its use as a low-boiling-point replacement solvent for butyl alcohols in the paint and lacquer industry.

It can be produced by the liquid-phase oxidation of *n*-butenes or from *sec*-butyl alcohol by catalytic dehydrogenation in either the liquid or vapour phase. Commercially, the vapour-phase dehydrogenation of *sec*-butyl alcohol is the dominant route. Methyl ethyl ketone is available as a by-product of acetic acid manufacture from butane. The attractiveness of recovering the methyl ethyl ketone depends on the demand for acetic acid as it can be recycled for further conversion.

Currently, most methyl ethyl ketone is produced from butylene but Hoechst Celanese obtain the ketone as a by-product of its acetic acid manufacture.

Capacities range from 10 000 to 125 000 tonnes per year.

PROCESSES

From butyl alcohol by dehydrogenation

sec-Butyl alcohol vapour, preheated by passing through a heat exchanger, is fed into a multi-tubular reactor containing a catalyst consisting of zinc or copper oxides. The reactor tubes are heated by oil or high-pressure steam and the reaction takes place at 400–500°C and at low pressures of less than 4 bar. (*See Figure 86*)

The exit gases are cooled by passage through a heat exchanger and condensed before entering a separator. Hydrogen is flashed off overhead, scrubbed to remove any residual methyl ethyl ketone and used as fuel. The condensate is dehydrated by fractionation. The methyl ethyl ketone phase separates from the water-ketone azeotrope obtained and is combined with the liquid stream from the base of the

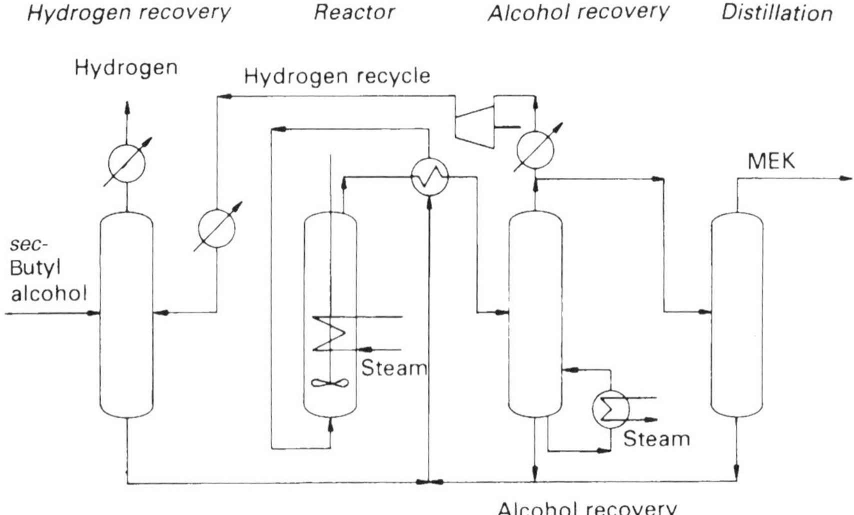

FIGURE 86 Methyl ethyl ketone from butyl alcohol by dehydrogenation

dehydrator. It is distilled and pure product collected overhead. Any unreacted *sec*-butyl alcohol is recovered from the residues and recycled.

The reaction can be carried out in the liquid phase using Raney nickel or copper chromate at a temperature of 150°C. Hydrogen and methyl ethyl ketone are separated in the vapour state as above.

Reaction

$$CH_3CH(OH)C_2H_5 \rightarrow CH_3COC_2H_5 + H_2$$

Raw material requirements and yield

Raw materials required per tonne of methyl ethyl ketone:

sec-Butyl alcohol	1170kg

Yield 95%

OTHER PROCESSES

A new process has been developed in which *n*-butene is directly oxidized in a solution of palladium and cupric chlorides which act as catalysts. The reaction is carried out at 120°C and a pressure of 10–20 bar. The methyl ethyl ketone formed is separated and purified by distillation. Yields of 88% are claimed.

Reaction

$$2CH_2{=}CHCH_2CH_3 + O_2 \rightarrow 2CH_3COCH_2CH_3$$

PROPERTIES

Colourless, mobile, flammable liquid with an acetone like odour. Soluble in ethyl alcohol, benzene, ether, and acetone.

Molecular Weight	72.10
Density at 20°C	0.806
Melting Point	–86.3°C
Boiling Point	79.6°C
Autoignition Temperature	516°C
Explosive Limits in air	
lower	1.4vol%
upper	7.5vol%
Flash Point Closed Cup	–6.6°C
Vapour Density (air = 1)	2.5
Exposure Limit COSHH	300ppm 15 minutes 200ppm 8 hour TWA
Exposure Limit ACGIH	300ppm 15 minutes TLV-STEL 200ppm 8 hour TLV-TWA

GRADES

Technical 99%.

INTERNATIONAL CLASSIFICATIONS

UN No.	1193
CAS Reg No.	78–93–3
EINECS No.	201–159–0
EC No.	606–002–00–3
Description	Flammable liquid
Packing Group	II
Emergency Action Code	2YE
HI (Kemler Code)	33

APPLICATIONS

The major outlet of methyl ethyl ketone is as a low-boiling solvent for nitrocellulose, acrylic and vinyl surface coatings. Its advantage is that low-viscosity solutions can be obtained, having a high solids content, without affecting the film properties. These lacquers are widely used in the automotive, electrical goods and furniture industries. In the US, about 63% of the ketone goes into coatings but some methyl ethyl ketone based formulations are being replaced.

The fast evaporation rate of methyl ethyl ketone makes it popular for rubber-based industrial cements, low-temperature bonding agents and printing inks. Niche outlets with rising demand are adhesives and in VCR tapes.

Other openings are as a solvent for paint removers and degreasing. Methyl ethyl ketone is used as a dewaxing agent in the refining of lubricating oils, for the extraction of vegetable oils, and in the manufacture of photographic film and artificial leather. It is also employed as an intermediate in the preparation of catalysts, flavours and perfumes. A derivative, methyl ethyl ketone peroxide, is used as a hardening agent in reinforced polyester fibre glass manufacture.

Attempts to reduce solvent emissions are affecting future demand for methyl ethyl ketone, and world demand is decreasing at around 0.5% per year.

HEALTH AND HANDLING

Methyl ethyl ketone vapour is irritating to the eyes and throat leading to coughing, headaches and nausea. The liquid is absorbed through the skin and this can lead to cracking. Protective clothing should be worn together with gloves and goggles to prevent skin and eye contact.

Storage containers can be made of iron, mild steel, copper or aluminium as methyl ethyl ketone is not corrosive to metals to any extent. They should be closed and stored in a well-ventilated area away from sources of ignition. Equipment should be spark-proof and earthed to prevent static build-up.

If spills occur, evacuate the area and extinguish all forms of ignition. Use non-sparking tools to clean up the sand or vermiculite used to absorb the liquid. Place the waste in containers for disposal by burning in an approved incinerator. The liquid must not be flushed into sewers or waterways. Methyl ethyl ketone is highly flammable and fires should be tackled with carbon dioxide, dry chemical or foam. Firefighters and clean-up staff must wear protective clothing to prevent skin contact and breathing equipment against inhalation of the vapour.

The movement of methyl ethyl ketone, or thinners containing it, is subject to transport regulations because of its flammability.

MAJOR PLANTS

Plants with capacities greater than 40 000 tonnes per year:

Elf Atochem	La Chambre	France
CONDEA Chemical	Moers	Germany
Shell Nederland Chemie	Pernis	Netherlands
Exxon Chemical	Fawley	UK
Exxon Chemical	Bayway	US
Hoechst Celanese	Pampa	US
Shell Chemical	Norco	US
SASOL	Sasolburg	South Africa
Idemitsu Kosan	Tokuyama	Japan
Maruzen Petrochemical	Ichiara	Japan
Tonen Chemical	Kawasaki	Japan
Yukong	Ulsan	South Korea
TASCO Chemical	Linyuan	Taiwan

MAJOR LICENSORS

ABB Lummus Global/Petroquisa
Aldehyd
Arco Technology
Edeleanu
Hoechst/Wacker Chemie
Idemitsu Petrochemical
IFP
Maruzen
Texaco Development
Uhde

Methyl Isobutyl Ketone (MIBK)

$$(CH_3)_2CHCH_2-\underset{\underset{O}{\|}}{C}-CH_3$$

SYNONYMS

METHYL ISOBUTYL KETONE hexanone, 4-methyl-2-pentanone, 4-methylpentan-2-one, iso-propylacetone, MIBK

Methyl isobutyl ketone's use as a solvent has resulted in this product's commercial significance. It is made from acetone in a three-step low-temperature process via diacetone alcohol and mesityl oxide. Methyl isobutyl ketone can be produced by the two-step catalysed high-temperature condensation of acetone in the liquid phase, but this route is only of importance where by-product isophorone is required as yields are lower than for the conventional low-pressure process.

A newer direct liquid-phase route (involving condensation–dehydration–hydrogenation of acetone) has been developed but suffers from low conversion rates and difficulties in purification due to the number of by-products formed. Another process which is not of commercial significance is the catalytic liquid-phase condensation of acetone with hydrogen directly to the ketone.

Methyl isobutyl ketone can be obtained as a by-product of the dehydrogenation of isopropyl alcohol to acetone but this process is not commercially important.

The consumption of methyl isobutyl ketone has been limited by restrictions on its use due to air pollution control standards. This has resulted in many formulation changes and substitution by other solvents.

Capacities range from 10 000 to 50 000 tonnes per year.

PROCESSES

1. From acetone via diacetone alcohol and mesityl oxide

Acetone containing less than 0.5wt% of water is fed into a column containing a fixed-bed alkali catalyst at a temperature between 10 and 20°C. The catalyst can be either solid barium or calcium hydroxide impregnated on a suitable carrier, or dilute sodium hydroxide. Heat from the exothermic reaction is removed continuously by interstage cooling. When the reaction is complete, diacetone alcohol is recovered under low pressure. If water is present it must be removed prior to the recovery. (*See Figure 87*)

The catalyst, 0.05wt% of phosphoric acid, is added to the diacetone alcohol formed; dehydration to mesityl oxide takes place at 90–130°C. The resultant mixture is distilled and acetone and mesityl oxide–water azeotrope are continuously recovered overhead. Mesityl oxide is separated from any unreacted acetone which is recycled.

Mesityl oxide and hydrogen vapours are passed over a catalyst consisting of a fixed bed of copper–chromium, nickel or palladium on alumina. The hydrogenation reaction can be carried out in either the liquid or vapour phase. The liquid-phase reaction takes place at 80–130°C and 3–30 bar pressure. Reaction temperature for the vapour-phase reaction is 150–170°C at atmospheric pressure. Whichever route is employed, the methyl isobutyl ketone produced is separated and purified by two-stage distillation. Acetone and isopropyl alcohol are formed as by-products.

Reaction

$$2CH_3COCH_3 \rightarrow (CH_3)_2C(OH)CH_2COCH_3$$

$$(CH_3)_2C(OH)CH_2COCH_3 \rightarrow (CH_3)_2C=CHCOCH_3 + H_2O$$

$$(CH_3)_2C=CHCOCH_3 + H_2 \rightarrow (CH_3)_2CHCH_2COCH_3$$

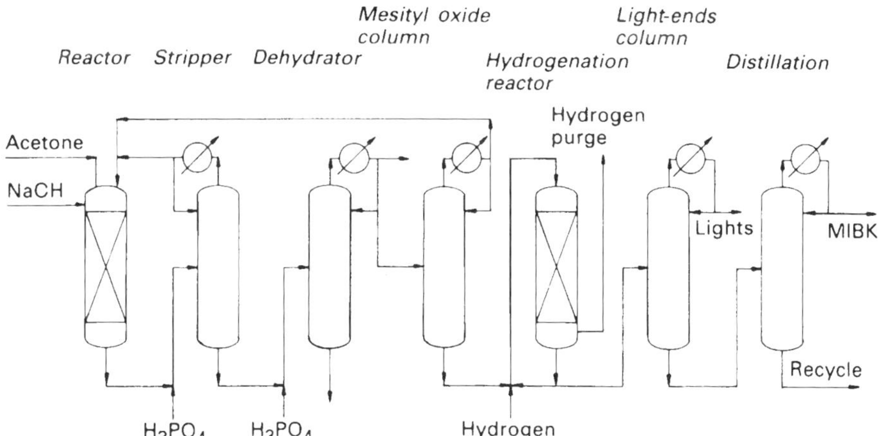

FIGURE 87 Methyl isobutyl ketone from acetone via diacetone alcohol and mesityl oxide

Raw material requirements and yield

Raw materials required per tonne of methyl isobutyl ketone:

Acetone	1150kg
Hydrogen	230m^3

Yield 75%

2. From acetone by direct hydrogenation

In a process developed in the 1980s, methyl isobutyl ketone can be produced by the one-step hydrogenation of acetone. Acetone and hydrogen in a molar ratio of 1:0.2–2.0 are preheated before being fed into a tubular reactor packed with a catalyst consisting of 0.5wt% metallic palladium dispersed on a sulphonated polystyrene–divinyl benzene cation exchange resin. The hydrogenation reaction takes place at 80–130°C and 50–60 bar. Heat from the highly exothermic reaction is removed by water-cooling jackets around the tubes. As the catalyst rapidly loses its activity above 150°C, accurate temperature control is vital.

Hydrogen is recovered from the top of the reactor and the reaction mixture containing acetone, methyl isobutyl ketone, diisobutyl ketone, isopropyl alcohol and water is distilled. Low-boiling compounds and unconverted acetone are removed overhead. The remaining aqueous crude methyl isobutyl ketone separates into two layers. The organic phase is distilled to remove isopropyl alcohol and any residual water overhead. In the third column, methyl isobutyl ketone is separated by fractionation from the heavy by-products. The quantities of by-products formed makes their separation difficult. (*See Figure 88*)

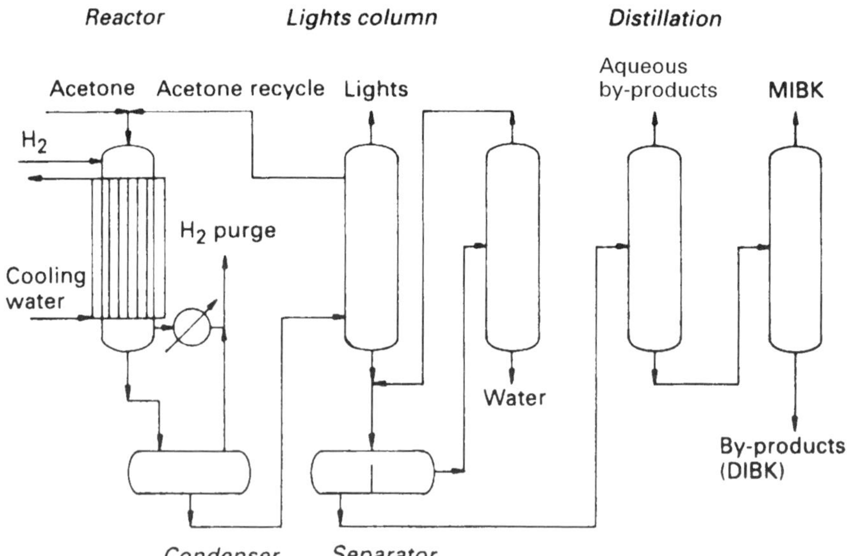

FIGURE 88 Methyl isobutyl ketone from acetone by direct hydrogenation

Reaction

$2CH_3COCH_3 + H_2 \rightarrow (CH_3)_2CHCH_2COCH_3 + H_2O$

Raw material requirements

Raw materials required per tonne of methyl isobutyl ketone (theoretical):

Acetone	1160kg

PROPERTIES

Colourless, flammable liquid with a characteristic ketone odour. Slightly soluble in water. Miscible with most organic solvents.

Molecular Weight	100.16
Density at 20°C	0.804
Melting Point	–80.2°C
Boiling Point	116.2°C
Autoignition Temperature	475°C
Explosive Limits in air	
lower	1.4vol%
upper	7.5vol%
Flash Point Closed Cup	14°C
Vapour Density (air = 1)	3.5
Exposure Limit COSHH	100ppm 15 minutes 50ppm 8 hour TWA
Exposure Limit ACGIH	75ppm 15 minutes TLV-STEL 50ppm 8 hour TLV-TWA

GRADES

Technical 99%.

INTERNATIONAL CLASSIFICATIONS

UN No.	1245
CAS Reg No.	108–10–1
EINECS No.	203–550–1
EC No.	606–004–00–4
Description	Flammable liquid
Packing Group	II
Emergency Action Code	3YE
HI (Kemler Code)	33

APPLICATIONS

The main outlet for methyl isobutyl ketone is as a solvent (either alone or blended with other aromatics) for surface-coating resins such as nitrocellulose, cellulose acetate–butyrate, oil-modified polyesters, acrylics, alkyds, and vinyls. It is also used as a solvent for adhesives based on nitrile rubber, for acrylics and pesticides.

Methyl isobutyl ketone can be used in the solvent extraction of rare metals, usually as their complexes, and the separation of mineral oils, pharmaceuticals, and butyl alcohols. It is employed as a denaturant for ethyl alcohol. Its phenylenediamine derivatives are utilized as antioxidants for rubbers.

Because of pollution controls, the use of methyl isobutyl ketone has been curtailed in several outlets.

HEALTH AND HANDLING

Methyl isobutyl ketone vapour is irritating to the skin, eyes and respiratory tract, and in high concentrations leads to nausea, headaches, dizziness and unconsciousness. Contact lenses concentrate the vapour and must be avoided. Protective clothing and goggles must be worn when handling the product to avoid skin and eye contact.

In closed containers, methyl isobutyl ketone is stable at room temperature. It should be stored in a well-ventilated, explosion-proof area away from strong oxidizing agents. Because of its flammability, all equipment must be explosion-proof and earthed to prevent static build-up. Methyl isobutyl ketone can attack some rubbers, plastics and resins and care is needed in the choice of valves and flanges.

In the event of spills, all sources of ignition must be extinguished and clean-up personnel should wear protective clothing. Small spills can be absorbed with vermiculite, which should be placed in containers, using non-sparking tools, for disposal by incineration. Although water can be used to clear methyl isobutyl ketone from hazardous areas, care must be taken to avoid discharge into waterways and sewers.

Methyl isobutyl ketone forms explosive mixtures with air; fires should be extinguished with carbon dioxide, dry chemical or foam. Water jets can scatter the fire. As the vapour is heavier than air, flashback is a danger. Firefighters should wear protective clothing and self-contained breathing apparatus.

MAJOR PLANTS

Plants with capacities greater than 15 000 tonnes per year:

Shell Chimie	Berre	France
Shell Nederland Chemie	Pernis	Netherlands
Acetica Sintetica	Pioltello	Italy
Eastman Chemical	Kingsport	US
Shell Chemical	Deer Park	US
Union Carbide	Institute	US
Kyowa Yuka	Yokkaichi	Japan
Mitsubishi Chemical	Kurashiki	Japan
Mitsui Petrochemical Industries	Otake	Japan

MAJOR LICENSORS

Edeleanu
Huels
Texaco Development
Toyo Engineering

Methyl Methacrylate (MMA)

$CH_2=C(CH_3)COOCH_3$

SYNONYMS

METHYL METHACRYLATE 2-methylacrylic acid methyl ester, methyl-2-propenoate, MMA

Methyl methacrylate was first produced by ICI in the mid-1930s; following further development, this technology is still the basis of the acetone cyanohydrin process used today. Most producers have access to a captive supply of hydrogen cyanide, the other raw material used in the process, either as a by-product of acrylonitrile manufacture or produced directly from natural gas.

Several companies have carried out research to find alternative routes to methyl methacrylate avoiding the use of hydrogen cyanide. Many processes from ethylene via methyl propionate as an intermediate have been proposed, but conversions are low and none is yet economic.

Work has been carried out on the carbonylation of propylene followed by dehydrogenation to methacrylic acid but difficult separation and low yields have made these routes unattractive.

Japan has experienced problems with the availability of hydrogen cyanide due to:

- a shipping ban because of its hazardous properties;
- lower yields per unit of acrylonitrile with newer catalysts;
- uneconomic direct production from natural gas.

This shortfall has led to the development of new processes based on C_4 hydrocarbons. These utilize *tert*-butyl alcohol or isobutylene as feedstock, which is oxidized in the vapour phase in two stages. One company, Asahi Chemical, employs mixed technology, producing methacrylonitrile from isobutylene which is then converted to methacrylamide sulphate, an intermediate in the acetone cyanohydrin process. The company plans to replace this plant with one based on the direct oxidative esterification of methacrolein obtained from isobutylene.

Unlike in Japan, isobutylene–*tert*-butyl alcohol is used as a feedstock for octane-enhancing of motor fuels in Western Europe and the US. Isobutane, which is cheaper than isobutylene, would be more attractive as a feedstock in these countries.

Shell has developed a process to make methyl methacrylate from propylene feedstocks, using a palladium catalyst. In this process, methyl acetylene is formed which reacts with carbon monoxide and methyl alcohol to give methyl methacrylate. Exclusive rights worldwide to this process have been acquired by ICI.

All production of methyl methacrylate is based on acetone cyanohydrin technology in Western Europe and the US. C_4-based routes in commercial operation in Japan are likely to be chosen for new large plants in other countries as concern increases over the potential environmental hazards of hydrogen cyanide and the disposal of spent sulphuric acid and ammonium sulphate by-products. Where sources of hydrogen cyanide and acetone are available, feedstock prices will still be the determinant.

Globally, most methyl methacrylate is produced from acetone cyanohydrin with some from isobutylene and a little based on methacrylonitrile. Small amounts of methyl methacrylate are obtained from the reprocessing of polymethylmethacrylate scrap back to the monomer.

Capacities range from 20 000 to 360 000 tonnes per year.

PROCESSES

1. From acetone and hydrogen cyanide (acetone cyanohydrin process)

Dry hydrogen cyanide is reacted with acetone in the presence of an alkali catalyst, usually a solution of caustic soda. The temperature is kept at 15–25°C by external water cooling to prevent the formation of by-products. To ensure maximum utilization of hydrogen cyanide, acetone is present in excess. The reaction products are immediately neutralized with acid and filtered to remove any insoluble salts before being distilled. Water and excess acetone are removed and the acetone cyanohydrin is fed into a reactor where 98% sulphuric acid is added. With the temperature kept below 40°C, methacrylamide sulphate is formed. Water and excess acid are flashed off before the sulphate is esterified with excess methyl alcohol in the third stage of the process. (*See Figure 89*)

The esterified product is stripped with steam to remove methyl methacrylate and methyl alcohol which are collected overhead. Ammonium salts are obtained from the base of the stripper column.

The methyl methacrylate–methyl alcohol mixture is washed with alkali to neutralize any residual acid, condensed and excess water removed. The organic layer is distilled to give pure methyl methacrylate, and methyl alcohol is recovered from the aqueous layer. Inhibitors are added both to the reaction stages and in storage

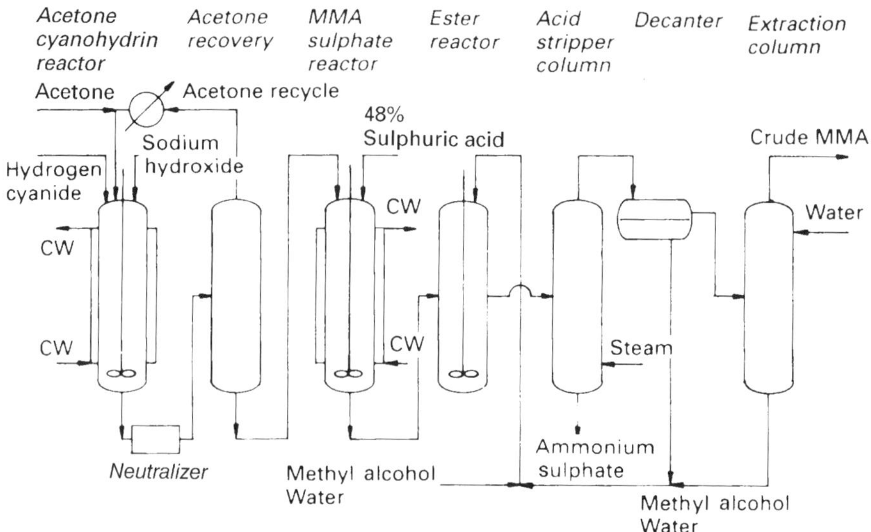

FIGURE 89 Methyl methacrylate from acetone and hydrogen cyanide (acetone cyanohydrin process)

to prevent polymerization. Typical inhibitors used are hydroquinone and derivatives, mono-methyl ether, pyrogallol, and phenolic compounds.

Reaction

$$CH_3COCH_3 + HCN \rightarrow (CH_3)_2COHCN$$

$$(CH_3)_2COHCN + H_2SO_4 \rightarrow CH_2{=}C(CH_3)CONH_3HSO_4$$

$$CH_2{=}C(CH_3)CONH_3HSO_4 + CH_3OH \rightarrow CH_2{=}C(CH_3)COOCH_3 + NH_4HSO_4$$

Raw material requirements and yield

Raw materials required per tonne of methyl methacrylate:

Acetone	665kg
Hydrogen cyanide	320kg
Methyl alcohol	350kg
Sulphuric acid (98%)	1150kg

Yield 85%

2. From isobutylene or *tert*-butyl alcohol

Isobutylene or *tert*-butyl alcohol is mixed with air and steam before entering a multi-tubular, water-cooled reactor. Isobutylene is separated from C_4 streams by hydration to *tert*-butyl alcohol. The reactor contains a fixed-bed catalyst composed of a multi-component metal oxide system usually molybdenum–bismuth–antimony or molybdenum–tungsten–tellurium. The vapour-phase reaction takes place at 300–400°C. (*See Figure 90*)

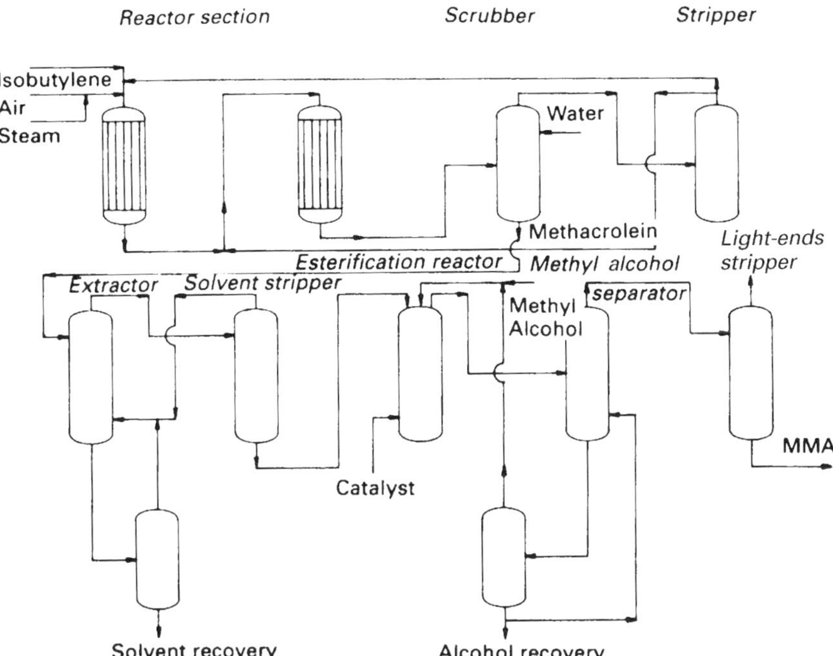

FIGURE 90 Methyl methacrylate from isobutylene

The methacrolein formed is immediately passed to a second reactor, without separation, where at 270–350°C and in the presence of a phosphorus–molybdenum catalyst, methacrylic acid is formed. The reaction gases are quenched, scrubbed with water and the aqueous solution is extracted with an organic solvent.

Any unreacted methacrolein is separated from the off-gases and recycled to the second reactor. Part of the remaining off-gas is used as a diluent to the feed gases to improve selectivity. The remainder passes through a combustion system before being purged as waste gas. The solvent is stripped from the methacrylic acid before its esterification with methyl alcohol in the liquid phase in the presence of an acid catalyst.

Crude methyl methacrylate is purified by distillation to remove excess methyl alcohol and low boilers. To prevent polymerization, inhibitors are added at various stages during the process.

Reaction

$$C_4H_8 + CH_3OH + 1\tfrac{1}{2}O_2 \rightarrow CH_2 = C(CH_3)COOCH_3 + 2H_2O$$

Raw material requirements and yield

Raw materials required per tonne of methyl methacrylate:

Isobutylene	860kg
Methyl alcohol	340kg

Yield 68%

PROPERTIES

Colourless, volatile, flammable liquid. Slightly soluble in water but soluble in most organic solvents.

Molecular Weight	100.1
Density at 20°C	0.938
Freezing Point	–48.2°C
Boiling Point	100.1°C
Autoignition Temperature	421°C
Explosive Limits in air	
lower	2.1vol%
upper	12.5vol%
Flash Point Open Cup	10°C
Vapour Density (air = 1)	3.45
Exposure Limit COSHH	100ppm 15 minutes
	50ppm 8 hour TWA
Exposure Limit ACGIH	100ppm 8 hour TLV-TWA

GRADES

Technical >99% (inhibited).

INTERNATIONAL CLASSIFICATIONS

UN No.	1247
CAS Reg No.	80–62–6
EINECS No.	201–297–1
EC No.	607–035–00–6
Description	Flammable liquid
Packing Group	II
Emergency Action Code	3YE
HI (Kemler Code)	339

APPLICATIONS

Almost all methyl methacrylate is polymerized to form homopolymers or copolymers with a wide range of mixed monomers.

Methyl methacrylate resins used in the automotive and construction industries account for almost 50% of world demand for the monomers.

The major outlet for polymethylmethacrylate is as sheet, divided between cast and moulded types. Its clarity, weather-resistance and light weight make it a good substitute for glass in light fittings, glazing panels and illuminated light displays.

Surface coatings and impregnation resins are the second most important use, where colour fastness and weather-resistance properties are required, for example in latex paints, lacquer resins and stoving enamels. Coatings incorporating methacrylates are used on paper, a steadily increasing market, fabrics and leather. Moulding and extrusion resins (used for small instrument cases, light fittings, car light covers, knobs, and decorative motifs) are large consumers of methyl methacrylate.

The monomer can be added as a modifier to rigid polyvinyl chloride (PVC) plastics to improve impact resistance, or combined with butadiene and styrene to give a new range of plastics. A growing market for methyl methacrylate is in partial replacement of styrene in unsaturated polyester resins to give better weather resistance and longer life.

Numerous minor outlets include adhesives, sealants, inks and floor polishes. Higher esters of methacrylic acid are used as viscosity index improvers in lubricating oils and pour-point depressants.

Although methyl methacrylate is finding new outlets in video disks, large-sized TV screens, and PC VDU anti-glare covers and optical fibres, demand has been hit by rising acetone and methyl alcohol prices and by-product substitution from lower priced materials particularly polycarbonates.

Future demand for methyl methacrylate is expected to grow at 2–3% per year to 2000. With new investments coming on-stream during 1999 and 2000, some European capacity will have to close. The largest market for methyl methacrylate is North America with 35% of total world consumption, followed by Western Europe with 25%. The other important and growing markets are Japan and Asia with 17% and 16% respectively.

HEALTH AND HANDLING

Methyl methacrylate vapour is irritating to the eyes, nose and throat. Over exposure can lead to headaches, drowsiness and nausea. Skin contact with the liquid can lead to sensitivity and dermatitis, and this should be avoided by wearing protective clothing and eye protection.

Store in closed containers, earthed to prevent static build-up, in a well-ventilated, cool area, out of sunlight and away from amines, halogens and oxidizing agents. Methyl methacrylate will polymerize easily especially when heated. When inhibited it is stable at room temperature but the storage life of the product is closely linked to the temperature. For this reason the inhibitor level must be monitored regularly and oxygen must be present for it to be effective.

In the event of leaks, evacuate the area and extinguish all forms of ignition. Absorb with sand or vermiculite and keep the liquid away from sewers and waterways. Scoop up the waste with non-sparking tools, and burn in an approved incinerator. Clean-up staff must wear protection against skin contact and vapour inhalation.

Fires should be extinguished with carbon dioxide, dry chemical or foam. Methyl methacrylate will polymerize at high temperatures and its heavy vapour can roll long distances. Flashback is a hazard. Because carbon monoxide is produced on burning, firefighting staff must wear protective clothing and self-contained breathing equipment.

MAJOR PLANTS

Plants with capacities greater than 60 000 tonnes per year:

Elf Atochem	Saint Avold	France
Rohm & Haas	Worms	Germany
Degussa	Wessling	Germany
Elf Atochem	Rho	Italy
ICI	Billingham	UK
CYRO Industries	New Orleans	US
ICI	Memphis	US
Rohm & Haas	Deer Park	US
Asahi Chemical	Kawasaki	Japan
Mitsubishi Rayon	Otake	Japan
KMC	Hsian	Taiwan

MAJOR LICENSORS

ICI
KMC
Mitsubishi Chemical
Nippon Shokubai
Rohm & Haas
Shell Development
Sumitomo Chemical

Methyl tert-*Butyl Ether (MTBE)*

CH_3
|
$H_3C - C = O$
|
CH_3

SYNONYMS

METHYL TERT-BUTYL ETHER MTBE, 2-methoxy-2-methyl propane

First synthesized in 1904, methyl *tert*-butyl ether (MTBE) was introduced as an oxygen enhancer for gasoline in 1973. Since then, the growth in demand for lead-free gasoline for vehicles in many of the developed countries of the world as concern about pollution has grown, has resulted in an upsurge in consumption of MTBE.

MTBE has many advantages over methyl alcohol as an octane enhancer for gasoline. It has good blending properties and can be shipped easily. MTBE is especially important for refiners with limited reforming capacity, or who wish to replace toluene in gasoline outlets so that it can be used for petrochemicals.

MTBE is manufactured by the catalytic addition of methyl alcohol to isobutylene obtained from a C_4 stream from which butadiene has been removed. Isobutylene can be obtained from steam cracker raffinate, in the form of butene–butane fractions from fluid catalytic crackers, or by the dehydrogenation of isobutane from refineries.

The catalyst can be an acidic ion exchange resin or sulphuric acid. An alternative source of isobutylene is *n*-butane in liquefied petroleum gas (LPG) which can be isomerized to isobutane and then dehydrogenated to isobutylene. The reaction can be carried out adiabatically in a fixed bed or by reaction distillation.

MTBE can be obtained as a by-product of the manufacture of propylene oxide by the Halcon process, whereby the *tert*-butyl alcohol produced is dehydrated to isobutylene and used as the feedstock source (see Propylene Oxide). Currently, around 48% of MTBE is obtained from isobutane, 33% from raffinate and 19% from *tert*-butyl alcohol.

As the cost of raw materials is critical to MTBE profitability, new plants are being sited in countries having cheap sources of methyl alcohol and C_4s. With so many plants having been built or coming on-stream, MTBE has been one of the fastest growing chemicals in the world. Recently, health and environmental concerns have been expressed about MTBE's usage in gasoline.

Capacities range from 15 000 to 700 000 tonnes per year.

PROCESSES

1. From isobutylene and methyl alcohol

A C_4 fraction (free from butadiene) and methyl alcohol in a 1:1 molar ratio are preheated to 70°C before being sent to a reactor. The reaction takes place either in the liquid phase or in a mixed gas–liquid phase, in the presence of an acidic ion exchange resin at a temperature of 50–90°C and under a pressure of 20 bar. Heat produced by the slightly exothermic reaction is removed by means of cooling jackets so as to maintain the temperature below 120°C. (*See Figure 91*)

The reaction mixture from the top of the reactor is distilled and any unreacted C_4s and methyl alcohol are removed overhead. The bottoms contain high-purity MTBE.

Additional methyl alcohol is added to the distillate before the mixture passes to a second reactor. Products from the second-stage reactor are cooled and extracted with water to remove any residual methyl alcohol. The water–methyl alcohol solution is distilled and the recovered methyl alcohol is recycled to the reactor.

Bottoms from the base of the distillation column containing MTBE, methyl alcohol and a little C_4s are recycled to the first reactor.

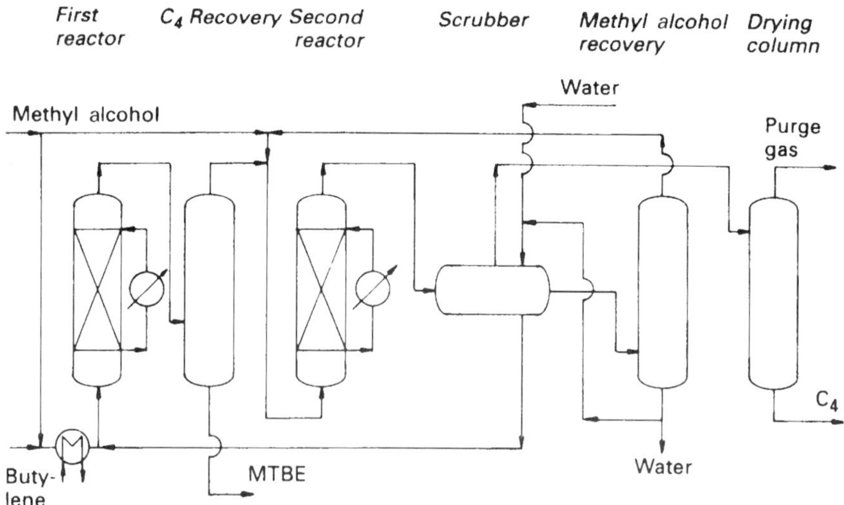

FIGURE 91 MTBE from isobutylene and methyl alcohol

In the single-stage process, a molar excess of 10% methyl alcohol is usually employed as it not only increases the conversion of isobutylene but also increases catalyst life. When excess methyl alcohol is used, traces of the alcohol remain in the MTBE which can be removed by azeotropic distillation. If the MTBE is to be used for gasoline, then the removal of by-products is unnecessary.

A process which uses 100% sulphuric acid as the catalyst at a concentration of 5wt% in a mixed butenes feed was developed in the 1980s. A 90% conversion of isobutylene in the first reactor, rising to 98% in the second, is claimed.

Reaction

$$CH_3\,C(CH_3) = CH_2 + CH_3OH \rightarrow CH_3OC(CH_3)_3$$

Raw material requirements and yield

Raw materials required per tonne of MTBE:

Isobutylene	653kg
Methyl alcohol	370kg

Yield 98%

PROPERTIES

Colourless, mobile liquid with a characteristic ether odour. Highly flammable and can form peroxides on storage. Miscible with water.

Molecular Weight	88.15
Density at 20°C	0.741
Melting Point	–108.6°C
Boiling Point	55.3°C
Autoignition Temperature	460°C
Flammability Limits in air	
lower	1.65vol%
upper	8.4vol%
Flash Point Closed Cup	–28°C

Exposure Limit COSHH Non established
The Working Group on the Assessment of Toxic Chemicals (WATCH) is recommending an Occupational Exposure Standard to the Advisory Committee on Toxic Substances (ACTS).
Exposure Limit ACGIH 40ppm proposed
Classified A3, animal carcinogen.

GRADES

Technical >99%, gasoline blending 60%.

INTERNATIONAL CLASSIFICATIONS

UN No.	2398
CAS Reg No.	1634–04–4
EINECS No.	216–768–7
EC No.	Not allocated
Description	Flammable liquid
Packing Group	II
Emergency Action Code	3YE
HI (Kemler Code)	33

APPLICATIONS

Around 95% of MTBE produced is used as an octane booster especially in mixtures with other alcohols such as methyl alcohol, ethyl alcohol, or *sec*-butyl alcohol. One of its major advantages is its good blending properties and it is excellent for increasing the octane levels of lead-free gasoline. Pressures in Europe and the US to reduce lead emissions into the atmosphere led to MTBE's rapid growth in demand. However, health and environmental concerns have been expressed about the use of MTBE in gasoline, and consumption has been limited by restrictions on its use due to air pollution control standards.

MTBE is also employed in the petrochemicals industry to produce pure isobutylene from C_4 streams by reversing its formation reaction. It is a good solvent and extractant. Other outlets include the production of methacrolein and methacrylic acid. MTBE is expected to grow at 3.5% per year to the end of the 20th century.

HEALTH AND HANDLING

MTBE vapour is irritating to the eyes, nose and throat, and at high concentrations rapidly leads to unconsciousness. The liquid is absorbed by the skin and causes smarting and irritation. Good ventilation is essential when handling the product.

MTBE should be stored in closed air-tight stainless steel or aluminium tanks under a nitrogen atmosphere in a well-ventilated, cool area, away from strong oxidizing materials. Attention should be paid to materials used for valves and flanges as MTBE can attack some plastics and rubbers. Peroxides can be formed in the presence of air on long storage or in sunlight; thus any product which has been stored should be tested for purity before use. Containers and handling equipment must be earthed to prevent static build-up which could result in a vapour explosion.

Although MTBE is miscible with water, at high concentrations it will form an air–vapour explosive mixture above the water which can be ignited by sparks or contact with hot surfaces.

Leaks are highly dangerous due to the rapid formation of explosive mixtures in air. Evacuate personnel, extinguish all forms of ignition and stop the source of emission if possible. The liquid should be absorbed with sand or vermiculite, collected with non-sparking tools and disposed of promptly by controlled burning in an approved incinerator. Clean-up staff and staff handling MTBE should wear protective clothing, eye protection and a respirator. Any contaminated clothing must be washed before reuse.

MTBE is a dangerous fire and explosion hazard. Its vapours will travel long distances along the ground and can lead to flashback. Use dry chemical, carbon dioxide or alcohol foam to extinguish fires. Toxic gases are given off during burning and all firefighting staff must wear self-contained breathing apparatus.

MAJOR PLANTS

Plants with capacities greater than 400 000 tonnes per year:

ARCO Chimie	Fos sur Mer	France
ARCO Chemie Nederland	Botlek	Netherlands
Complex	Mazheikiai	Russia
Belvieu Environmental Fuels	Mount Belvieu	US
Enron	La Porte	US
Exxon	Baton Rouge	US
	Baytown	US
Global Octanes	Deer Park	US
Huntsman Corp.	Port Neches	US
Lyondell	Channel View	US
	Corpus Christi	US
Oxyfuel	Beaumont	US
Shell Chemical	Norco	US
Texas Petrochemicals	Houston	US
Valero Refining	Corpus Christi	US
Alberta Envirofuels	Edmonton	Canada
Petrobras	Camaraci	Brazil
Pequiven	Jose	Venezuela
Saudi European Petrochemical	Al Jubail	Saudi Arabia
Qatar Fuels	Umm Said	Qatar
Singapore Refining	Pulau Merlimau	Singapore

MAJOR LICENSORS

Arco Technology
BP Chemicals/ABB Lummus Global
CDTECH
Edeleanu
Exxon
Phillips Petroleum
Shell Development
Snamprogetti
Sumitomo Chemical
Texaco Development
UOP

Nitrobenzene

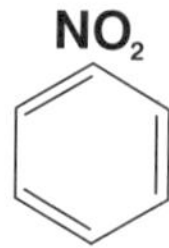

SYNONYMS

NITROBENZENE nitrobenzol, oil of bitter almonds, oil of mirbane

Nitrobenzene is an important precursor for the dyestuffs industry. Its commercial manufacture is by the nitration of benzene in either a batch or continuous process, the latter being the route of choice. Tubular reactors are normally employed in continuous nitration plants because of the higher conversion rates achieved.

Capacities range from 5000 to 315 000 tonnes per year.

PROCESSES

From benzene by nitration

The nitrating acid (composed of 27–32wt% nitric acid, 56–60wt% sulphuric acid and 8–17wt% water) is fed into a series of agitated nitrators containing benzene. The ratio of acid to benzene is maintained at a level sufficient to ensure that all the nitric acid is consumed. The nitration temperature is 50°C and heat from the reaction is removed by cooling coils in the reactor. Where agitation is provided by turbulent flow, reaction time is shorter, the temperature can be controlled more effectively, and the smaller volumes of reactants present at any one time reduce the risk of explosion. Many plants use nitrogen blanketing to improve safety. (*See Figure 92*)

On completion, any spent acid and nitrobenzene are removed from the reactor into a separator, where they form two layers. The crude nitrobenzene passes through a series of separators where it is washed with water, then sodium carbonate to remove any residual acid, and finally with water again. Final purification is by distillation.

If nitration is carried out at a temperature of 120–160°C, excess water can be removed from the reaction products as an azeotrope with benzene. The spent acid is reacted with fresh benzene to remove any residual nitric acid and nitrobenzene. Waste gases are scrubbed prior to incineration.

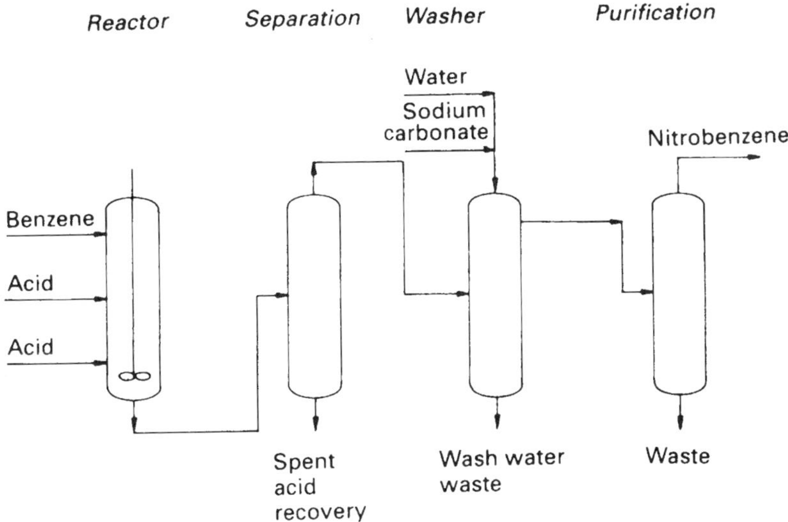

FIGURE 92 Nitrobenzene from benzene by nitration

American Cyanamid has developed an adiabatic process which utilizes the heat of the reaction. The feed stream enters a tubular reactor where it is vigorously agitated at a temperature of 60–80°C and has a residence time of 4 minutes. The exit temperature is 120°C.

The process is economical because weaker sulphuric acid streams can be employed and the heat of reaction is almost sufficient to evaporate the nitrobenzene formed, resulting in a considerable energy saving.

Water and benzene are removed as an azeotrope in a still and the benzene is recycled. Final purification of the nitrobenzene is carried out by vacuum distillation.

Reaction

$C_6H_6 + HNO_3 \rightarrow C_6H_5NO_2 + H_2O$

Raw material requirements and yield

Raw materials required per tonne of nitrobenzene:

Benzene	660kg
Nitric acid	535kg
Sulphuric acid	790kg
Sodium carbonate	10kg

Yield 95%

PROPERTIES

Pale yellow oil with a bitter almond odour, which is miscible with most organic solvents. Only slightly soluble in water.

Molecular Weight	123.1
Density at 20°C	1.204
Melting Point	5.8°C
Boiling Point	210.9°C
Autoignition Temperature	482°C
Flammability Limits in air	
lower	1.8vol%
upper	40vol%
Flash Point Closed Cup	88°C
Vapour Density (air = 1)	4.1
Exposure Limit COSHH	2ppm 15 minutes 1ppm 8 hour TWA
Exposure Limit ACGIH	1ppm TLV-TWA (skin)

GRADES

Technical, distilled, 97% (mirbane oil).

INTERNATIONAL CLASSIFICATIONS

UN No.	1662
CAS Reg. No.	98–95–3
EINECS No.	202–716–0
EC No.	609–003–00–7
Description	Toxic substance
Packing Group	II
Emergency Action Code	2X
HI (Kemler Code)	60

APPLICATIONS

The most important use of nitrobenzene is in the manufacture of aniline which is used in the dyestuffs, pharmaceuticals and rubber chemicals industries. Hence aniline producers have their associated nitrobenzene plants.

Nitrobenzene is also used for the manufacture of explosives, as an inert solvent in Friedel–Crafts reactions, in the depolymerization of rubber and as a constituent of shoe and metal polishes.

World demand is increasing at 4–6% per year.

HEALTH AND HANDLING

Nitrobenzene is an eye irritant and is readily absorbed by inhalation and through the skin. The first effects of exposure are a blue discoloration to the lips and skin followed by headaches, giddiness and nausea. It is a powerful methaemglobin former, which can lead to liver and spleen damage and injury to the central nervous system. Staff handling nitrobenzene should be instructed to recognize the signs and symptoms of over-exposure so that rapid treatment can be given.

Nitrobenzene is stable when stored in closed containers, which should be protected from physical damage and stored separately in a well-ventilated area. Nitrobenzene forms explosive mixtures with aluminium chloride, aniline and glycerine, and reacts with tin, zinc and alkalis.

Spills must be treated as an emergency and advance planning procedures should be in place to deal with them. All personnel must be evacuated and any forms of ignition extinguished. Clean-up staff, wearing protective equipment and clothing against inhalation of vapour and skin contact, should contain and scoop up the nitrobenzene using non-sparking tools. The waste should be placed in containers for disposal in accordance with local regulations. Nitrobenzene is harmful to aquatic life and the liquid must be prevented from entering streams or watercourses.

Fires should be extinguished with carbon dioxide, water, dry chemical or foam. Firefighting personnel must wear protective clothing and self-contained breathing apparatus because of the poisonous vapours given off during burning.

MAJOR PLANTS

Plants with capacities greater than 100 000 tonnes per year:

BASF	Antwerp	Belgium
Bayer	Antwerp	Belgium
Bayer	Krefeld	Germany
Anilina do Portugal	Estareja	Portugal
ICI	Huddersfield	UK
Moravske Chemical	Ostrava	Czech Republic
BASF	Geismar	US
Du Pont	Beaumont	US
First Chemical	Pascagoula	US
Rubicon	Geismar	US
Sumitomo Chemical	Niihama	Japan

MAJOR LICENSORS

AgriEvo
Bayer
ICI
Sumitomo Chemical

Perchloroethylene

$Cl_2C{=}CCl_2$

SYNONYMS

PERCHLOROETHYLENE ethylene tetrachloride, tetrachloroethylene, carbon dichloride, perc

First obtained by heating hexachloroethane, perchloroethylene was manufactured initially from acetylene. Although direct chlorination is possible, commercial processes proceed via trichloroethylene.

In most countries acetylene has been replaced by a variety of feedstocks: by ethylene or ethylene dichloride, by C_1–C_3 hydrocarbons or by chlorinated hydrocarbon wastes. High-temperature chlorination processes have been developed and, with increasing environmental pressures on the disposal of chlorine-containing wastes, they have become the major source for the production of perchloroethylene. The reaction can be carried out at high temperature, at either low or high pressure. The commercialization of oxychlorination processes have opened up a route which enables by-product hydrogen chloride to be utilized.

Capacities range from 10 000 to 150 000 tonnes per year.

PROCESSES

1. From ethylene dichloride

Perchloroethylene and trichloroethylene are co-produced by the chlorinated–oxychlorination of ethylene dichloride. (*See Figure 93*)

In the chlorinated–oxychlorination route, ethylene dichloride (produced by the additive chlorination of ethylene or chlorinated residues) is fed with chlorine and oxygen into a fluidized-bed reactor containing a catalyst consisting of a mixture of cupric and potassium chloride on silica or fuller's earth. The feed is cracked at a temperature of 420–440°C under slight pressure to yield a mixture of trichloroethylene and perchloroethylene. Temperature control is essential to obtain optimum

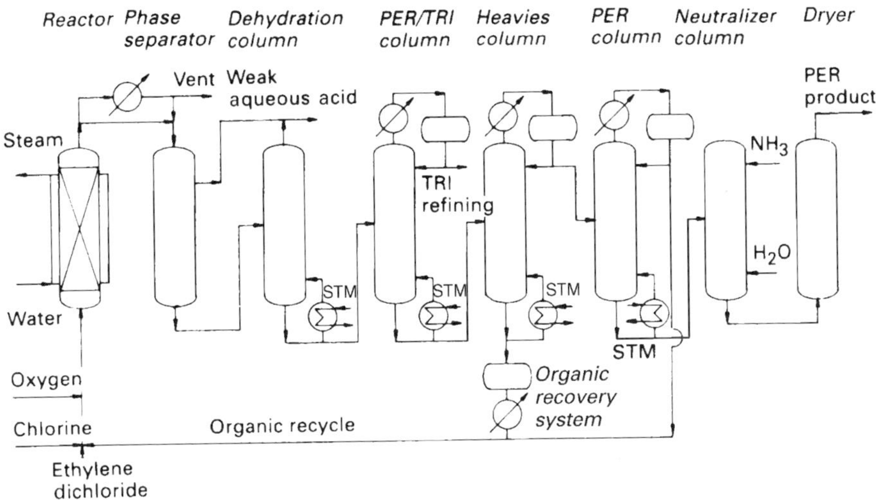

FIGURE 93 Perchloroethylene from ethylene dichloride

yields. The ratio of trichloroethylene to perchloroethylene can be varied by altering the chlorine–ethylene dichloride ratios.

The exit gases are scrubbed with water and the chloroethylenes are recovered by azeotropic distillation. Perchloroethylene and trichloroethylene are separated by distillation, and any light and heavy fractions remaining are recycled. Tars and heavy residues are burnt. The major advantage of the oxychlorination route is that minimal amounts of by-product aqueous hydrogen chloride are produced.

Reaction

$$2C_2H_4Cl_2 + 5Cl_2 \rightarrow C_2H_2Cl_4 + C_2HCl_5 + 5HCl$$

$$C_2H_2Cl_4 + C_2HCl_5 \rightarrow C_2HCl_3 + 2HCl + C_2Cl_4$$

$$2C_2H_4Cl_2 + 1\tfrac{1}{2}Cl_2 + 1\tfrac{3}{4}O_2 \rightarrow C_2HCl_3 + C_2Cl_4 + 3\tfrac{1}{2}H_2O$$

Raw material requirements and yield

Raw materials required per tonne of perchloroethylene (790kg trichloroethylene is co-produced):

Ethylene dichloride	1190kg
Chlorine	640kg
Oxygen	385kg

Yield 90%

2. From acetylene by chlorination

Trichloroethylene produced by the chlorination of acetylene is reacted with chlorine in the liquid phase at 80–110°C. The catalyst used consists of a 1wt% of ferric chloride. The pentachloroethane formed is cracked at 180–330°C to perchloroethylene. In an alternative liquid-phase process, perchloroethylene is obtained as an overhead distillate from the reaction between a 10% suspension of calcium hydroxide and a counterflow of pentachloroethane at a temperature of 80–120°C (see Trichloroethylene).

Reaction

$$C_2H_2 + 2Cl_2 \rightarrow C_2H_2Cl_4$$

$$2C_2H_2Cl_4 + Ca(OH)_2 \rightarrow 2CHCl{=}CCl_2 + CaCl_2 + 2H_2O$$

$$CHCl{=}CCl_2 + Cl_2 \rightarrow C_2HCl_5$$

$$2C_2HCl_5 + Ca(OH)_2 \rightarrow 2CCl_2{=}CCl_2 + CaCl_2 + 2H_2O$$

Raw material requirements and yield

Raw materials required per tonne of perchloroethylene:

Acetylene	170kg
Chlorine	1450kg

Yield 85%

OTHER PROCESSES

From hydrocarbons or chlorinated wastes by chlorinolysis

Perchloroethylene can be produced from C_1–C_3 hydrocarbons or chlorinated hydrocarbon wastes. The process can be carried out by three routes:

- high-temperature chlorination;
- low-pressure chlorinolysis;
- high-pressure chlorinolysis.

The low-pressure route is the most popular because it facilitates the purification of the hydrogen chloride formed.

Propane was originally the feedstock but this hydrocarbon has largely been replaced by cheaper ethane or chlorinated residues. The principal product is perchloroethylene with carbon tetrachloride and hydrogen chloride as by-products.

Propane, ethane or chlorinated hydrocarbons residues are preheated, mixed with gaseous chlorine and introduced into a tubular or fluidized-bed reactor. The reaction takes place at 600–800°C and 2–10 bar pressure. The exit gases are quenched rapidly to reduce hexachloroethane formation, and the chlorinated products are separated from chlorine and hydrogen chloride which are collected overhead.

Hydrogen chloride is condensed and the remaining chlorine is absorbed in carbon tetrachloride. After stripping, the chlorine is recycled.

The hydrocarbon mixture is fractionated and crude perchloroethylene and carbon tetrachloride are condensed and purified by distillation.

Reaction

$C_2Cl_6 \rightarrow C_2Cl_4 + Cl_2$

$2CCl_4 \rightarrow C_2Cl_4 + 2Cl_2$

PROPERTIES

Clear, colourless liquid with an ethereal odour. Non flammable. Soluble in ethyl alcohol and ether.

Molecular Weight	165.83
Density at 20°C	1.623
Melting Point	–22.4°C
Boiling Point	121°C
Vapour Density (air = 1)	5.83
Exposure Limit COSHH	100ppm 15 minutes
	50ppm 8 hour TWA
Being reviewed under the Existing Substances Regulations.	
Exposure Limit ACGIH	100ppm TLV-STEL
	25ppm TLV-TWA
Classified A3, animal carcinogen.	

GRADES

Technical 99%.

INTERNATIONAL CLASSIFICATIONS

UN No.	1897
CAS Reg No.	127–18–4
EINECS No.	204–823–8
EC No.	602–028–00–4
Description	Toxic substance
Packing Group	III
Emergency Action Code	2Z
HI (Kemler Code)	60

APPLICATIONS

Perchloroethylene is an important solvent because of its stability and non-flammability. It has been largely replaced as a dry cleaning solvent for clothes by less toxic compounds, but problems have been experienced with colour bleeding and interaction with some synthetic fibres. It is also used in the textile industry, as a scourer, dye carrier, and for sizing and finishing of cloth.

It is an excellent metal degreaser especially for aluminium. As a solvent, it can be used for the extraction of fats, dissolving rubber, paint removal and sulphur recovery. Other minor outlets include the manufacture of fluorocarbons and trichloroacetic acid. Consumption is declining due to recycling of the product in closed systems and product replacement.

HEALTH AND HANDLING

Perchloroethylene vapour affects the central nervous system causing headaches, dizziness and finally unconsciousness. Gloves, aprons of polyvinyl alcohol or neoprene, boots, goggles and faceguards must be worn to prevent skin contact when handling the product. Contact lenses will concentrate the vapour and should be avoided.

Store in closed containers in a cool, well-ventilated area away from sunlight. It is incompatible with sodium hydroxide, barium, beryllium and lithium. Perchloroethylene is normally inhibited to prevent decomposition; levels should be monitored regularly and good stock rotation practised.

Contain spills and either absorb with dry sand, earth or vermiculite, or collect with non-sparking tools, and place in containers for disposal according to local regulations. Care must be taken to keep the liquid away from sewers or waterways. All clean-up staff must be protected against skin contact and inhalation of the vapour. Any contaminated clothing must be laundered prior to reuse.

Perchloroethylene does not burn but will degrade at high temperatures with the evolution of phosgene and hydrogen chloride. The vapour being heavier than air can collect in sumps and wells and self-contained breathing apparatus must be worn when entering such areas.

MAJOR PLANTS

Plants with capacities greater than 40 000 tonnes per year:

Dow Chemical	Stade	Germany
Huels	Marl	Germany
Wacker Chemie	Burghausen	Germany
EniChem	Assemini	Italy
	Porto Marghera	Italy
ICI	Runcorn	UK
Dow Chemical	Plaquemine	US
PPG Industries	Lake Charles	US
Vulcan Chemicals	Geismar	US

MAJOR LICENSORS

ABB Lummus Global
Diamond Shamrock
Elf Atochem
Hoechst
Huels
PPG
Stauffer Chemical
Toagosei Chemical
Vulcan Materials

Phenol

SYNONYMS

PHENOL carbolic acid, hydrobenzene, oxybenzene, phenic acid, phenyl hydrate, phenyl hydroxide, phenylic acid, phenyl alcohol

Phenol was isolated from coal tar produced in the coking of coal over a hundred years ago, but the first commercial process was the sulphonation of benzene and subsequent fusion with caustic soda. The demand for phenol rose rapidly with the discovery that trinitrophenol was a powerful explosive for military use.

There are three other synthetic routes to phenol. Phenol can be produced by chlorobenzene–caustic soda hydrolysis but this process no longer has any economic importance.

The toluene oxidation route, originally developed by Dow Chemical, has not gained wide acceptance. The major disadvantages of the toluene-based process are:

- loss of the methyl group as carbon dioxide and water;
- major waste disposal problems with the copper-containing tar residues which require special incineration equipment;
- complex separation procedures.

In view of these difficulties and the need for corrosion-resistant construction materials, this process currently accounts for around 4% of world capacity.

Cumene is the dominant raw material for phenol manufacture, accounting for most of the world's capacity, because of its ready availability, and the value of by-product acetone. It is the most competitive process for the manufacture of phenol, supported by the demand for acetone. The remainder is obtained from coal tar, petroleum distillates, or chlorobenzene.

With the over-capacity of acetone, there is still a strong commercial interest in finding alternative routes to phenol, which like the benzene chlorination and toluene processes, do not produce acetone by-product. A process in which benzene is oxidized in the presence of acetic acid to phenyl acetate, followed by hydrolysis to phenol, has been developed by Mitsui Petrochemical, who has a plant using this route in Japan. Monsanto has also been working on a one-step benzene-to-phenol route.

A new process developed by Solutia, due to be commercialized by 2000, is the direct catalytic conversion of benzene to phenol avoiding cumene as the intermediate and eliminating acetone production. Any move to alternative feedstocks for acetone derivatives, which lead to a reduction in demand, will continue to provide interest in non-acetone routes.

Capacities range from 30 000 to 500 000 tonnes per year.

PROCESSES

1. From cumene by peroxidation

In the peroxidation process, phenol is obtained by the decomposition of cumene hydroperoxide, derived from cumene, which in its turn is produced from benzene and propylene (see Cumene). (*See Figure 94*)

In the two-stage process, pure cumene is fed into an oxidation vessel where it is mixed with a dilute solution of sodium carbonate. Air is introduced and the mixture, at a temperature of 110–130°C, is left in contact until 25–30% of the cumene is oxidized to the hydroperoxide. If the conversion is allowed to proceed beyond this

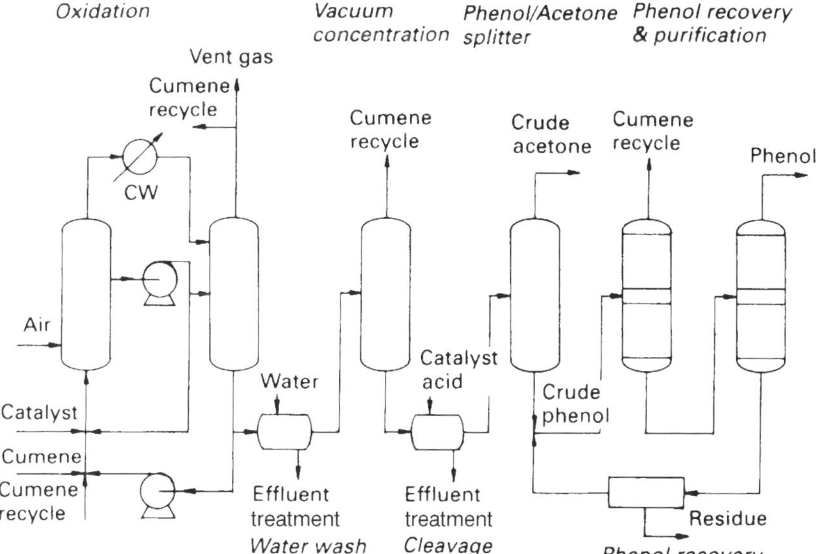

FIGURE 94 Phenol from cumene by peroxidation

point there is an increased risk of by-product formation. Lower operating temperatures favour the yield of hydroperoxide but conversion is lower.

In the second stage, the crude mixture is concentrated to about 80% cumene hydroperoxide before being fed into a cleavage reactor. The reaction is carried out at a temperature of 70–80°C and a pressure of 0.3 bar in the presence of a dilute acid, such as a 10% solution of sulphuric acid. Both the oxidation and cleavage reactions are very exothermic. Temperature control can be effected by evaporation of water which is present or by using cooled mixture to dilute the hydroperoxide. The major reaction products are phenol and acetone, with small quantities of acetophenone, α-methyl styrene and cumene.

The reaction products are separated by distillation. Acetone is flashed from the top of the first column and purified by distillation. The bottoms from the column are further distilled to remove any unreacted cumene and α-methyl styrene. By-product α-methyl styrene can be recovered and sold or converted to cumene by catalytic hydrogenation before being recycled to the first-stage reactor. Further distillation separates by-product acetophenone from phenol which is recovered overhead.

Improvements by Dow Chemical and Allied Chemical have reduced the amount of by-products, especially di-methyl phenol carbinol, by converting it to α-methyl styrene which can be hydrogenated to cumene and recycled.

Reaction

$$C_6H_5CH(CH_3)_2 + O_2 \rightarrow C_6H_5C(CH_3)_2OOH$$

$$C_6H_5C(CH_3)_2OOH \rightarrow C_6H_5OH + CH_3COCH_3$$

Raw material requirements and yield

Raw materials required per tonne of phenol (610 kg of acetone is co-produced):

Cumene	1340kg
Air	1400m^3
Sulphuric acid	Small
Sodium Carbonate	Small

Yield 90–95%

Yield 95–98% if by-product α-methyl styrene is recycled

2. From toluene by oxidation

The liquid-phase air oxidation of toluene takes place in two steps, starting with the oxidation of toluene to benzoic acid which is then further oxidized to phenol. (*See Figure 95*)

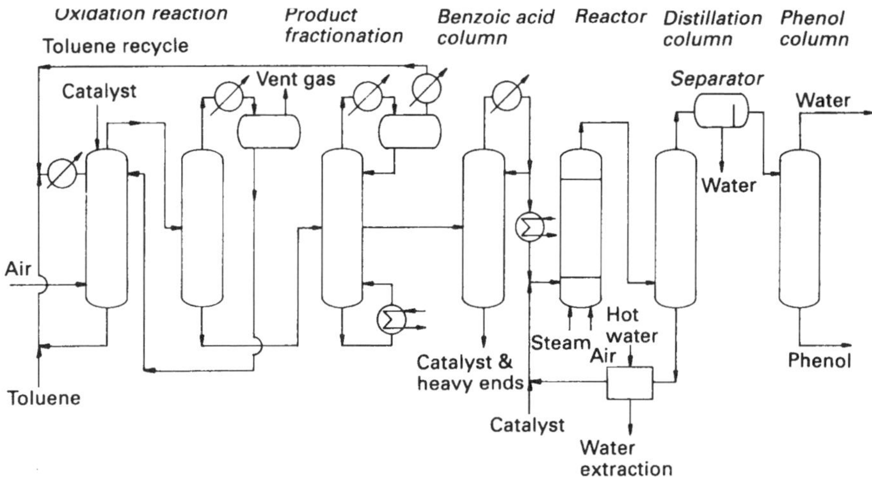

FIGURE 95 Phenol from toluene by oxidation

The first reaction is carried out at 160–170°C over a cobalt naphthenate catalyst in the presence of air or oxygen at a pressure of 8–10 bar. The solvent which can be toluene, water, or excess product benzoic acid, contains cobalt salts (either bromide or naphthenate) as a catalyst. When a 40% conversion has been achieved, the solution is sent to a distillation column where unreacted toluene and light ends are stripped from the reaction mixture. Water is removed before the toluene is recycled. Many by-products are formed and benzoic acid is recovered either by distillation as a pure overhead product or by crystallization from water (see Benzoic Acid).

In the second stage, purified benzoic acid in the molten form is mixed with a catalyst consisting of cupric salts promoted by manganese salts, and fed into a reactor. Air and steam, which must be present in excess, are sparged into the melt. The phenol formed is kept in contact with the copper salts as briefly as possible by continuous removal in order to reduce the formation of polymeric tar materials. The reaction takes place at 220–245°C at atmospheric or slightly above atmospheric pressure. Some benzene and diphenyl ether are produced as by-products.

The reaction products are separated in a series of distillations. Phenol and water are removed overhead in the first distillation column, and the bottoms are extracted with hot water to remove any benzoic acid. The benzoic acid solution is concentrated in evaporators before being recycled.

The phenol–water mixture separates into two layers; any water in the lower phenol-rich layer is removed as an azeotrope by further distillation. Phenol is withdrawn from the bottom. Phenol is recovered from the overhead water streams from the first and phenol distillation columns by distillation. The water-insoluble tar products which contain copper salts are specially incinerated to insure that pollutants do not reach the atmosphere.

ABB Lummus Global have developed a vapour-phase process which does not produce tars or non-volatile residues by operating at very high space velocities per pass. Operating temperatures and pressure are higher than in the liquid-phase process.

Reaction

$$C_6H_5CH_3 + 1\tfrac{1}{2}O_2 \rightarrow C_6H_5COOH + H_2O$$

$$C_6H_5COOH + \tfrac{1}{2}O_2 \rightarrow C_6H_5OH + CO_2$$

Raw material requirements and yield

Raw materials required per tonne of phenol:

Toluene	1250kg
Air	1050m^3
Copper catalyst	Small

Yield 80–85%

OTHER PROCESSES

Mitsui Petrochemical has introduced a process based on benzene. In this route, benzene is partially hydrogenated to cyclohexene followed by conversion to cyclohexanol and then to phenol by dehydrogenation, thus avoiding the production of acetone. Solutia has also developed a direct catalytic benzene to phenol process which does not produce cumene as an intermediate. It is due to be commercialized in 2000.

PROPERTIES

White crystals, with a carbolic odour, which turn pink on exposure to light and air. Strongly hygroscopic, liquefying in moist air. Soluble in water, ethyl alcohol, ether and chlorinated hydrocarbons. Forms salts with aqueous solutions of alkalis. Corrosive poison.

Molecular Weight	94.12
Density at 20°C	1.071
Melting Point	41°C
Boiling Point	181.8°C
Autoignition Temperature	715°C
Flammability Limits in air	
lower	1.5vol%
upper	8.6vol%
Flash Point Closed Cup	81°C
Vapour Density (air = 1)	3.24

Exposure Limit COSHH 10ppm 15 minutes
5ppm 8 hour TWA

Substance on the current Advisory Committee on Toxic Substances (ACTS) and the Working Group on the Assessment of Toxic Chemicals (WATCH) work programme.

Exposure Limit ACGIH 10ppm TLV-STEL
5ppm TLV-TWA

GRADES

Solid 98%, liquid commercial 92% and 84%.

INTERNATIONAL CLASSIFICATIONS

UN No.	
solid	1671
molten	2312
CAS Reg No.	108–95–2
EINECS No.	203–632–7
EC No.	604–001–00–5
Description	Toxic substance
Packing Group	II
Emergency Action Code	2X
HI (Kemler Code)	60

UN No.	2821 (solution)
CAS Reg No.	108–95–2
EINECS No.	203–632–7
EC No.	604–001–00–5
Description	Toxic substance
Packing Group	II or III
Emergency Action Code	2X
HI (Kemler Code)	60

APPLICATIONS

In the US, rapid growth in demand for bisphenol A, used in the manufacture of polycarbonates and epoxy resins, has resulted in this outlet consuming 37% of total phenol output to move into first position.

Phenolic resins are still a major outlet for phenol accounting for around 34% of total demand. They are used for the manufacture of adhesives and laminated boards used in the building, electrical, appliances, and car industries, and as a binder for foundry sands.

Around 15% of phenol is used for the textile intermediate, caprolactam. Other minor uses include the manufacture of alkyl phenols for non-ionic detergents, 7%, and aniline for dyestuffs which accounts for a further 5%. The remainder is used in the manufacture of chlorinated phenols for wood preservatives and herbicides, diphenols, pyrocatechol, hydroquinone, vanillin, and in the pharmaceutical industry for acetylsalicylic acid and paracetamol,

Due to the rise in the market for bisphenol A, global phenol consumption is estimated to grow at 4% per year. Several new plants are due to come on-stream in the US and Asia by 2000 and an over-supply position is forecast to 2000. Some old plants are expected to close.

HEALTH AND HANDLING

Phenol vapours or liquid are rapidly absorbed through the skin or mucous membranes. Liquid phenol causes severe burns in contact with the skin and acute poisoning can occur from repeated low concentrations from either liquid or vapour. Phenol is not sufficiently volatile to be a respiratory hazard under normal conditions. Protective clothing to prevent skin contact and eye protection must be worn. Any contaminated clothing must be laundered before reuse.

Phenol can be stored at room temperature in tightly closed containers to prevent discoloration and absorption of water. The area should be cool and well ventilated, well away from strong oxidizing agents and halogens. Hot phenol is corrosive to aluminium, lead, magnesium and zinc and care must be taken to ensure that there is no contact with these metals during storage, handling and transportation. All containers must be marked with poison labels.

Leaks should be closed off if possible and the area cleared of personnel and all lights extinguished. Spills, if small, should be absorbed with paper, vermiculite or dry sand. After containment, large spills can be allowed to solidify before removal. Phenol can be recovered from the waste by steam stripping. The area should be flushed with water to remove any residues but the washings must be collected and not allowed to enter watercourses or sewers.

Phenol emits toxic, sooty fumes on burning. Carbon dioxide, dry chemical or alcohol foam will extinguish fires but water tends to scatter the flames. Staff handling phenol, fighting fires or on clean-up duty must wear full protection against skin and eye contact and inhalation of fumes. Training should be given to all workers on the correct procedures to be followed prior to being allowed to handle phenol.

MAJOR PLANTS

Plants with capacities greater than 180 000 tonnes per year:

Phenolchemie	Antwerp	Belgium
Rhone-Poulenc	Rousillon	France
Phenolchemie	Gladbeck	Germany
EniChem	Mantua	Italy
Allied Signal	Frankford	US
Aristech Chemical	Haverhill	US
Dow Chemical	Freeport	US
Georgia Gulf	Plaquemine	US
GE Plastics	Mount Vernon	US
Shell Chemical	Deer Park	US
Chiba Phenol	Ichihava	Japan
Mitsubishi Chemical	Kashima	Japan
Mitsui Petrochemical Industries	Chiba	Japan
Mitsui Toatsu Chemical	Senboku	Japan

MAJOR LICENSORS

Cumene	*ABB Lummus Global*
	BP Chemicals/Hercules
	Hoechst/Uhde
	Kellogg
	Mitsui Petrochemical
	Monsanto
	Rhone-Poulenc
Toluene	*Dow Chemical*
	Stamicarbon
Chlorobenzene	*Hooker Chemicals*
Cresylic acid	*Hydrocarbon Research*
Benzene	*Mitsui Petrochemical*
	Solutia

Phthalic Anhydride

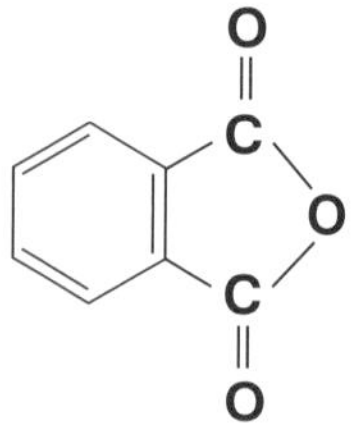

SYNONYMS

PHTHALIC ANHYDRIDE phthalic acid anhydride,
1,2-benzene dicarboxylic acid anhydride, phthalandione,
1,3-isobenzofurandione, 1,3-dioxophthalan

Initially produced from naphthalene in West Germany over one hundred years ago, phthalic anhydride is now a chemical of major importance.

The historical route, the oxidation of naphthalene in concentrated sulphuric acid in the presence of mercury sulphate, was replaced by the catalytic vapour-phase oxidation of naphthalene in air in the presence of a vanadium oxide catalyst.

Research has centred on new catalysts, mainly potassium-modified vanadium pentoxides. Fluid-bed systems were replacing fixed-bed systems until BASF improved the fixed-bed process and brought it back into favour.

Most new plants use *o*-xylene instead of naphthalene because stoichiometrically it is a more favourable raw material. The availability of *o*-xylene in commercial quantities at low cost has added impetus to the change. Many plants operate on either feedstock because naphthalene is available captively.

A range of catalysts is employed depending on the feedstock used. All are based on vanadium pentoxide with titanium dioxide and antimony trioxide, or molybdenum trioxide and calcium oxide, or manganese oxides, or similar catalyst mixtures. Varying amounts of maleic anhydride, benzoic acid and citraconic anhydride are formed as by-products.

Although the processes used today appear to be little changed over the years, catalysts have a longer life, yields have increased, reactors are smaller and the number of operatives has been reduced.

Approximately, 90% of phthalic anhydride is produced from *o*-xylene and 10% from naphthalene. Many plants will operate on either feedstock.

Capacity ranges from 10 000 to 160 000 tonnes per year.

PROCESSES

1. From *o*-xylene

Preheated compressed air is mixed with *o*-xylene in a carburettor to give an air–*o*-xylene weight ratio of 20:1. The air–*o*-xylene mixture passes to a multi-tubular reactor containing vanadium pentoxide with titanium dioxide–antimony trioxide catalyst. The catalyst life can be prolonged by the addition of sulphur dioxide. (*See Figure 96*)

The reaction takes place at a temperature of 375–425°C and a pressure below 1 bar. The temperature is maintained by circulating molten salt around the tubes which removes the excess heat generated by the exothermic reaction. The heat of reaction is used to generate steam, part of which is used in the plant to heat up the air stream.

The effluent gases are cooled in a heat exchanger before entering switch condensers where the phthalic anhydride collects on the walls as a solid. The product is recovered by sublimation and collected in a storage tank. One switch condenser is in use while phthalic anhydride is sublimed from the other.

Purification is carried out in a continuous thermal system. Firstly, any by-product phthalic acid is converted to the anhydride in a precomposer. Then the phthalic anhydride is distilled under vacuum in a two-stage system and stored either in its molten state or bagged as flakes. The waste gases are scrubbed with water to remove any residual phthalic anhydride and either the by-products are incinerated, or fumaric and maleic anhydride can be recovered economically from the aqueous solution. A catalyst life of more than 3 years is claimed.

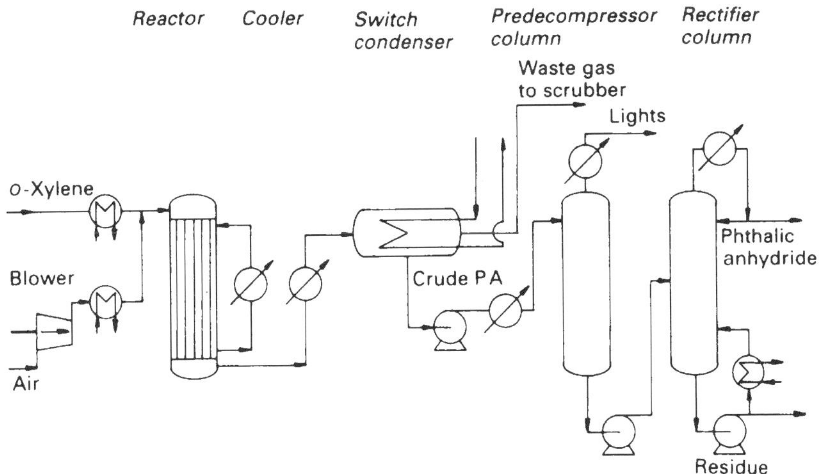

FIGURE 96 Phthalic anhydride from *o*-xylene

A recent development has been the low air ratio (LAR) process, where the air–*o*-xylene weight ratio has been reduced to 9.5:1. The decrease in the volume of air per unit of production has resulted in a reduction in capital costs, 60% energy saving, and increased catalyst productivity of 40%, as more heat is removed from the reactor by molten salt circulation.

After cooling, the increased concentration of phthalic anhydride in the exit gases from the reactor results in about half the crude product being recovered as a liquid in an after-cooler. As a result, the size of the switch condensers can be reduced.

Reaction

$$C_6H_4(CH_3)_2 + 3O_2 \rightarrow C_6H_4(CO)_2O + 3H_2O$$

Raw material requirements and yield

Raw materials required per tonne of phthalic anhydride:

	Conventional process	*LAR process*
o-Xylene	950kg	925kg
Air:hydrocarbon ratio	20:1	9.5:1

Yield 85% (Conventional process)
Yield 91% (LAR process)

2. From naphthalene

Naphthalene is vaporized by bubbling preheated air through the molten material in a vaporizer. The naphthalene vapour is mixed with more air until the air–naphthalene weight ratio reaches 20:1. The mixture passes into a multi-tubular reactor containing a modified vanadium oxide on silica gel catalyst. The reaction takes place at 390°C with a contact time of 4–5 seconds. Heat from the highly exothermic reaction is removed by molten salt or mercury circulating in a jacket around the tubes. (*See Figure 97*)

The reaction gases are cooled to around 135°C and the crude phthalic anhydride is condensed and collected. Pretreatment of the crude phthalic anhydride is carried out with chemicals to remove by-product naphthaquinone, prior to purification as in the *o*-xylene process.

Reaction

$$C_{10}H_8 + 4O_2 \rightarrow C_6H_4(CO)_2O + 2CO_2 + 2H_2O$$

Raw material requirements and yield

Raw materials required per tonne of phthalic anhydride:

Naphthalene	1075kg

Yield 80%

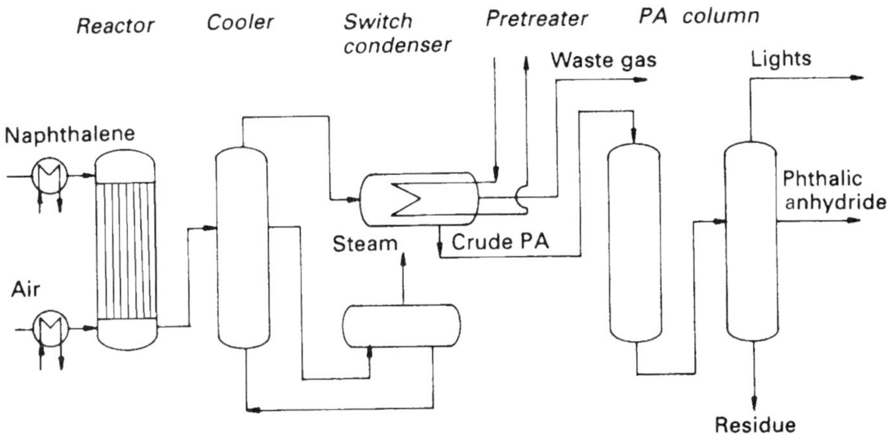

FIGURE 97 Phthalic anhydride from naphthalene

PROPERTIES

White crystalline solid or clear molten liquid, with irritating odour. Reacts with moisture to form phthalic acid.

Molecular Weight	148.1
Density at 20°C	1.527
Melting Point	131°C
Boiling Point	295°C (at 1 atm.)
	284°C sublimes
Autoignition Temperature	580°C
Flash Point Closed Cup	152°C
Flammability Limits in air	
lower	1.7vol%
upper	10.4vol%
Vapour Density (air = 1)	5.1
Exposure Limit COSHH	12mgm^3 15 minutes (Maximum Exposure Limit)
	4mgm^3 TWA 8 hour (Maximum Exposure Limit)
Lower dust explosion limit	25gm^3
Exposure Limit ACGIH	1ppm TLV-TWA

GRADES

Technical 99.5–99.8%, pure 99.9%.

INTERNATIONAL CLASSIFICATIONS

UN No.	2214
CAS Reg No.	85–44–9
EINECS No.	201–607–5
EC No.	607–009–00–4
Description	Corrosive substance
Packing Group	III
Emergency Action Code	2X
HI (Kemler Code)	80

APPLICATIONS

The largest outlet for phthalic anhydride, accounting for around 49% of demand, is for the manufacture of plasticizers (such as di-2-ethylhexyl phthalate) used in polyvinyl chloride (PVC) polymers and copolymers. Unsaturated polyester resins consume 22% of total phthalic anhydride production and alkyd resins a further 18%.

Other minor uses are in the manufacture of dyes, magnesium peroxyphthalate for detergents, phenolphthalein, fire retardants, polyester resin cross-linking agents and drying oil modifiers.

Future growth is expected to be in polyester resins, around 4–5% per year. As new plants come on-stream in the Far East, exports of phthalic anhydride from Europe is decreasing which has led to the closure of some old uneconomic plants.

HEALTH AND HANDLING

Phthalic anhydride is irritating to the eyes, skin, nose and throat and excessive exposure can lead to sensitization and asthma. Molten phthalic anhydride will cause burns in contact with skin. Goggles, rubber or vinyl gloves, and protective clothing to prevent contact must be worn when handling the product. Contact lenses should not be worn as they tend to concentrate any vapours and increase irritation to the eyes.

Store in closed containers in a cool area well away from heat, strong oxidizing agents and alkalis. Good explosion-proof ventilation is needed to prevent dust build-up which is potentially explosive.

Phthalic anhydride is stable but reacts with moisture to give an acid which can corrode metals, giving off nitrogen. Storage tanks are normally blanketed with dry inert gas to prevent corrosion and earthed to avoid static sparks when being handled. Molten phthalic anhydride must be stored at temperatures below 148.9°C under an inert gas.

Spills should be contained; small quantities of the anhydride can be neutralized with aqueous sodium bicarbonate prior to disposal according to local regulations. Carbon dioxide or dry chemicals are used to treat fires, but water sprays are employed to contain flake or dust.

MAJOR PLANTS

Plants with capacities greater than 74 000 tonnes per year:

Europhthal	Ostend	Belgium
Elf Atochem	Chauny	France
BASF	Ludwigshafen	Germany
Lonza	Scanzorosciate	Italy
BP Chemicals	Hull	UK
Aristech Chemical	Pasadena	US
BASF/Sterling Chemicals	Texas City	US
Exxon Chemical	Baton Rouge	US
Koppers	Ciccero	US
Stephan	Millsdale	US
Japan Phthalic	Mizushima	Japan
Kawasaki Kasei Chemicals	Kawasaki	Japan
Samkyung Chemical	Ulsan	South Korea
Union Petrochemical	Linyuan	Taiwan

MAJOR LICENSORS

o-Xylene	*Alusuisse Italia*
	BASF
	Huels
	Lonza
	Lurgi
	Rhone-Poulenc
	Scientific Design
	Wacker Chemie/Von Heyden
Naphthalene	*Badger*
	Lonza
	Lurgi
	Nippon Shokubai
	Rhone-Poulenc
	Scientific Design
	Wacker Chemie/Von Heyden

Polyethylene High Density (HDPE) & Polyethylene Linear Low Density (LLDPE)

$[-CH_2-]_n$

The discovery by Ziegler and Natta in the 1950s of catalysts capable of polymerizing ethylene at lower pressures and temperatures than that used previously, revolutionized the production of polyolefins. These catalysts consist of the derivatives of transition metals titanium, vanadium or zirconium with organo-aluminium compounds. The polymer formed by this new process had a more crystalline structure and higher density due to its linearity with only a few short-chain branches. In order to differentiate it from low-density polyethylene (LDPE), it was called high-density polyethylene (HDPE). About the same time, researchers in the US discovered that catalysts based on chromium or molybdenum oxides on silica or silica–alumina were also capable of polymerizing ethylene at lower temperatures and pressures.

Commercial production began in the late 1950s using all of these new catalyst systems. The exact composition of the proprietary catalysts used are kept secret by the companies concerned although the major constituents are known. The method of catalyst production determines not only their efficiency but also controls the molecular weight of the polymer formed. Hydrogen is likewise employed as a molecular weight regulator.

Since the introduction of the original Ziegler catalysts, high-activity catalysts based on magnesium dichloride-supported titanium have been developed. These permit the control of physical properties, molecular weight, stereospecificity and the degree of copolymerization.

The addition of alpha olefins during the polymerization of ethylene results in a polymer which has a density similar to that of LDPE, but with the linearity of HDPE. This polymer is called linear, low-density polyethylene (LLDPE). In the 1980s, zirconium catalysts (composed of bis-cyclopentatienyl zirconium dichloride) were developed which maintained their activity over a long period. By employing this catalyst for the production of LLDPE, resins with a narrow molecular weight range can be obtained.

Suspension (slurry), solution and gas-phase polymerization can be used for the production of HDPE. Solution, slurry or gas-phase routes are employed for LLDPE. LLDPE is made in the same way as HDPE except for the choice of comonomer and catalyst.

Suspension was the original technology used because of its flexibility. Solution processes are employed where low molecular weight polymers are required. The most recent and advanced technology, first introduced by Union Carbide and later by other companies, is gas-phase polymerization. In this process a solid catalyst is employed, the bed consisting of polyethylene granules. The reaction, carried out at low pressure, can produce a wide range of polyethylene grades in a single reactor.

The current trend is towards raw material and energy-saving processes and speciality products with high added value. In the 1960s and 1970s, slurry was the dominant process but now gas-phase processes have gained ground at the expense of other routes, and in the 1990s 70% of new capacity was gas-phase swing plants capable of producing HDPE and LLDPE. The gas-phase process produces polymers which can be used for all applications while slurry-based materials mainly find outlets in blow moulding, pipe and film. By 2005 it is estimated that the polyolefins market will be made up of 44% HDPE and 32% LLDPE.

Metallocene resins, manufactured using single-site catalysts which produce very uniform chains with narrow molecular weight distribution, are forecast to account for 10% of production by the end of the first decade of the 21st century. Their range of new properties will strongly impact the polyolefin market.

As ethylene is the biggest factor in the price of polyethylene, future plants will tend to be built in lower feedstock cost areas. In Japan, Europe and the US, mergers and alliances will continue in order to help reduce overheads and fixed costs. The two dominant companies are Union Carbide who, with its own production and that of its licensees, has 18% of the market and BP Chemicals with 9%.

Capacities range from 20 000 to 1 300 000 tonnes per year.

PROCESSES

1. From ethylene by gaseous polymerization

Polymer-grade ethylene, comonomer and hydrogen are fed via a distributor plate to ensure even gas distribution into the base of a fluid-bed reactor into which a high-activity catalyst (based on titanium and magnesium chlorides) is injected directly into the reactor. The comonomers used are 4-methyl-pentene-1, hexene, propylene, or octene, depending on the physical properties required. Polymerization takes place at 60–100°C and 22 bar pressure. Accurate temperature control is maintained by regulating the amount of catalyst and by heat transfer to prevent a runaway reaction and fusion of the catalyst particles. (*See Figure 98*)

The polymer particles formed are kept in the fluidized state by the gas stream which prevents agglomerization and ensures homogeneity. The polymer leaves

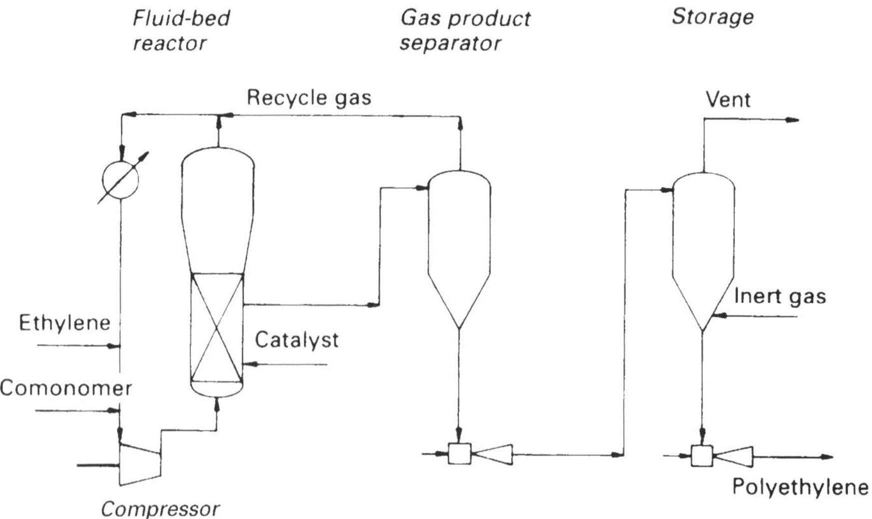

FIGURE 98 Polyethylene high density from ethylene by gaseous polymerization

the reactor in the form of a white powder and is fed into a degasser where any unreacted olefins are removed by a nitrogen gas flow. The powder is transported pneumatically to the finishing area where additives and stabilizers are incorporated prior to storage.

Unconverted overhead gases from the reactor are cooled in a heat exchanger, mixed with additional ethylene and comonomers to maintain the required composition, compressed and recycled.

The advantages of the gas-phase process are: lower temperature operation, lower operating costs, and the ability to produce a wide range of products from a single reactor.

Raw material requirements

Raw materials required per tonne of HDPE:

Ethylene and comonomer	1010kg
Nitrogen	5Nm^3

2. From ethylene by suspension (slurry) polymerization

Polymer-grade ethylene, diluent and catalyst are continuously fed into a reactor, the type being dependent on the catalyst used. Fluidized reactors are employed with chromium-based catalysts, while loop reactors are associated with Ziegler–Natta types. A low-boiling-point hydrocarbon such as hexane is frequently used as the diluent, but isobutane or isopentane are alternatives. Polymerization takes place at a temperature of 85–100°C and a pressure of 5–10 bar. High-activity

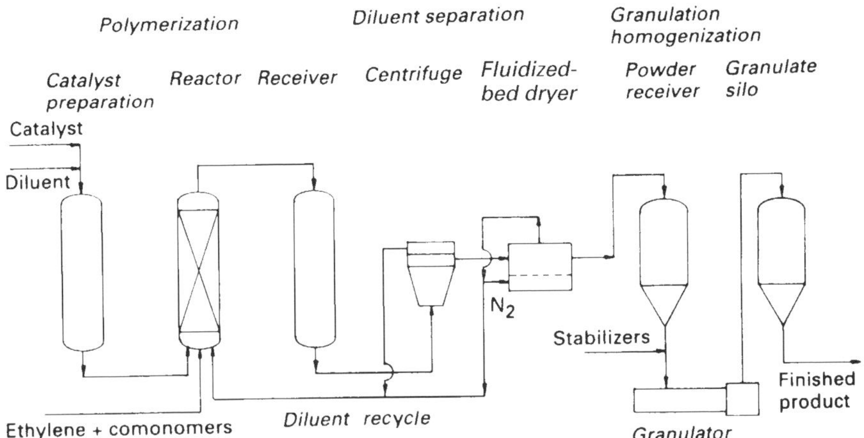

FIGURE 99 Polyethylene high density from ethylene by suspension polymerization

catalysts based on titanium or chromium are normally employed so that the catalyst removal steps can be eliminated. The slurried catalyst and polymer particles formed are circulated at high speed through the reactor to enable excess heat to be transferred efficiently by external cooling and to prevent fouling. (*See Figure 99*)

When about 60wt% of the polymer has been achieved, the suspension is withdrawn from the reactor and centrifuged to remove the diluent. As much as 90% of the diluent can be recovered and recycled without further treatment. The polymer is steam stripped to remove any residual diluent and dried with a nitrogen gas stream before being pelletized.

Although the catalyst systems used control the molecular weight of the polymer formed, the addition of hydrogen will influence the range obtained. Alpha olefin comonomers can be incorporated if desired.

Raw material requirements

Raw materials required per tonne of HDPE:

Ethylene	953kg
Comonomer	57kg
Nitrogen	30Nm^3

3. From ethylene by solution polymerization

Ethylene, dissolved in a diluent such cyclohexane, catalyst and hydrogen are fed into a reactor at a pressure of 80 bar. The catalyst, slurried with the diluent prior to entering the reactor, is fed in at the rate necessary to maintain the polymerization rate required. A range of catalyst systems can be employed including titanium tetrachloride, vanadium oxychloride and tributylaluminium, or titanium tetrachloride and triisobutylaluminium modified with ammonia, or chromium acetylaceonate and triisobutylaluminium on a silica support. The polymerization takes place at 200–300°C with a residence time of around 2 minutes. The polyethylene produced dissolves in the cyclohexane and the reaction is allowed to continue up to a polymer concentration of 35% depending on the viscosity of the mixture formed.

After leaving the reactor, unreacted monomer and solvent are flashed off the hot polymer and any residual monomer is removed in a devolatilizer. The polymer is processed as in the suspension process.

In a variation of the final stages of the solution process introduced by Du Pont, the catalyst is removed from the solution leaving the reactor by treatment with a deactivating agent. The mixture then passes through an alumina bed where deactivated catalyst residues are absorbed prior to the volatilization of diluent and unreacted monomer.

The advantages of the solution process are that a wide range of comonomer types and product densities can be produced depending on the catalyst used.

OTHER PROCESSES

The latest innovation is the introduction of metallocene single-site catalysts. They provide a method of introducing long-chain branching into polymer chains, thereby reducing the molecular weight distribution, giving increased clarity and toughness and improved film products processability.

The catalysts can be based on metallic elements from Group IV and V, such as zirconium, vanadium and titanium, or Group VIII, for example, nickel and palladium. The most commonly investigated metallocene catalysts are those based on biscyclopentadienyl. Although more expensive than conventional catalysts, metallocenes are more reactive. Effective removal of the heat generated in the reactor is achieved by liquid hydrocarbon injected directly into the fluid bed. This also helps to disperse the catalyst within the bed.

The process was first introduced by Dow Chemical in a converted LLDPE plant in Spain. Several companies have collaborated in order to reduce research and development costs.

PROPERTIES

Colourless, translucent solid with a dull surface. Very resistant to weak acids, alkalis, and inorganic chemicals.

	HDPE	*LLDPE*
Density at 20°C	0.94–0.97	0.92–0.94
Vicat Softening Point	127°C	101°C
Melt Index g/10 minutes	0.1–3.5	0.13–1.2
Tensile Strength MPa	26.5	10–11
Tensile Elongation at rupture %	906	811
Hardness (Rockwell D)	60–70	50–60

GRADES

HDPE	Homopolymers, copolymers.
LLDPE	Copolymers.

INTERNATIONAL CLASSIFICATIONS

UN No.	Not allocated
CAS Reg No.	9002–88–4

APPLICATIONS

The major outlet for HDPE is in blow moulding which accounts for over 27% of total world demand for the polymer. Included in this outlet are milk bottles, packaging containers, drums, fuel tanks for automobiles, toys and housewares. Film and sheet, which consume a further 20% of production, are used in a wide range of applications including wrapping, refuse sacks, carrier bags and industrial liners.

The wide range of articles made by injection moulding includes crates, pallets, packaging containers and caps, paint cans, housewares and toys which together require about another 18% of total demand. Extruded pipe for water, gas and irrigation, conduit, wire coating and cable insulation require around 7% of HDPE production.

The largest outlet for LLDPE, consuming 77% of total demand, is in film applications where it is replacing LDPE. Injection moulding and wire and cable account for the remainder.

World growth rate for HDPE to 2005 is expected to be around 5.5–6.0% per year, while LLDPE is forecast to grow at 9.5% per year during the same period. Metallocene-based resins will have a large impact on the polyolefins market especially if they can compete on price. The biggest effect will be demonstrated by LLDPE as it continues its substitution of LDPE. Around 10% of polyethylenes are expected to be metallocene based by the middle of the first decade of the 21st century. In order to get access to Far Eastern markets, further alliances between the large US and European companies and Asian producers can be expected.

HEALTH AND HANDLING

HDPE dust can cause eye and skin irritation. Personnel handling the product should wear protective clothing, boots, goggles and gloves to prevent contact. Contact lenses should be avoided. HDPE is listed as a possible carcinogen.

Store in a cool, well-ventilated area away from strong oxidizing agents. The product can be disposed of by landfill or incineration. Recycling is becoming more important as the cost of waste disposal increases. Polyethylenes can be incinerated and the heat recovered.

LLDPE should not be exposed to flames as it gives off dense smoke on burning. Carbon dioxide or water fog should be used to contain and extinguish fires. Flammable mixtures can be formed in air during a fire. If fillers, additives or colorants are present then their effects must be examined separately, and the manufacturers should be consulted for guidance. Firefighters must wear protective clothing and breathing apparatus.

MAJOR PLANTS

HDPE

Plants with capacities greater than 250 000 tonnes per year:

Fina Oil & Chemicals	Antwerp	Belgium
Elanac	Wesseling	Germany
DSM	Geleen	Netherlands
Repsol Quimica	Puertollano	Spain
BP Chemicals	Grangemouth	UK
Chevron Chemicals	Orange	US
Eastman Chemical	Longview	US
Exxon Chemical	Mont Belvieu	US
Formosa Plastics	Point Comfort	US
Lyondell	Bay City	US
Phillips Petroleum	Pasadena	US
Solvay	Deer Park	US
Ipiranga Petroquimica	Triunfo	Brazil
Korea Petrochemical	Ulsan	South Korea

Swing plants (HDPE & LLDPE)

Plants with capacities greater than 250 000 tonnes per year:

Dow Chemical	Terneuzen	Netherlands
Orgsyntez	Kazan	Russia
Chevron Chemicals	Orange	US
Dow Chemical	Freeport	US
	Plaquemine	US
Millenium Petrochemicals	Deer Park	US
	Morris	US
Mobil Oil	Beaumont	US
Union Carbide	Seadrift	US
	Taft	US
Dow Canada	Fort Saskatchewan	Canada
Imperial Oil (Exxon)	Sarnia	Canada
NOVA Chemicals	Joffre	Canada
Petromont	Montreal East	Canada
Eastern Color & Chemical	Al Jubail	Saudi Arabia
Saudi Yanpet	Yanbu	Saudi Arabia
Sharq (Sabic/Mitsubishi)	Al Jubail	Saudi Arabia
PT Peri	Merok	Java, Indonesia

MAJOR LICENSORS

HDPE

Amoco
Asahi Chemical
BP Chemicals
Chisso
Dow Chemical
Du Pont
Hoechst
Huels
Mitsubishi Chemical
Mitsui Petrochemical
Nissan Chemical
Phillips Petroleum
Snamprogetti
Solvay
Stamicarbon
Union Carbide
Veba Chemie

LLDPE

BP Chemicals
CDTECH
Dow Chemical
Du Pont
EniChem
Exxon
Himont
Mitsui Petrochemical
NOVA Chemical
Phillips Petroleum
Showa Denko
Stamicarbon
Union Carbide

Polyethylene Low Density (LDPE)

$[-CH_2-]_n$

Low-density polyethylene (LDPE) was first produced in the laboratories of ICI; due to its versatility, it is now one of the major polymers in the world. Large-scale production, which began in the 1950s, is carried out in a tubular or stirred autoclave reactor using a free radical initiator and chain transfer agent. A wide range of initiator systems can be used, most of the commercial ones being proprietary. They consist of one or two peroxides with or without oxygen which operate at low or high temperatures.

The large amount of heat created by the reaction is removed by external cooling jackets or, in the case of a stirred autoclave, by cold ethylene feed absorption. The heat is used to generate low-pressure steam. Because temperature control is poor, conversion rates obtained from the autoclave processes are generally lower than those gained by the tubular route. However, savings can be made on operating and equipment costs due to the lower pressures employed. In the US, around 53% of LDPE is produced by the autoclave process compared to 47% by the tubular route. Many producers use both routes.

Recent emphasis has been on raw material and energy-saving processes and this trend can be seen in the move from high-pressure LDPE to low-pressure linear low-density polyethylene (LLDPE). With the production of competitively priced LLDPE in Saudi Arabia due to lower-cost ethylene, this polymer offers a cost-performance advantage over LDPE which it is replacing, especially in films.

The advent of metallocene LLDPE resins has helped to increase their penetration into the LDPE market. Olefins which had been used to produce LDPE are being diverted to LLDPE production. This trend is expected to continue (see Polyethylene High Density).

Increasing use is being made of LDPE grades containing either higher olefin comonomers such as hexene and octene or polar comonomers which extend the range of properties. Examples of these comonomers are vinyl acetate, butyl acetate and acrylic acid.

Capacities range from 30 000–570 000 tonnes per year.

PROCESSES

1. From ethylene by polymerization in a tubular reactor

Polymer-grade ethylene is dried and compressed in stages, firstly in a precompressor to 250–300 bar and then secondly in a hypercompressor up to 3500 bar. Recycle ethylene and oxygen are added before the hypercompressor stage. Peroxide initiators, usually at a concentration level of 0.5–1g per kg of polyethylene, are injected into the feedstock just prior to entry into the reactor. The most popular initiator systems consist of a mixture of low-temperature peroxides, such as *tert*-butyl or amyl peroxypivalate, with a high-temperature peroxide like di-*tert*-butyl hydroperoxide. Alternatives include di(3,5,5-trimethylhexanoyl) peroxide, *tert*-butyl peroxybenzoate or *tert*-butyl peroxy-2-ethyl hexanoate. (*See Figure 100*)

The compressed ethylene is heated and introduced into a tubular reactor which consists of a number of sections in the form of a long coil. Polymerization is initiated at 140–180°C, depending on the initiator used, the temperature rising rapidly to 300°C as the reaction proceeds. In multi-zone tubular reactors, the reaction mixture is cooled slightly, more peroxide is added and the polymerization is reinitiated. Conversions of up to 38% can be achieved, which is higher than that obtained in an autoclave reactor.

In order to obtain the high velocity required for effective reaction-heat transfer to the external cooling jacket, the exit pressure is intermittently reduced. This action also helps to control the reaction, reduces reactor fouling and improves the polymer properties.

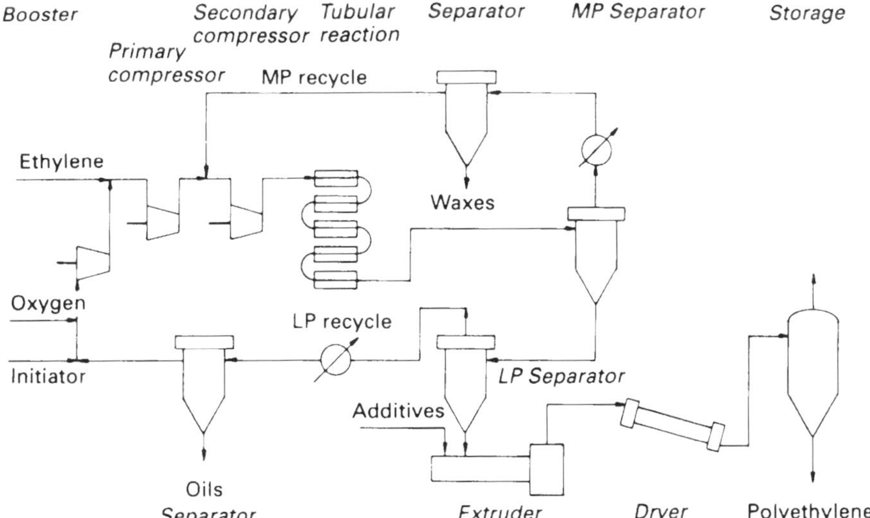

FIGURE 100 Polyethylene low density from ethylene by polymerization in a tubular reactor

Molten polymer is taken from the reactor and passes to a separator where any unreacted monomer is removed by degassing at two separate pressure stages. The molten polymer is fed to an extruder and pelletized under water. The resin is dried and sent to storage.

The physical properties of the polymer formed can be controlled by initiator selection, its concentration and the volume of comonomer added. In order to tailor-make products with a distinct molecular weight distribution and short- and long-chain branching, it is necessary to optimize the reactor design and reaction conditions to meet the required criteria.

Although oxygen can be used as the primary initiator in tubular reactors, newer processes tend to employ organic peroxides instead. *tert*-Amyl peresters or per-neodecanoates are becoming increasingly important due to their higher activity which allows lower temperatures to be used.

Raw material requirements

Raw materials required per tonne of LDPE:

Ethylene	1005kg

Conversion up to 38%.

2. From ethylene by polymerization in a stirred autoclave

These processes utilize either a single-stage autoclave with back mixing or a multi-stage autoclave in which case no mixing is used between the stages.

Purified ethylene is fed at low temperature directly into the autoclave containing the hot reaction mixture. The temperature of the gas is raised while that of the reaction mixture is lowered. By regulating the rate of flow of the incoming ethylene, the rate of polymerization can be controlled. A peroxide, such as di(3,5,5-trimethylhexanoyl) peroxide, is added and the mixture is stirred mechanically. In autoclave reactors, initiator levels of 1–2g per kg of polyethylene are used. The reaction takes place at a temperature of 200–300°C and at pressures around 1700 bar.

The polymer formed separates out and is removed and degassed as for the tubular process. Because temperature control is poorer, conversion rates obtained from the autoclave processes are generally lower than those gained by the tubular route. However, savings can be made on operating and equipment costs due to the lower pressures employed.

Raw material requirements

Raw materials required per tonne of LDPE:

Ethylene	1010kg

PROPERTIES

Translucent solid with a waxy surface. Resistant to weak acids and alkalis.

Molecular Weight	90,000
Density at 20°C	0.92–0.93
Vicat Softening Point	93°C
Melt Index g/10 minutes	0.2–3.5
Tensile Strength MPa	12.4
Tensile Elongation at rupture %	653
Hardness (Rockwell D)	41–46

GRADES

Homopolymer.

INTERNATIONAL CLASSIFICATIONS

UN No.	Not allocated
CAS Reg No.	9002–88–4

APPLICATIONS

Film is the largest application for LDPE, consuming 50% of total world demand. It is used for food packaging, industrial liners, pallet and shrink wrapping, overwrapping, heavy-duty sacks and bags. Non-packaging outlets for film include household wrapping, refuse bags, bin liners, storage bags, industrial sheeting, agricultural film and backing for disposable nappies.

The second important outlet is sheathing for electrical and communications cables which takes 10% of LDPE production.

The third important application consuming 8% of LDPE output is extrusion of which the extrusion coating of paper and board is the most important. These products are used for the packaging of liquids like milk, fruit juices and soft drinks, and in moisture barrier applications. Foil is also coated with LDPE in multi-layer structures as it can be heat sealed.

Another large segment is injection moulding for the production of toys, housewares, lids, caps and closures of all kinds. Around 4% of total demand goes into this segment, and a further 2% is consumed by blow moulding.

LLDPE is replacing LDPE in many of its traditional film outlets. As metallocene polymers become more widely available at competitive prices, LLDPE will play a bigger role in the market, substituting much of the traditional LDPE. However, there are some LDPE applications for which LLDPE resin properties are unsuitable.

World growth is expected be around 1% by 2002. By 2005, LLDPE is expected to have 24% of the global polyolefins market.

HEALTH AND HANDLING

Dust can cause irritation to the nose and throat. In a confined space, good explosion-proof ventilation must be provided to remove dust, and personnel should wear dust masks to prevent inhalation as well as eye protection. Contact lenses should be avoided. LDPE is a potential carcinogen. Store at room temperature in a cool, dry area.

The product can be disposed of by landfill or incineration. Recycling is becoming more important as the cost of waste disposal increases. LDPE should not be exposed to flames as it gives off smoke on burning. Carbon dioxide or water fog should be used to extinguish fires. If fillers, additives or colorants are present then their effects must be examined separately, and the manufacturers should be consulted for guidance.

MAJOR PLANTS

Plants with capacities greater than 200 000 tonnes per year:

Petrochemie Danubia	Schwechat	Austria
Exxon Chemical	Antwerp	Belgium
	Meerhout	Belgium
Elf Atochem	Carling	France
Elanco	Wesseling	Germany
Erdoelchemie	Cologne	Germany
Dow Chemical	Terneuzen	Netherlands
DSM	Geleen	Netherlands
Polimeri Europa	Gela	Italy
Repsol Quimica	Puertollano	Spain
Borealis	Porvoo	Finland
Azot	Severodonezk	Russia
	Sumgait	Russia
Chevron Chemicals	Cedar Bayou	US
Dow Chemical	Freeport	US
Du Pont	Orange	US
Eastman Chemical	Longview	US

Exxon Chemical	Baton Rouge	US
Lyondell	Clinton	US
	Deer Park	US
	Morris	US
Mobil Oil	Beaumont	US
Union Carbide	Seadrift	US
Westlake Polymers	Lake Charles	US
Pemex	La Grangrejera	Mexico
Polyolefin Co.	Palau Ayer	Singapore

MAJOR LICENSORS

Arco Technology
Dow Chemical
Du Pont
Elf Atochem
EniChem
Exxon
Gulf Oil Chemicals
ICI/BASF
Imhausen
Mitsui Petrochemical
Snamprogetti
Stamicarbon
Sumitomo Chemical
Union Carbide

Polypropylene (PP)

$$\left[-CH_2-\underset{\displaystyle CH_3}{\underset{|}{CH}}- \right]_n$$

Polypropylene (PP) was first produced commercially in the 1950s following the discovery of Ziegler–Natta catalysts based on titanium chlorides and organo-aluminium compounds.

Polypropylene can exist in three stereoisomeric forms:

- isotactic — the methyl groups are attached in the same plane;
- syndiotactic — the methyl groups are alternately distributed in the same plane;
- atactic — the methyl groups are randomly distributed.

Isotactic polymers have the high crystallinity and stereoregularity required for plastics and fibre processing. For this reason the amount of atactic production is usually kept below 5%. It is removed from isotactic polymers by dissolving the atactic material in a hydrocarbon or chlorinated hydrocarbon solvent and separating the slurry by centrifugation. The incorporation of small amounts of other monomers, such as ethylene, to extend the properties of polypropylene, is being increasingly employed.

Polymer processes used for the production of polypropylene can be divided into three main types:

- slurry or suspension;
- bulk;
- gas phase.

Traditionally, polypropylene was made by slurry polymerization using an aluminium alkyl and titanium trioxide catalyst in a hydrocarbon diluent. The process suffered from two major defects; firstly, the catalyst had to be treated with alcohol to deactivate and extract it; secondly, unwanted atactic polymer had to be extracted and removed. Concentrations of up to 300ppm of catalyst could still be entrained in the final polymer.

Bulk technologies were developed in which the hydrocarbon diluent was replaced by liquid propylene. The polymer is continuously withdrawn from the solution and any unreacted monomer is flashed off.

In the 1970s, considerable effort was spent in improving the process and high-activity catalysts (consisting of titanium trichloride on magnesium dichloride with modified aluminium alkyl activators) were developed. Their performance and stereospecificity has made catalyst and atactic polymer removal unnecessary. Moreover, these catalysts enabled the production of differentiated polypropylene grades with specific application properties. They also permitted the introduction of gas-phase stirred or fluidized-bed processes operating with lower purity propylene – 95% chemical grade instead of the 99.5% polymerization grade – which has considerably improved the economics of the polymerization process. The reduction in the amount of catalyst lost due to entrainment in the polymer, together with the removal of several extraction steps, have substantially reduced production costs too. Another advantage is that polypropylene is recovered as a dry powder. Himont's Spheripol, Union Carbide/Shell's Unipol and BASF's Novolen processes dominate current world technology.

Random copolymers produced from propylene and ethylene can be easily produced by the gas-phase polymerization process because the presence of ethylene increases the rate of reaction.

Metallocenes, single-site catalysts which produce uniform polymer chains with a narrow molecular weight distribution, are a major new development. Although more expensive than conventional catalysts, metallocenes open up the possibility of tailoring polymer properties which could generate new markets for polypropylene. BASF is investigating their use to make impact- and fibre-grade material.

The emergence of these new processes has led to the rapid growth in polypropylene production in recent years and the replacement of old plants by those employing the new generation of catalysts. Although slurry processes dominated total installed polypropylene capacity, this is decreasing due to the growing importance of the gas-phase route.

Capacities range from 10 000 to 720 000 tonnes per year. Montell (Shell) is the world's largest producer of polypropylene.

PROCESSES

1. From propylene by gas-phase polymerization

Liquid propylene is passed through a molecular sieve dryer to remove any moisture and impurities before being pumped into a reactor. The catalyst, consisting of a suspension of titanium chloride and diethylaluminium chloride in a solvent, is added. The polymerization is carried out at 70–90°C and a pressure of 25–35 bar with a catalyst concentration of 200ppm with respect to the feed. Evaporation of the liquid propylene cools the reactor and helps to maintain the reaction temperature as well as aerating the stirred catalyst bed. Alternatively a fluidized bed can be used.

The gases from the reactor are cooled externally in a heat exchanger, liquefied and recycled. The molecular weight of the polymer formed is controlled by the continuous addition of hydrogen, and ethylene can be added if random polymers are required. A mixture of powdered polymer and propylene gas is carried via a tube in the reactor to a cyclone where the powder is separated. The powder is purged with nitrogen to remove any residual propylene, which can be recovered or burnt, before being sent to a storage silo.

If high-impact polymers are required, the powder is fed into a second reactor where further polymerization takes place with some of the propylene and ethylene feed. The reactor temperature is 50–70°C with a pressure of 11–25 bar. The pressure in the second reactor is always lower, so that powder is transferred from the first by the pressure difference between them. Sufficient catalyst is present to initiate the reaction. A separate hydrogen feed is used to control the molecular weight of the copolymer. The polymer powder is blended with additives and fed to an extruder. The extruded polymers are cut into pellets, dried and steam stripped to remove any residual solvent before bagging.

The major advantages of the gas-phase process are: firstly, dry polymer is produced and the new stereospecific catalysts give higher yields and greater selectivity without the need for their removal; secondly, the molecular weight of the polymer produced and its comonomer content can be controlled easily to give a wide range of molecular weight materials; thirdly, the product composition can be changed rapidly by altering the gas make-up to the reactor.

Raw material requirements

Raw materials required per tonne of polypropylene:

	Homopolymers	*Impact polymers*
Propylene	1020kg	920kg
Ethylene	60kg	100kg

2. From propylene by slurry polymerization

Propylene, of 99.5% purity, is continuously fed into a polymerization reactor containing catalyst and a diluent. Copolymer ethylene can also be present. The catalyst consists of aluminium alkyls with titanium trichloride in solution. Hexane or heptane can be used as the diluent. In a few instances, liquid propylene is used instead. (*See Figure 101*)

The reaction takes place in the liquid phase at a temperature of 65–70°C and a pressure of 5–30 bar. The polymer particles are suspended in the diluent forming a slurry. Any atactic polymer formed dissolves in the diluent. Further diluent may be added as the polymer concentration increases, to prevent fouling.

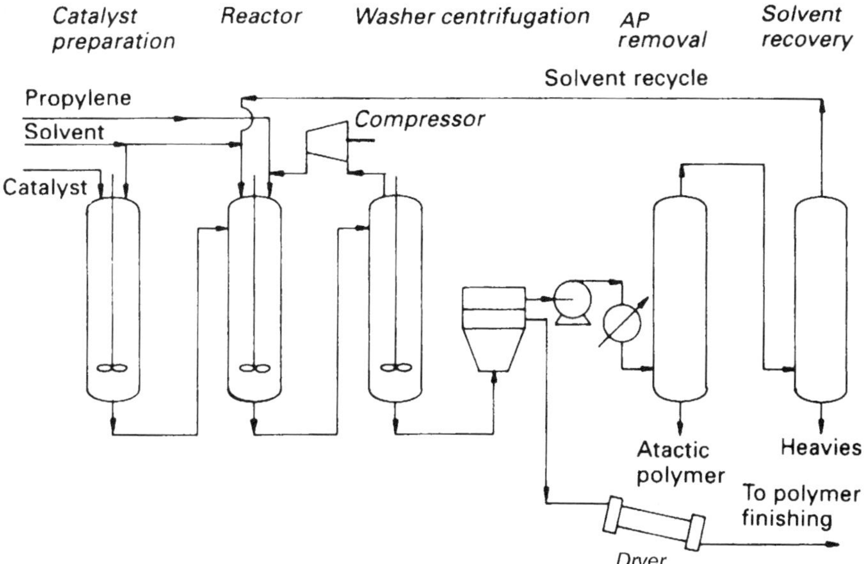

FIGURE 101 Polypropylene from propylene by slurry polymerization

At the end of the reaction, the mixture is passed to a heated tank where the pressure is released and unreacted propylene is flashed off, cooled and recycled. The product stream, consisting of a polymer slurry, is washed with ethyl alcohol or water to deactivate and remove the catalyst and unwanted atactic polymer. After centrifuging to remove the solvent, polypropylene in the form of fine granules is washed with acetone or isopropyl alcohol to remove all solubilized catalyst. The polypropylene leaving the dryer at 100–120°C can, after stabilization, be either sold as a powder or fed into a pelletizing extruder and converted to granules.

In some processes, the polymer is separated from the catalyst residues by extraction which avoids the need for centrifuges. The use of high-activity catalysts, which give low levels of atactic polymer, eliminates the need for the extraction step.

Raw material requirements

Raw materials required per tonne of polypropylene:

Propylene	1035kg

OTHER PROCESSES

In the bulk process, liquid propylene is used as the diluent instead of an inert liquid using the same catalyst system as that employed in the suspension route. The polymer slurry formed is withdrawn from the tubular reactor and any unreacted monomer is flashed off. Catalyst deactivation and subsequent processing are carried out as for the slurry process.

PROPERTIES

Colourless, translucent to transparent solid with a glossy surface. Very good resistance to acids, alkalis, and inorganic chemicals. Except for hydrocarbons and chlorinated compounds, polypropylene has a good resistance to organics.

Density at 20°C	0.88–0.91
Vicat Softening Point	154°C
Melting Temperature	170–172°C
Tensile Strength MPa	29.3–38.6
Tensile Elongation at rupture %	20–600
Impact Strength (notched Izod)	21

GRADES

Homopolymers, random copolymers, medium- and high-impact copolymers.

INTERNATIONAL CLASSIFICATIONS

UN No.	Not allocated
CAS Reg No.	9003–07–0

APPLICATIONS

Polypropylene has improved impact strength, a higher softening point, lower density, better stress cracking and more scratch resistance than the polyolefins. The one disadvantage is its brittleness below 0°C.

Injection moulding is the most important outlet for polypropylene accounting for nearly half of total demand. Moulded articles used in packaging consume 10% of polypropylene usage; electronics and small and large electrical appliances take 11% and furniture a further 6%. Other injection moulded outlets include housewares, toys, and luggage. Due to its strength and lightness, around 11% of polypropylene is being used in motor vehicles for battery cases, ducting, interior trim, heating and air-conditioning equipment.

The polymer can be made into fibres and its combined strength and resistance has led to the replacement of sisal and jute in ropes, twine and string. It can be made into film and sheet. Film tape derived from polypropylene film is used in carpet backing and woven sacks. Around 26% of polypropylene is used for the manufacture of fibre and filament, and 19% goes into film and sheet.

Polypropylene can be extruded and used for such outlets as pipe and conduit, wire and cable. Another important area of packaging where polypropylene is used, is for containers and closures. Containers made by blow moulding are used extensively in medical and consumer outlets. Atactic polypropylene finds outlets in paper laminating, sealants and adhesives.

Since the early 1990s, over 50% of polypropylene's growth has come by substitution of polyvinyl chloride (PVC), styrenics and polyamides, 13% by replacement of conventional materials such as paper, glass and metal, and 3% from new applications. The introduction of metallocene catalyst-based polymers will open up new opportunities, especially for film and fibre applications, without changing the bulk of the polypropylene market.

A world growth rate of 6% for polypropylene is predicted for the period 1997–2000. Its high growth is due to polypropylene's versatility and the lower processing costs of the newest processes.

HEALTH AND HANDLING

With normal handling, polypropylene does not present any risk to skin. Good ventilation is required to keep dust levels to a minimum especially in confined areas. Staff should wear dust masks and eye protection to prevent irritation. Contact lenses are not advised. Polypropylene is a potential carcinogen. Handling equipment should be spark-proof and earthed to prevent the risk of explosion.

Waste polymer can be disposed of by landfill or by incineration. The polymer should not be exposed to flames as it gives off smoke on burning. Water spray should be used to extinguish fires. If fillers, additives or colorants are present then their effects must be examined separately. Manufacturers should be consulted for guidance.

MAJOR PLANTS

Plants with capacities greater than 300 000 tonnes per year:

Amoco	Geel	Belgium
Borealis	Beringen	Belgium
PetroFina	Feluy	Belgium
ROW	Wesseling	Germany
DSM	Limbourg	Netherlands
Montell	Brindisi	Italy

Amoco	Chocolate Bayou	US
Epsilon Products	Marcus Hook	US
Exxon Chemical	Baytown	US
Fina Oil & Chemicals	La Porte	US
Huntsman Chemical	Longview	US
Montell	Bayport	US
	Lake Charles	US
Phillips Petroleum	Pasadena	US
Solvay	Deer Park	US
OPP	Paulmau	Brazil
Daelim Industrial	Yeochon	South Korea
Korea Petrochemical	Ulsan	South Korea
Tri Polyta	Cilegan	Indonesia

MAJOR LICENSORS

Amoco
BASF
Borealis
Eastman Chemical
Elf Atochem
El Paso
EniChem
Exxon
Hercules
Himont
Hoechst
Mitsubishi Chemical
Mitsui Petrochemical
Montell Technology
Phillips Petroleum
Solvay
Sumitomo Chemical
Tokuyama Soda
Uhde
Union Carbide

Polystyrene & Expanded Polystyrene

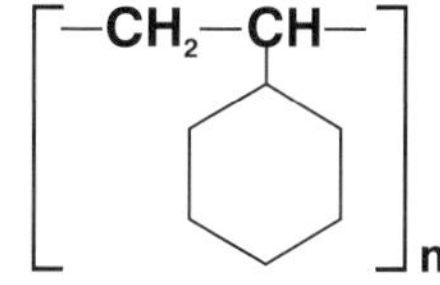

SYNONYMS

POLYSTYRENE PS

EXPANDED POLYSTYRENE expanded beads EPS

Commercial production of polystyrene began in the 1930s, with capacity increasing rapidly during World War II to supply plants producing synthetic rubber. The ready availability of styrene has ensured its continued growth.

Although styrene will polymerize spontaneously on heating in an oxygen-free atmosphere, catalysts are added in order to obtain complete polymerization at lower temperatures. These can be cationic or anionic compounds, Ziegler or free radicals such as peroxide or azo compounds. Free radicals are the catalysts of major commercial importance. Difunctional radical initiators are employed to reduce the polymerization time and give more effective molecular weight control. High polymerization rates can be achieved with anionic catalysts such as organo-metallic compounds.

Emulsion, aqueous suspension, solution and bulk polymerization processes have been developed to aid heat transfer from the exothermic reaction, which if uncontrolled leads to the formation of low molecular weight polymers. Each process produces different types of polymers. The two main types are crystal, a clear amorphous resin with good stiffness and electrical properties, and impact, containing varying levels of polybutadiene, which improves its toughness and impact resistance.

The main advantages of the mass process are the clarity and excellent colour of the resins produced compared to the suspension process which can easily yield polymers of different molecular weights but lacking the clarity.

Expanded or foam polystyrene (EPS) is a rigid cellular form of the polymer. It is manufactured by adding an expanding agent such as iso- or *n*-pentane to styrene prior to carrying out the suspension polymerization. Vaporization of the pentene causes the beads to expand and produce the cellular form. Stabilizers are added to the reaction mixture to produce beads of a larger size than those normally obtained which will retain the expanding agent.

Idemitsu and Dow Chemical have researched a process for the production of a syndiotactic polystyrene obtained by catalysis with metallocenes. Unlike conventional polystyrene which has an atactic structure, these materials are engineering plastics which melt at 270°C and have a high density of 1.27.

Continuous solution and suspension processes are most widely used for the production of polystyrene. The advantages of the solution route are low residual monomer content and high-purity polymers. However, it does suffer from heat dispersal problems because of the high viscosity of the resultant solution. The suspension route is popular for making specialist grades of crystal and high-impact grades of polystyrene.

Capacities range from 40 000 to 315 000 tonnes per year.

PROCESSES

1. From styrene by suspension polymerization

Styrene and demineralized water in a molar ratio of 1:1–2 are fed into a jacketed reactor fitted with a stirrer. An initiator such as benzoyl peroxide is added. After agitation to disperse the styrene, suspension agents such as polyvinyl alcohol or celluloses are fed in, in a concentration of 0.5wt% with respect to styrene, to stabilize the aqueous phase. The mixture is heated to 80–90°C to start the reaction. Excess heat generated by the reaction is removed by the aqueous phase in order to keep the temperature below 140°C.

The suspending agent and continuous agitation disperse the monomer to form beads. Pentene, butane or other blowing agents are introduced into the reactor and the polymerization continued. Additional initiator can be added towards the end of the polymerization to increase the reaction rate in the final stages. (*See Figure 102*)

After cooling, the slurry of polymer beads formed is discharged into a tank where any remaining unreacted monomer is flashed off. After the addition of acid to destroy the stabilizer, the slurry is washed, centrifuged and dried. The beads are screened before blending with lubricants.

Raw material requirements

Raw materials required per tonne of expanded polystyrene:

Styrene	1005kg
Demineralized water	1000–2000kg
Pentane	1000kg

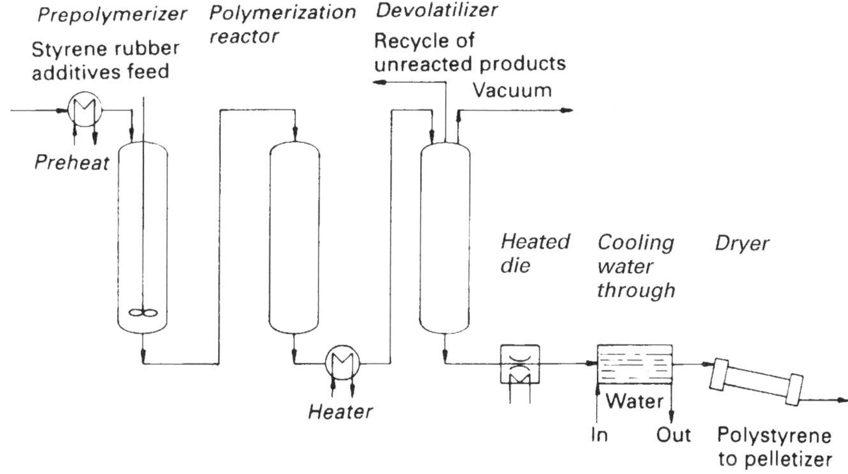

FIGURE 102 Polystyrene from styrene by suspension polymerization

2. From styrene by solution polymerization

Solution polymerization, which can be carried out either continuously or in batches, is conducted in diluent such as ethylbenzene. This enables the polymerization to be controlled, which is necessary when anionic initiators are used because the rate of reaction is so rapid that without a diluent an explosion could result.

Styrene and ethylbenzene are fed continuously into the first of three stirred tank reactors used sequentially and the initiator is introduced. The polymerization takes place at 90–170°C at a pressure of 0.5–2 bar with a conversion of 60–90%. Heat given out by the reaction is removed by a cooling agent which circulates around the reactors. Unconverted monomer and the diluent are vaporized under vacuum. After further degassing, the polymer melt leaving the reactor can be mixed with stabilizers and dyes before being extruded under water. The styrene–solvent mixture is condensed and recycled to the reactor, a small volume normally being removed to prevent the build-up of contaminants.

Raw material requirements

Raw materials required per tonne of polystyrene:

Styrene	1010kg

3. From styrene by mass polymerization

Styrene and a diluent are preheated before being pumped into a prepolymerizer. When 30–40% conversion has taken place, the viscous mass is fed into a polymerization tower. Heat is generated by the reaction and the temperature is controlled by means of internal cooling coils. Liquid polystyrene removed from the base of the tower is passed to a devolatilizer where any unconverted styrene is removed under vacuum and recycled. The liquid mass is extruded through a water bath into a pelletizer. The pellets are sent to bulk storage. Mass polymerization is used for copolymerization of styrene with other monomers.

Raw material requirements

Raw materials required per tonne of polystyrene:

Styrene	1015kg

OTHER PROCESSES

Emulsion polymerization is carried out using a water-soluble initiator, with stabilization effected by the use of a detergent. The polymerization begins at a soap micelle and forms polystyrene with high molecular weights. These tend to have poorer clarity than polymers made by other routes.

PROPERTIES

Clear, crystalline resin. Soluble in cyclohexane, ethylbenzene, ethyl acetate and carbon disulphide. Insoluble in ether, acetone, phenol and saturated hydrocarbons. Burns with a sooty flame.

Density at 20°C	1.05
Vicat Softening Point	96°C
Tensile Strength MPa	42
Elongation at rupture %	1.8
Modulus MPa	3170
Impact Strength (notched Izod)	21

GRADES

General-purpose, high-impact, expanded.

INTERNATIONAL CLASSIFICATIONS

UN No.	2211 (polystyrene beads expanded)
CAS Reg No.	9003–53–6

APPLICATIONS

The largest proportion of world polystyrene output goes into packaging which accounts for around 31% of total demand. Included in packaging are food and dairy containers, closures, lids, tumblers, glasses, yogurt cups, produce baskets, egg trays, vending cups and fast food containers. Durable goods, including housewares and furniture made of polystyrene, take another 19%.

Polystyrene is used in a wide range of appliances such as refrigerators, freezers, air conditioners and small electrical goods, and these outlets consume 8% of polystyrene production. A further 11% goes into audio-visual outlets, especially radios, TV cabinets, video cassette holders and compact disks.

Approximately 20% goes to produce expanded polystyrene which finds outlets in building, construction and insulation while most of the remainder is used in packaging.

Polystyrene is a mature product and its consumption tends to fluctuate with the level of economic activity. The growth rate is expected to be around 3.6% per year until 2000.

HEALTH AND HANDLING

Polystyrene powder can cause irritation by inhalation and skin contact. Personnel handling the powder should wear goggles and dust masks. Store in a cool, dry, well-ventilated place away from strong oxidizing agents. Spills should be collected, care being taken not to generate dust, and placed in containers for disposal according to local regulations.

Polystyrene depolymerizes when heated above 300°C and burns with a smoky flame. It may form flammable and explosive mixtures in air. Carbon dioxide or water can be used to extinguish fires. Firefighting staff must wear protective clothing and breathing apparatus.

MAJOR PLANTS

Plants with capacities greater than 160 000 tonnes per year.

BASF	Antwerp	Belgium
Dow Chemical	Tessenderloo	Belgium
BP Chemicals	Wingles	France
BASF	Ludwigshafen	Germany
BP Chemicals	Marl	Germany
Bredase (Shell)	Breda	Netherlands

EniChem/BP Chemicals	Mantua	Italy
Complex	Shevchenko	Russia
BASF	Joilet	US
Chevron Chemicals	Marietta	US
Dow Chemical	Midland	US
Fina Oil & Chemicals	Carville	US
Huntsman Chemical	Belpre	US
	Chesapeake	US
Nova Chemicals	Monaca	US
Dow Chemical	Sarnia	Canada
Asahi Chemical	Sodegaura	Japan
Denki Kagaku	Chiba	Japan
Mitsubishi Chemical	Yokkaichi	Japan
Chi Mei Industrial	Tainan	Taiwan
Dow Chemical	Hong Kong	Hong Kong

MAJOR LICENSORS

ABB Lummus Global
Badger
BASF
Cosden Technology
Dow Chemical
Elf Atochem
EniChem
Fina Technology
Hoechst/Uhde
Huels
Huntsman Chemical
Idemitsu Petrochemical
Mitsui Petrochemical
Petrocarbon Developments
Toyo Engineering

Polyvinyl Chloride (PVC)

$\left[-CH_2-CH- \right]_n$

SYNONYMS

POLYVINYL CHLORIDE PVC, vinyl chloride homopolymer, chloroethylene polymer

After the polyolefins, polyvinyl chloride (PVC) is the most widely used of the thermoplastic polymers. There are two main types of homopolymer available: rigid or unplasticized; and flexible which contains plasticizer. Its chains are mainly atactic with a small amount of chain branching.

Polymerization processes used for the production of PVC are:

- suspension;
- emulsion;
- mass or bulk;
- solution.

All employ free radical initiators which are either soluble in the monomer (as in suspension and mass polymerization) or in the aqueous solution (as in the emulsion route). The greatest proportion of capacity is based on the suspension route because these resins are the most versatile and suitable for a wide range of applications. Mass polymers can also be employed for a multitude of uses. Emulsion-based resins are better suited for pastes while solution-produced resins tend to be limited to surface coatings.

Because of PVC's rigidity and limited heat stability, additives are incorporated into the resin prior to processing. They can be blended with the polymer which is then heated to 100°C followed by cooling, or compounded by passing the polymer and additives through heated rollers which knead the resultant mass. Plasticizers are added to make the polymer more flexible and easier to process at lower temperatures. Phthalate esters are most commonly used in the ratio of 40–60 parts of plasticizer to 100 parts of resin. The monomer, vinyl chloride will polymerize with a range of comonomers thus extending its range of properties. Comonomers used include vinyl acetate, vinylidene chloride, acrylonitrile, acrylic esters and ethylene or propylene.

The technology is continuously evolving and operating cost, reliability and safety are key factors. Flexibility is becoming more important because it allows producers to reduce the quantity of polymer stored.

Considerable regrouping has occurred amongst polymer producers particularly in Europe during the 1980s and 1990s.

Capacities range from 35 000 to 1 360 000 tonnes per year.

PROCESSES

1. From vinyl chloride by suspension polymerization

Demineralized water and a dispersing agent, such as polyvinyl alcohol or a cellulose derivative, are fed into the reactor. As some hydrochloric acid is formed during the polymerization process, the acidity is buffered by the addition of an aqueous solution of sodium or magnesium hydroxide. (*See Figure 103*)

The reactor is sealed and evacuated and vinyl chloride is introduced under pressure. The vinyl chloride is dispersed by mechanical agitation, and an initiator which is soluble in the monomer is added. Free radical initiators (for example azobisisobutylonitrile, dibutyl peroxide carbonate or benzoyl peroxide) are commonly employed. The mixture is heated to the reaction temperature of 60–70°C, and the polymerization is initiated by the free radicals formed in the vinyl chloride

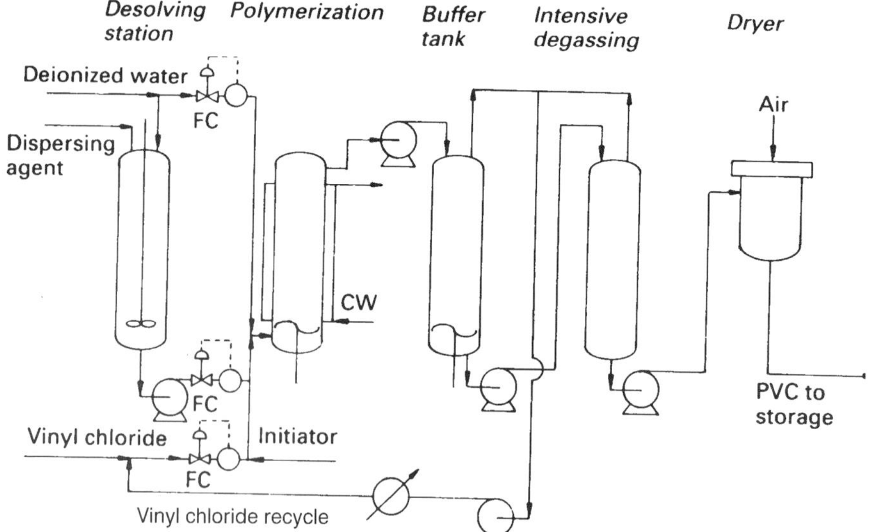

FIGURE 103 Polyvinylchloride from vinyl chloride by suspension polymerization

droplets. The suspension agent prevents the formation of polymer lumps which could impede the cooling process. Excess heat generated by the polymerization is removed by means of a water-cooled jacket around the reactor.

When around 90% of monomer conversion has been reached, the reaction is halted by discharging the slurry into a degasser where any unreacted monomer is removed by steam stripping. The resin is filtered, centrifuged to remove any remaining water and dried prior to transfer to the storage area.

Raw material requirements

Raw materials required per tonne of PVC:

Vinyl chloride	1005kg

2. From vinyl chloride by mass polymerization

The mass polymerization reaction is usually carried out in two stages in order to improve heat control and to overcome problems of polymer grain size as the resin is grown firstly in solution and secondly in the dry phase. (*See Figure 104*)

In the first reactor, polymerization takes place in liquid vinyl chloride in the presence of dibutyl peroxydicarbonate which acts as the initiator. By controlling the amount of initiator, polymerization is stopped at 10% conversion. The solution is sent to a second reactor where more initiator and vinyl chloride are added and the mixture is heated. During the second polymerization, the polymer suspension is

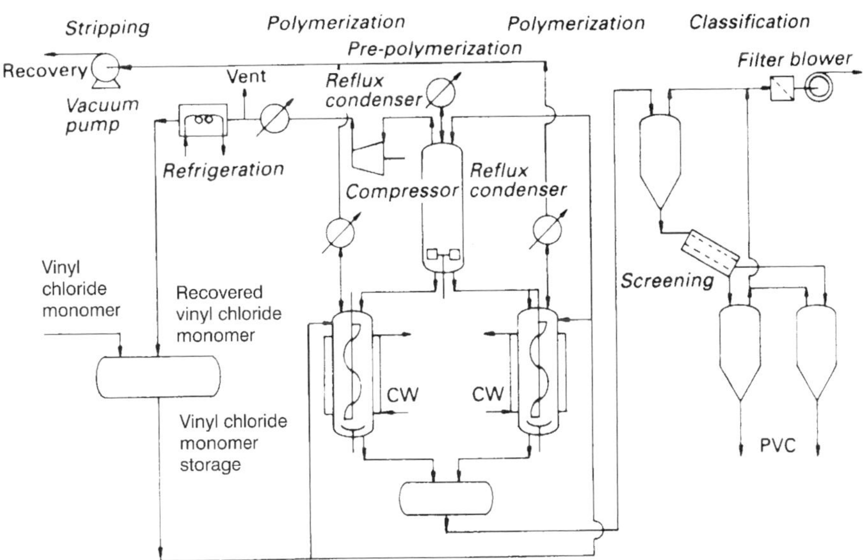

FIGURE 104 Polyvinylchloride from vinyl chloride by mass polymerization

converted to a powder. The reaction is allowed to proceed to 80% conversion; any excess monomer is removed by degassing and the passage of an inert gas through the polymer. The resin is conveyed pneumatically to storage silos.

Raw material requirements

Raw materials required per tonne of PVC:

Vinyl chloride	1007kg

OTHER PROCESSES

Emulsion polymerization

In emulsion polymerization, the monomer is dispersed in the aqueous solution and the initiators are water soluble. Emulsifiers (such as *sec*-alkyl sulphonates or sulphosuccinates with ammonium persulphate, sodium bisulphate and ferrous sulphate initiators) are frequently used. Polymerization takes place in the micelles formed by the emulsifiers. The polymer migrates to form PVC particles in water stabilized by the emulsifiers present. Reaction conditions are 40–65°C and a pressure of 0.05 bar. The reaction is stopped when 80% of the monomer has been converted. The PVC latex formed is spray dried to give spherical beads.

Solution polymerization

Solution polymerization is carried out in a solvent, with benzoyl peroxide as the initiator. As the process is more expensive to operate, because of the use of a solvent instead of water, it is limited to the manufacture of surface-coating polymers.

PROPERTIES

White, amorphous, odourless powder. Soluble in nitrobenzene, cyclohexanone, and tetrahydrofuran but insoluble in vinyl chloride. Resistant to dilute alkalis and acids but is attacked by concentrated nitric and chromic acids. Flame retarding, but plasticizers reduce its resistance to burning. Reacts violently with fluorine.

Density at 20°C	1.35
Melting Temperature	100°C
Tensile Strength MPa	55
Thermal Expansion 10^{-5}/°C	7
Elongation %	15

Exposure Limit COSHH	
Total inhalable dust	10mg/m^3 8 hour TWA
Respirable dust	4mg/m^3 8 hour TWA

GRADES

Rigid, flexible (plasticized).

INTERNATIONAL CLASSIFICATIONS

UN No.	Not allocated
CAS Reg No.	9002–86–2

APPLICATIONS

Two types of the homopolymer are produced: rigid which accounts for around 65% of PVC demand and flexible accounting for the remainder. Products made from rigid resin are inflexible and hard while flexible resins, which contain a large proportion of plasticizer, are soft and can be stretched.

Items made from rigid PVC include pipe and conduit, fittings, automobiles, blow moulding and roofing tiles. Pipe and conduit, used in the building and construction industry for water, gas and drainage are the largest consumer of PVC, accounting for around 32% of total demand. Rigid profiles consume a further 17%.

Flexible PVC finds outlets in calendered sheet, wire and cable coating, flooring, coated fabrics, shower curtains automobile upholstery and furniture. Film and sheet consume around 13% of PVC production, wire and cable coating 9% and coated fabrics and flooring a further 5% each.

Global growth increased by nearly 6% in 1997, with the highest rise in the Middle and Far East.

HEALTH AND HANDLING

PVC has limited stability; above 80°C the resin discolours and evolves hydrogen chloride. It should be stored in closed containers in a well-ventilated, dry, cool area away from sunlight, heat, and oxidizing agents. Inhalation of PVC causes damage to the lungs and liver. Repeated exposure to PVC dust can lead to skin irritation and dermatitis and conjuctivitis in the eyes. Contact lenses must not be worn.

Freshly manufactured PVC can contain traces of toxic vinyl chloride monomer, a confirmed human carcinogen. The foam may ignite spontaneously if not allowed to cool before stacking.

If a leak occurs, the area must be evacuated and non-sparking tools used to scoop up the powder, which should be binned and disposed of in accordance with local regulations. Clean-up staff should wear protective clothing to prevent skin and eye contact and masks against inhalation. Contaminated clothing must be laundered before reuse and boots cleaned to remove traces of PVC.

Waste PVC can be disposed of by landfill; incineration gives rise to hydrogen chloride gases which cause severe pollution problems. Dust is a nuisance and, although it is not an explosion hazard due to the high amount of energy required to ignite PVC, effort should be made to keep it to minimum levels.

Although PVC has limited flammability because of the presence of the chlorine molecule, it will produce copious smoke dispersed with fine solids and liquid particles in a fire started by other materials. The resin drips on burning causing the fire to spread. Carbon dioxide or water are used to extinguish fires. Toxic fumes (such as carbon monoxide, hydrochloric acid and phosgene) are given off and personnel must wear self-contained breathing apparatus when fighting fires where PVC is present.

MAJOR PLANTS

Plants with capacities greater than 245 000 tonnes per year:

Solvay	Tavaux	France
BSL Olefinverbund	Schkopau	Germany
EVC	Wilhelmshafen	Germany
Hoechst	Ludwigshafen	Germany
Vestolit	Marl	Germany
Rovin	Pernis	Netherlands
Borden	Addis	US
CONDEA Vista	Aberdeen	US
Formosa Plastics	Baton Rouge	US
	Point Comfort	US
Georgia Gulf	Plaquemine	US
Occidental Chemical	Pasadena	US
Shintech	Freeport	US
CPC	Camacari	Brazil
PQ Columbia	Cartagena	Colombia
National Plastic	Al Jubail	Saudi Arabia
Reliance Industries	Gujarat	India
Shin-Etsu Chemical	Kashima	Japan
LG Chemical	Yeochon	South Korea
Hanwha Chemical	Ulsan	South Korea
	Yeochon	South Korea
Formosa Plastics	Jen Wu	Taiwan
	Linyuan	Taiwan

MAJOR LICENSORS

BASF
Chisso
Diamond Shamrock
Dynamit Nobel
Elf Atochem
EVC
Geon
Goodrich
Hoechst
Huels
Kanegafuchi Chemical
Kema Nord
Kureha Chemical
Mitsubishi Chemical
Monsanto
Mitsui Petrochemical
Shin-Etsu Chemical
Solvay/ICI
Stauffer Chemical
Sumitomo Chemical
Texaco Development
Veba Chemie
Wacker Chemie

Propylene $CH_2{=}CHCH_3$

SYNONYMS

PROPYLENE propene, methylethene, methyl ethylene

Originally a by-product of ethylene production, propylene has become increasingly important as a chemical intermediate, which has led to a major increase in its price relative to ethylene. There has been a rapid growth in polypropylene production since the early 1980s.

There are three major sources of propylene:

- by-product of the manufacture of gasoline from crude oil;
- co-product with ethylene from naphtha cracking;
- cracking of propane or butane.

Refinery gas streams (produced by crude oil distillation, catalytic cracking and catalytic reforming) have provided important sources of propylene especially in the US. In Europe, demand for petrochemical feedstocks favoured naphtha as the prime source of olefins. Although most crackers were operated to maximize ethylene output, some producers have changed feedstocks and cracking conditions to alleviate 'tightness' of supply at certain sites. The ethylene:propylene ratio can be varied from 0.4:1 to 0.75:1 depending on the cracking severity, with propylene production being greater at low severity. In the 1990s, there has been a trend towards lower-severity cracking to increase propylene formation in new crackers to meet the rising demand for propylene derivatives.

Propane and butane are the preferred feedstocks if only propylene is required. Sources of these gases are refinery off-gases, natural gasolines and wet natural gas. Propylene is also obtained as a by-product of the catalytic cracking of gas oils, and by the dehydrogenation of *n*-butane to butadiene.

With propylene demand growing faster than ethylene, efforts have been made to find alternative methods to maximize propylene output. As large surpluses of butadiene are available, some of the excess has been absorbed by the installation of partial or total hydrogenation units. Total hydrogenation recycles the butenes formed back to the cracker where they are converted to additional ethylene and propylene. The raw C_4 cut can be selectively dehydrogenated to maximize 2-butenes which are then, in the presence of ethylene, converted to propylene via the metathesis reaction.

Currently, 70% of world propylene production is obtained as a co-product from naphtha cracking, with 2% from propane and the remainder from refinery operations and dehydrogenation. In the US, only 55% of propylene is obtained from naphtha cracking due to the availability of refinery gases and high gasoline demand.

Capacities range from 30 000 to 1 400 000 tonnes per year.

PROCESSES

1. From hydrocarbons by steam cracking

Propylene is produced as a by-product of ethylene manufacture and is separated out from the C_3 stream (see Ethylene). (*See Figure 105*)

The C_3 fraction separated overhead from the depropanizer consists of a mixture of propane, propylene, propadiene and propyne with traces of C_2 and C_4 hydrocar-

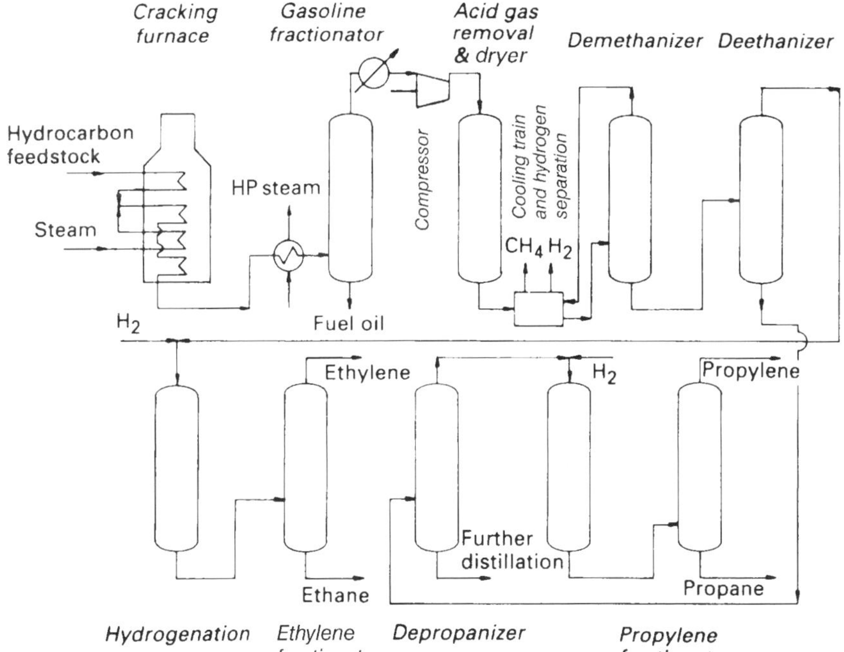

FIGURE 105 Propylene from hydrocarbons by steam cracking

bons. The mixed gases are fed into a selective dehydrogenation unit containing a palladium catalyst at a pressure of 18 bar. A measured amount of hydrogen is added, sufficient to convert propadiene and propyne to propylene with minimal conversion to propane being obtained.

The reaction can be carried out in either the gas phase at a temperature of 50–120°C, or in the liquid phase below 25°C. The exit gases are scrubbed to remove any oligomers formed prior to any unconsumed hydrogen being stripped off and recycled. The propane–propylene mixture can be used for chemical use. If further purification to polymer-grade propylene is required, the propane–propylene is separated in a single- or double-column process. Because of the close boiling points of propane and propylene, reflux ratios between 0.90 and 0.97 are normally employed.

Reaction

$$2C_3H_8 \rightarrow C_3H_6 + H_2 + C_2H_4 + CH_4$$

Raw material requirements and yield

Raw materials required per tonne of propylene and typical yields in wt% are given under Ethylene.

2. From propane by dehydrogenation

There are currently four technologies available for the dehydrogenation of propane obtained from liquefied petroleum gas (LPG) fields. These are: Catofin from ABB Lummus Global; Oleflex from UOP; Fluidized Bed Dehydrogenation (FED) from Snamprogetti; and Steam Active Reforming (STAR) from Phillips Petroleum. All differ in the type of catalyst used, operating temperature and pressure and the use of diluents.

Propane together with recycle propane and hydrogen are preheated to a temperature of 500–700°C and fed into a moving or fixed-bed reactor at near atmospheric pressure. The reaction is carried out in the presence of a noble metal catalyst such as platinum on activated alumina impregnated with 20% chromium. Operating either continuously or cyclically, the catalyst is regenerated by burning off any surface coke formed in air. The heat generated is used to heat the catalyst. Steam can be used as a diluent to maintain positive pressure in the reactor. The propylene formed is recovered from the C_3 stream.

Reaction

$$C_3H_8 \rightarrow C_3H_6 + H_2$$

Raw material requirements and yield

Raw materials required per tonne of propylene and typical yields in wt% are given under Ethylene.

3. From 2-butenes by metathesis

An equimolecular mixture of 2-butenes and ethylene are fed into a fixed-bed reactor, and a rhenium-based catalyst is introduced through the top. The reaction takes place at a temperature of 20–50°C with sufficient pressure to keep the reactants in the liquid phase. Alternatively, a tungsten-based catalyst can be used at a temperature of 300–400°C.

The liquid from the reactor is fed into a fractionation column where ethylene and propylene are recovered overhead. In a deethanizer, the unreacted ethylene is recovered and recycled, and propylene is recovered from the depropanizer. The residual C_4s are recovered in a second fractionation column and recycled. Some of the catalyst is withdrawn from the bottom of the reactor, regenerated by calcination at 500–600°C before being returned to the reactor. The conversion to propylene is favoured at low temperatures and the maximum achieved at 35°C is 63%.

Reaction

$$C_2H_4 + CH_3CH{=}CHCH_3 \rightarrow 2C_3H_6$$

OTHER PROCESSES

Propylene is obtained as a by-product of the fluid catalytic cracking of gas oils. The reaction is carried out in the presence of a zeolite catalyst at 450–580°C and 2.5–4 bar with a contact time of 5 seconds to 2 minutes. Yields of 2–5wt% are obtained depending on the type of feedstock used.

In refineries, cracked gases from the coking units contain 10–15 mole% C_3s, while in visbreaker units levels of 2–3% C_2–C_4 are obtained. After removal of the light components in a deethanizer, the C_3 fraction is separated in a depropanizer.

PROPERTIES

Colourless, flammable gas with a slightly sweetish odour. Slightly soluble in water. Soluble in ethyl alcohol and ether. It forms explosive mixtures with air.

Molecular Weight	42.08
Density of liquid at 20°C	0.508
Melting Point	–185.3°C
Boiling Point	–47.7°C
Autoignition Temperature	455°C
Flammability Limits in air	
lower	2.0vol%
upper	11.1vol%
Flash Point Closed Cup	–107.8°C
Vapour Density (air = 1)	1.49
Exposure Limit COSHH	Asphyxiant
Exposure Limit ACGIH	Asphyxiant

GRADES

Technical 93%, 99.5%, polymer 99.8%.

INTERNATIONAL CLASSIFICATIONS

UN No.	1077
CAS Reg No.	115–07–1
EINECS No.	204–062–1
EC No.	601–011–00–9
Description	Flammable gas
Packing Group	Not allocated
Emergency Action Code	2WE
HI (Kemler Code)	23

APPLICATIONS

Propylene is one of the three most important chemical intermediates used in the petrochemical industry. In the US, the major chemical outlet is for the manufacture of polypropylene which accounts for 46% of total production. Copolymerization with ethylene, with or without diene modifiers, results in ethylene–propylene rubbers.

Acrylonitrile and propylene oxide dominate the individual chemicals made from propylene, consuming 13% and 10% respectively. The next most important outlets are oxo alcohols consuming 10% and cumene consuming 6%.

Propylene is the starting material for a number of reactions, and other chemicals produced include acrylic acid (5%) and isopropanol needing (3%). Included under the remainder is the manufacture of propylene dimer for the detergents industry, heptenes and acrolein. Considerable quantities of propylene production are consumed for non-chemical uses such as gasoline alkylate.

World demand is growing at 5%.

HEALTH AND HANDLING

Propylene gas causes no hazard to the skin and eyes but is an asphyxiant in high concentrations. The liquid can cause burns if it comes into contact with the skin or eyes. Protective clothing, goggles and gloves are advised when handling the product, to avoid exposure to vapour or liquid.

Normally, propylene is stored as a liquid under a pressure of 20 bar in welded containers. It is non-corrosive and can be stored at atmospheric pressure. It must be kept away from strong oxidizing agents and it reacts dangerously with nitrogen dioxide. Because of its flammability, the area must be well ventilated, sources of ignition avoided and electrical apparatus earthed to prevent static build-up.

Propylene is a potential fire and explosion hazard. Although small fires can be extinguished with dry chemical, the source of the leak should be closed and the fire allowed to burn itself out. Any propylene leak must be regarded as an emergency because of the flammability of the gas. Liquid spills present a particular hazard, as the liquid, being heavier than air, collects in low-lying areas. Checks must be made for leaks after the fire has been extinguished because propylene–air mixtures can ignite without warning. Self-contained breathing equipment and protective clothing must be worn by firefighting staff.

When entering a propylene-rich-area, personnel should wear a safety line attached to a colleague with self-contained breathing apparatus so that rapid rescue can be effected in the event of an emergency.

MAJOR PLANTS

Plants with capacities greater than 450 000 tonnes per year:

Fina Olefins	Antwerp	Belgium
Naphtachimie	Lavera	France
Dow Chemical	Terneuzen	Netherlands
DSM	Geleen	Netherlands
Erdoelchemie	Dormagen	Germany
ROW	Wesseling	Germany
Vestolen	Gelsenkirchen	Germany
Chevron Chemicals	Port Arthur	US
Dow Chemical	Freeport	US
Exxon Chemical	Baton Rouge	US
	Baytown	US
Lyondell	Channelview	US
Phillips Petroleum	Sweeny	US
Quest Energy	Mont Belvieu	US
Shell Chemical	Deer Park	US
	Norco	US
Ultramar Diamond Shamrock	Mont Belvieu	US
Copene	Camacari	Brazil
Showa Denko	Oita	Japan
Tosoh Corp.	Yokkaichi	Japan

MAJOR LICENSORS

ABB Lummus Global
Air Products & Chemicals
Braun
CDTECH
IFP
Kellogg
Kinetic Technology Inc.
Linde
Lurgi
Phillips Petroleum
Snamprogetti
Stone & Webster
Technip
UOP

Propylene Glycol

$$CH_3-\underset{\displaystyle OH}{\underset{|}{CH}}-\underset{\displaystyle OH}{\underset{|}{CH_2}}$$

SYNONYMS

PROPYLENE GLYCOL 1,2-dihydroxypropane, 1,2-propanediol, methylethylene glycol, trimethyl glycol, methyl ethyl glycol

Propylene oxide reacts readily with water to yield propylene glycol. Although the reaction can be accelerated by the use of acids or alkalis, this makes the separation of the final products more difficult.

Capacities range from 15 000 to 250 000 tonnes per year.

PROCESSES

1. From propylene oxide by hydration

Propylene oxide and water are fed into a reaction tower, where, at a temperature of 200°C and a pressure of 12 bar, propylene glycol and higher glycols are obtained. (*See Figure 106*)

In order to limit the quantity of higher alcohols formed, the amount of water is carefully controlled to give a molar ratio of water to oxide of 15:1. This favours the formation of monopropylene glycol, the isomer in greatest demand. Heat from the highly exothermic reaction is removed by means of a water-cooled jacket around the reactor. The rate of feed and withdrawal of the reaction products are adjusted so that the composition of the mixture in the reactor remains constant.

The reaction mixture is dehydrated by evaporation under vacuum and the anhydrous glycols separated by distillation. Propylene glycol, dipropylene glycol and tripropylene glycols are recovered.

Reaction

$CH_3COHCH_2 + H_2O \rightarrow CH_3CH(OH)CH_2OH$

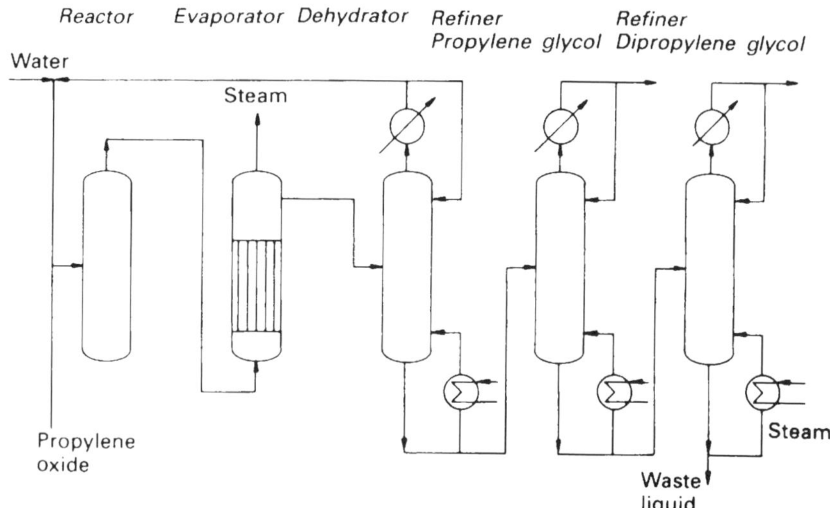

FIGURE 106 Propylene glycol from propylene oxide by hydration

Raw material requirements

Raw materials required per tonne of propylene glycol (theoretical):

Propylene oxide	760kg
Water	240kg

PROPERTIES

Colourless, tasteless, slightly viscous liquid with hardly any odour. Combustible. Miscible with water.

Molecular Weight	76.11
Density at 20°C	1.038
Melting Point	–54°C
Boiling Point	187.3°C
Autoignition Temperature	410°C
Flammability Limits	
lower	2.6vol%
upper	12.5vol%
Flash Point Open Cup	103°C
Exposure Limit COSHH	150ppm 8 hour TWA (vapour and particulates) 10mg/m ^{3} 8 hour TWA (particulates)
Exposure Limit ACGIH	None established

GRADES

Technical 99%, food 99+%.

INTERNATIONAL CLASSIFICATIONS

UN No.	Not allocated
CAS Reg No.	57–55–6
EINECS No.	200–338–0
EC No.	Not allocated

APPLICATIONS

The major outlet for monopropylene glycol, accounting for around 40% of total world demand, is in the production of unsaturated polyester resins which are used in surface coatings and glass fibre reinforced resins. The manufacture of plasticizers and hydraulic brake fluids consume considerable quantities of propylene oxide.

A rapidly growing market for propylene glycol, now amounting to 7% of consumption, is the manufacture of non-ionic detergents. These are utilized not only in the petroleum, sugar refining and paper-making industries, but also in the preparation of toiletries, antibiotics and liquid washing formulations. Around 5% is used as an antifreeze. Although higher in price, ecological pressure is leading to the replacement of ethylene glycol by propylene glycol in some specialist areas, for example de-icing aircraft and runways and as a coolant in the food industry.

Propylene glycol is an excellent solvent, and finds outlets in printing inks, alkyd resins and as an extractant. It is also widely used as a humectant in the pharmaceutical, cosmetics, animal foodstuffs and tobacco industries.

HEALTH AND HANDLING

Propylene glycol is not harmful but when handling the product goggles should be worn as the liquid is irritant if it comes into contact with the eyes.

As propylene glycol is not corrosive to metals, containers can be made of iron, mild steel, copper or aluminium. They should be stored in a well-ventilated area. Spills should be contained, absorbed with dry sand or vermiculite and the contaminated area flushed with water. The liquid must be kept away from water intakes.

Propylene glycol is combustible but not flammable at room temperature. Fires can be extinguished with water, carbon dioxide or alcohol foam.

MAJOR PLANTS

Plants with capacities greater than 35 000 tonnes per year:

Arco Chimie	Fos sur Mer	France
BASF	Ludwigshafen	Germany
Dow Chemical	Stade	Germany
Erdoelchemie	Cologne	Germany
Arco Chemie Nederland	Botlek	Netherlands
Dow Chemical	Freeport	US
	Plaquemine	US
Lyondell	Bayport	US
Dow Quimicas	Aratu	Brazil
Asahi Glass	Kashima	Japan
Nihon Oxirane	Chiba	Japan
Yukong Oxichemical	Ulsan	South Korea

MAJOR LICENSORS

CDTECH
IFP/Chinese Petroleum
Kinetic Technology Inc.
Phillips Petroleum/ABB Lummus Global
Shell Development
Union Carbide
UOP

Propylene Oxide

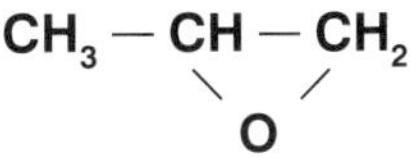

SYNONYMS

PROPYLENE OXIDE 1,2-epoxypropane, 1,2-propanediol, methyl oxirane, methyl ethylene oxide

Propylene oxide only became a chemical of importance after World War II, since which time capacity has grown rapidly.

It was the commercial development of ethylene oxide and its derivatives which led to the realization of the potential uses of propylene oxide as a chemical intermediate.

There are four major routes to propylene oxide:

- chlorohydrination of propylene;
- electrochemical integrated chlorohydrination of propylene;
- indirect oxidation of propylene;
- direct oxidation of propane.

The choice of route depends on the market utilization of the co-products as well as the availability and cost of raw materials.

The original process used for the production of propylene oxide was the dehydrochlorination of propylene chlorohydrin made from propylene and chlorine in aqueous solution, and this remained the basis of all manufacture for many years. Many plants were converted from ethylene oxide production when the more economic direct hydration of ethylene process was introduced.

The chlorohydrination process suffers from two major problems. Firstly, the expensive chlorine molecule is lost as calcium chloride, and secondly, the volume of dilute calcium chloride solution produced as a by-product (approximately 1.5 tonnes per tonne of propylene oxide) presents major disposal problems. In order to overcome this burden, calcium hydroxide was replaced by sodium hydroxide and the waste sodium chloride solution from the process was utilized, after concentration, in chloralkali electrolysis plants.

More recently, ABB Lummus Global have modified the process utilizing *tert*-butyl hypochlorite, as the chlorine-transfer agent, to replace the aqueous stream. This route can be integrated with a chlorine–caustic soda electrolysis unit where the caustic generated can be used to saponify the propylene chlorohydrin formed. After removal of organics, the resultant brine can be used as feed to the cells.

The major disadvantages of the Lummus process are its slow reaction rate and the complex range of co-products formed. Although many modifications to the traditional chlorohydrination process exist, none has substantially improved its economics and been used commercially.

Considerable effort has been spent on trying to develop a route based on the direct oxidation of propylene with air or oxygen over a silver catalyst. Unfortunately, gas-phase direct oxidation processes have not become popular because of difficulties in reactor temperature control, low yields, the number of by-products formed and the complexity of their separation.

Because of these problems, research has been concentrated on indirect oxidation routes using peroxygen carriers formed by the liquid-phase oxygen or air oxidation of the feedstock. In the process developed by Oxirane, a joint venture formed by Atlantic Richfield and Halcon International, propylene epoxidation is carried out by organic peroxides produced by the liquid-phase air oxidation of the parent feedstock. The two favoured feedstocks used industrially are isobutane and ethylbenzene which yield the co-products *tert*-butyl alcohol or 1-phenyl ethyl alcohol respectively. The 1-phenyl ethyl alcohol can be dehydrated to styrene.

Daicel operated a peracetic acid based propylene oxide plant which produced acetic acid as a co-product in the 1970s. Peroxidation was carried out in the presence of peracids, hydroperoxides or catalysts based on molybdenum, tungsten, chromium and vanadium.

The feature of these processes is that the quantity of co-product formed is greater and more valuable than propylene oxide. Thus, the choice of feedstock is heavily dependent on the value of the co-product. However, if demand for the co-product is not in balance, then propylene oxide production has to be limited.

In spite of the various routes which have been developed, the choice of process depends on the site integration and market utilization of the co-products as well as the availability and cost of raw materials.

Around 50% of propylene oxide capacity is based on the chlorohydrin route with epoxidation processes, based on *tert*-butyl alcohol and ethylbenzene, accounting for the remainder. Approximately 70% of world propylene oxide capacity is controlled by two producers, Dow Chemical and Arco.

Capacities range from 30 000 to 640 000 tonnes per year.

PROCESSES

1. From propylene by chlorohydrination

In the conventional two-stage chlorohydrination process, chlorine and water are fed into the base of a tower reactor where the hypochlorous acid formed reacts with an equimolecular weight of propylene which is introduced about halfway up the tower. The reaction, carried out at a pressure of 1–1.9 bar, is exothermic, the heat generated maintaining the temperature at 40–90°C. The effluent gas is scrubbed with sodium hydroxide before being returned to the reactor. (*See Figure 107*)

In a second reactor, the chlorohydrin solution formed is hydrolysed to propylene oxide by treatment with a 10–15% excess of calcium hydroxide. Excess water reduces the amount of co-product formation and also acts as a cooling agent. Liquid from the hydrolyser is distilled and propylene oxide is recovered overhead and purified by rectification. Calcium chloride is removed as an aqueous waste stream.

A portion of the recycle gases is purged to prevent the build-up of propane in the input feed and to prevent oxygen from the converted chlorine reaching explosive limits.

Because of the volume of waste water, calcium hydroxide can be replaced by sodium hydroxide and the resultant brine, after removal of organic impurities, used in an adjacent chlorine plant. Before it can be employed however, either expensive concentration has to be carried out or salt has to be added to bring the concentration up to the level required for the cell feed.

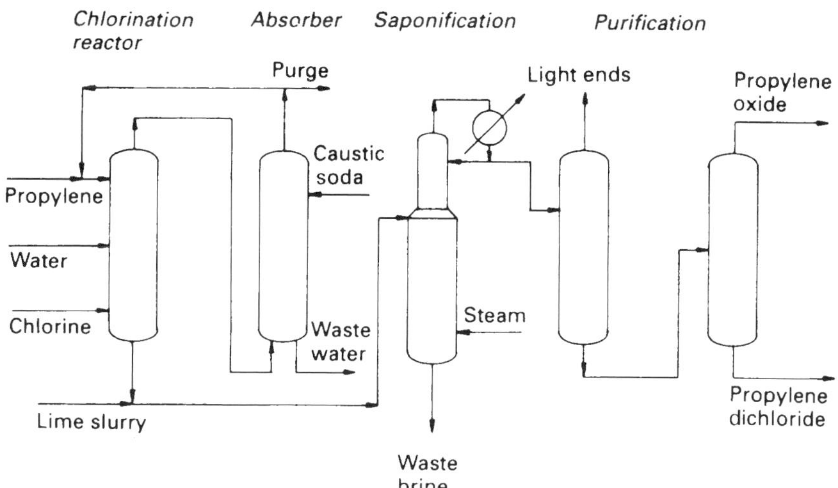

FIGURE 107 Propylene oxide from propylene by chlorohydrination

ABB Lummus Global has modified the chlorohydrin process so that the large by-product waste water stream is eliminated and only by-product hydrogen is formed which can be used in another chemical process or burnt as fuel.

In the Lummus route, chlorine is reacted with a *tert*-butyl alcohol–sodium hydroxide mixture in brine in a stirred tank to form *tert*-butyl hypochlorite. The effluent from the reactor is allowed to separate into two phases. The organic phase, containing *tert*-butyl hypochlorite together with added water, reacts with propylene to yield propylene chlorohydrin and *tert*-butyl alcohol. When propylene chlorohydrin is saponified with sodium hydroxide, propylene oxide is produced. To minimize by-product formation, crude propylene oxide is stripped from the alkaline brine solution. Purification of propylene oxide is carried out by distillation. After removal of light ends overhead, the bottoms go to the final column where pure propylene oxide is recovered.

Butyl alcohol is recovered from the bottom stream from the stripper and recycled. The remaining brine solution, after treatment to remove organics, is sufficiently concentrated for direct use in the cell feed.

Reaction

$$CH_2{=}CHCH_3 + HOCl \rightarrow CH_2ClCH(OH)CH_3$$

$$2CH_2ClCH(OH)CH_3 + Ca(OH)_2 \rightarrow 2CH_3CHOCH_2 + CaCl_2 + 2H_2O$$

Raw material requirements and yield

Raw materials required per tonne of propylene oxide:

Propylene (95%)	820kg
Chlorine	1390kg
Calcium hydroxide	730kg

Yield 95%

2. From propylene by epoxidation

In this two-step process, isobutane is oxidized with oxygen or air in the liquid phase at a temperature of 110–150°C and a pressure of 22–55 bar to *tert*-butyl hydroperoxide, some of which decomposes further to *tert*-butyl alcohol. Conversion of isobutane is kept at 15–25% to improve the selectivity to *tert*-butyl hydroperoxide, higher yields of peroxide being obtained at low conversion rates. The oxygen concentration is controlled in order to maximize the hydroperoxide yield and maintain a non-explosive condition within the reactor. (*See Figure 108*)

The reactor effluent is distilled and any unreacted isobutane is recycled. The *tert*-butyl hydroperoxide–butyl alcohol mixture is fed together with excess propylene into a series of reactors. The reaction, taking place at 80–130°C and a pressure of

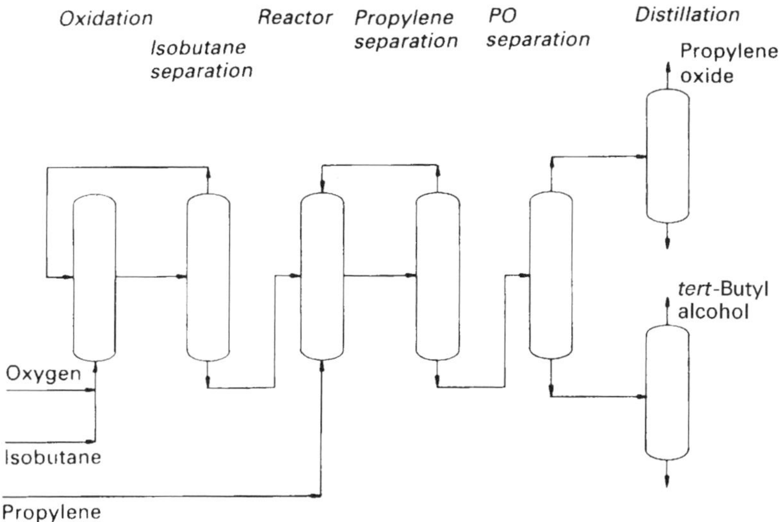

FIGURE 108 Propylene oxide from propylene by epoxidation

17–70 bar in the presence of a catalyst, forms propylene oxide and phenyl ethyl alcohol. A range of different catalyst systems has been proposed for the epoxidation but the most popular are molybdenum naphthenate, coordination compounds of molybdenum hexacarbonyl and oxyacetyllactonate, or the soluble salts of titanium, vanadium or tungsten.

The reaction mixture is separated by multi-distillation. Any unreacted propylene is removed overhead and recycled, before the propylene oxide is separated from the product stream. The crude propylene oxide is purified by further distillation. Catalyst is re-extracted from the remaining mixture before *tert*-butyl alcohol is recovered as an overhead stream. It can be sold or dehydrated to isobutene.

Reaction

$$4(CH_3)_3CH + 3O_2 \rightarrow 2(CH_3)_3COOH + 2(CH_3)_3COH$$

$$(CH_3)_3COOH + CH_3CH{=}CH_2 \rightarrow CH_2OCHCH_3 + (CH_3)_3COH$$

Raw material requirments and yield

Raw materials required per tonne of propylene oxide (2160kg of *tert*-butyl alcohol is formed as co-product):

Propylene	780kg
Isobutane	1700kg

Yield 90%

If ethylbenzene is used as the raw material instead of isobutane, then the co-product formed, phenyl ethyl alcohol, is dehydrated to styrene (2.2–2.5kg of styrene is formed as co-product to 1kg of propylene oxide) (see Styrene).

OTHER PROCESSES

Daicel used to produce propylene oxide by the epoxidation of propylene by peracetic acid at 50–80°C and 9–12 bar, no catalyst being required. In this route, peracetic acid is formed by the air oxidation of acetaldehyde in ethyl acetate as a solvent over a metal ion catalyst at 30–50°C and 25–40 bar. The resultant 30% peracetic acid solution in ethyl acetate is concentrated, and the mixture is reacted with propylene at 60–80°C and 10–12 bar to yield propylene oxide and acetic acid. The plant is no longer in operation.

In the Bayer–Degussa process, peracetic acid is produced by the reaction of acetic acid with hydrogen peroxide. The acetic acid co-product can be recovered and recycled. A process modification by EniChem, employs hydrogen peroxide in the liquid phase. The reaction takes place in the presence of a titanium silicate catalyst in a methyl alcohol–water mixture kept at 40–50°C and with propylene at a pressure of 4 bar. The major disadvantages of these routes are the hazards associated with the manufacture and handling of concentrated hydrogen peroxide, and the associated high plant costs.

Reaction

$$CH_3CHO + O_2 \rightarrow CH_3COOOH$$

$$CH_3COOOH + CH_2{=}CHCH_3 \rightarrow CH_2OCHCH_3 + CH_3COOH$$

Raw material requirements and yield

Raw materials required per tonne of propylene oxide (1300 kg of acetic acid is formed as co-product):

Propylene	960kg
Acetaldehyde	1100kg
Air	337kg

Yield 92%

PROPERTIES

Colourless, low boiling point, flammable liquid with an sweetish ethereal odour. Partially miscible with water and soluble in ethyl alcohol.

Molecular Weight	58.08
Density at 20°C	0.823

Melting Point	–111.9°C
Boiling Point	34.2°C
Autoignition Temperature	430°C
Explosive Limits in air	
lower	2.1vol%
upper	37.0vol%
Flash Point Open Cup	–37°C

Exposure Limit COSHH (Schedule 1) 5ppm 8 hour TWA (Maximum Exposure Limit)
Being reviewed under the Existing Substances Regulations.
Exposure Limit ACGIH 20ppm 8 hour TLV-TWA
Suspected possible human carcinogen.

GRADES

Technical 99.9% (must be acetylene free).

INTERNATIONAL CLASSIFICATIONS

UN No. (blanketed with nitrogen)	1280
CAS Reg No.	75–56–9
EINECS No.	200–879–2
EC No.	603–055–00–4
Description	Flammable liquid
Packing Group	I
Emergency Action Code	2WE
HI (Kemler Code)	339

APPLICATIONS

The largest outlet for propylene oxide, accounting for 63% of total consumption, is for polyether polyols which are used in the production of polyurethane foams. The most important of these, flexible foams, account for 40% of this total, the remainder going into rigid foams and polyurethane elastomers.

Of the other outlets, propylene glycol at 25% and polyethers at 5% are both increasing as they replace ethoxylated products in some applications, such as antifreeze for de-icing runways and aircraft, and as a solvent. Propylene oxide is employed as a raw material for the production of glycerine, solvents, and surfactants.

World growth rate is estimated at 4–5% in the period 1995–2000, with Western European levels being about half that.

HEALTH AND HANDLING

Propylene oxide vapour is irritating to the eyes, nose and throat. Inhalation of the vapour should be avoided as it causes nausea even some time after exposure. The liquid will cause burns in contact with the skin and protective clothing and goggles should be worn.

Because of its volatility, propylene oxide is normally stored under pressure in specially designed tanks made of iron, mild steel, copper or aluminium. Care must be taken with the choice of rubber or plastics used for seals and valve materials because many are attacked.

In the event of spills, the area should be evacuated and all sources of ignition extinguished. If possible the discharge should be contained and not allowed to get into sewers or waterways. Absorb the liquid with dry sand or earth using non-sparking tools, and dispose of in accordance with local regulations.

Combat fires from a safe distance using carbon dioxide, dry chemical or alcohol foam. Water may be ineffective on the fire but can be used to cool exposed containers. Because of the density of the vapour, flashback along a vapour trail can occur. All clean-up and firefighting staff must wear rubber clothing, goggles and self-contained breathing apparatus.

The movement of propylene oxide by road, rail or sea is governed by special regulations because of its flammability.

MAJOR PLANTS

Plants with capacities greater than 135 000 tonnes per year:

Arco Chimie	Fos-sur-Mer	France
Arco Chemie Nederland	Botlek	Netherlands
Shell Nederland Chemie	Moerdijk	Netherlands
Dow Chemical	Stade	Germany
Erdoelchemie	Cologne	Germany
Dow Chemical	Freeport	US
	Plaquemine	US
Lyondell	Bayport	US
	Channelview	US
Huntsman Corp.	Port Neches	US
Dow Quimicas	Aratu	Brazil
Nihon Oxirane	Chiba	Japan
Serayako Chemicals	Seraya	Singapore
Yukong Oxichemical	Ulsan	South Korea

MAJOR LICENSORS

ABB Lummus Global
Arco Technology
BASF
Bayer
Daicel
Halcon-Scientific Design
Huntsman Chemical
Mitsui Petrochemical
Shell Development
Union Carbide

Styrene

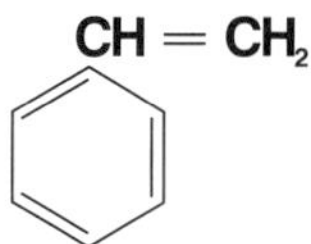

SYNONYMS

STYRENE cinnamol, phenylethylene, styrol, vinylbenzene, cinnamene, phenylethene, ethyenyl benzene, vinylbenzol cinnamenol

Ethylbenzene is the prime feedstock used for the production of styrene. Produced by the alkylation of benzene with ethylene, the reaction can be carried out either in the vapour or liquid phase. The ethylbenzene formed can be converted to styrene adiabatically or isothermally. Although both routes are in use, adiabatic dehydrogenation in multiple reactors or reactor beds is the preferred choice.

Ethylbenzene occurs in refinery xylene fractions but its separation for use in a styrene dehydrogenation plant requires fractionating columns having in excess of 100 theoretical plates and high-reflux ratios.

A more recent process, developed by Halcon, Arco and Shell, involves the oxidation of ethylbenzene to its hydroperoxide. This reacts with propylene to yield propylene oxide and methyl phenyl carbinol which is then dehydrated to styrene.

Because the dehydrogenation of ethylbenzene is an endothermic reaction and heat is needed for a commercially viable conversion, work has been carried out on a number of alternative processes. These include:

- direct oxidative coupling of benzene and ethylene;
- dimerization of butadiene followed by catalytic dehydrogenation;
- direct oxidative coupling of toluene followed by disproportionation of the stilbene formed to styrene and benzene;
- extractive distillation from pyrolysis gasoline with dimethyl formamide or dimethylacetamide.

None of these processes has gained any commercial importance.

More recently, catalysts have become more complex with several companies developing their own to meet specific needs as well as optimum yield at lowest utility cost.

About 87% of styrene is produced from ethylbenzene, the remainder being obtained via the propylene oxide route.

Capacities range from 30 000 to 950 000 tonnes per year.

PROCESSES

1. From ethylbenzene

Ethylbenzene is made by the catalytic alkylation of benzene with ethylene in the liquid or vapour phase. The catalyst is either aluminium chloride promoted by hydrogen chloride or ethyl chloride for the liquid-phase reaction, or crystalline aluminosilicate zeolite for the vapour-phase process (see Ethylbenzene). (*See Figure 109*)

The dehydrogenation of ethylbenzene is brought about either in an adiabatic reactor with interstage heating or in a pseudoisothermal reactor with heat added throughout the catalyst to maintain a constant temperature.

In the adiabatic process, purified ethylbenzene is preheated with steam before going through a heat exchanger to increase the temperature further. The vapours are mixed with superheated steam, in the ratio of 2.6:1 steam:ethylbenzene by weight. The gases are fed into a series of multiple beds containing a dehydrogenation catalyst consisting of potassium carbonate promoted iron–chromium oxides or zinc oxide promoted with alumina and chromates. Superheated steam

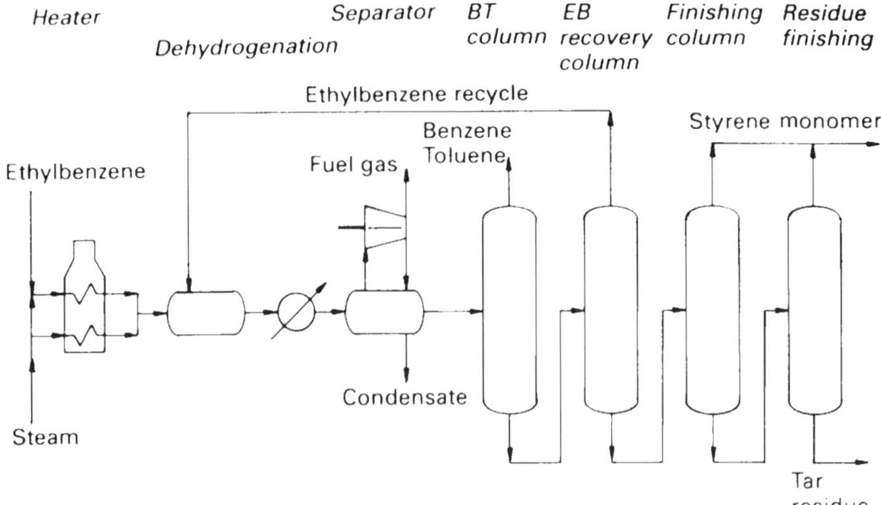

FIGURE 109 Styrene from ethylbenzene

increases the ethylene conversion (by reducing the partial pressures of the reactants) and decokes the catalyst. The reaction is carried out at a temperature of 690–700°C and below atmospheric pressure. Conversion levels of 50–70wt% are achieved with yields of 90–95mole%.

The gas stream from the reactor is cooled and the heat recovered is used to generate steam or preheat the reactor feed in order to minimize energy consumption. The crude product passes to the recovery section where any benzene and toluene generated by the reaction are removed by distillation under vacuum in the first column. The distillate is returned to the ethylbenzene unit where benzene is recovered.

Ethylbenzene is separated from styrene in the second column by distillation under vacuum and recycled to the reactor. Styrene distillation in the third column is carried out under vacuum and at as low a temperature as possible to reduce styrene polymerization. A non-sulphur inhibitor, such as phenylenediamine, dinitrophenols or dinitrocresols, is added to minimize polymer formation. The bottom product, consisting of tar and polymer, is burnt. Before storage, more inhibitor is added to the styrene.

In the isothermal route, the reactor consists of a series of tubes packed with catalyst through which the ethylbenzene–styrene mixture flows. The heat required for the reaction comes from the hot flue gas of the shell and tubes exchanger. The heating medium is a molten salt mixture consisting of sodium, lithium and potassium carbonates. With a steam:ethylbenzene ratio of 0.6–0.9, the reactor is operated under vacuum at around 600°C.

The advantages of the isothermal dehydrogenation are lower steam:hydrocarbon ratios, lower energy and cooling water requirements and a more favourable conversion equilibrium with a greater catalyst efficiency.

In spite of these advantages, the adiabatic route is the favoured one due to lower capital costs, greater economy of scale and the difficulties of loading catalyst into a tubular reactor.

A development to improve the production of styrene by removing the hydrogen produced has been marketed by UOP. Originally called the Styro-Plus process but now renamed the SMART SM process, it claims a styrene selectivity of greater than 96 mole% with an ethylbenzene conversion of 82%. This is achieved by the oxidative removal of hydrogen using a noble metal catalyst, with the reactor operating temperature kept at 620–645°C.

The major difference between this process and the adiabatic dehydrogenation route is the addition of oxygen between the stages of the adiabatic reactor. A demonstration unit has been operating successfully with a 30% increase in production and several licences have been sold.

Reaction

$C_6H_6 + CH_2{=}CH_2 \rightarrow C_6H_5CH_2CH_3$

$C_6H_5CH_2CH_3 \rightarrow C_6H_5CH{=}CH_2 + H_2$

Raw material requirements and yield

Raw materials required per tonne of styrene:

Benzene	780kg
Ethylene	284kg

Yield 90%

2. From ethylbenzene as a co-product of propylene oxide

Styrene can be produced as a co-product of the oxidation of ethylbenzene followed by the dehydration of the α-methylbenzyl alcohol formed. (*See Figure 110*)

In this two-step process, ethylbenzene is oxidized with air in the liquid phase at a temperature in the region of 130°C and a pressure of 22–55 bar to ethylbenzene hydroperoxide. The temperature is maintained by evaporation and condensation of the ethylbenzene in an external heat exchanger. Around 20–25% of ethylbenzene is converted per pass, higher yields of the peroxide being obtained at low conversion rates. The oxygen concentration is controlled in order to maximize the hydroperoxide yield and maintain a non-explosive condition within the reactor. Acetophenone and α-methylbenzyl alcohol are formed as by-products.

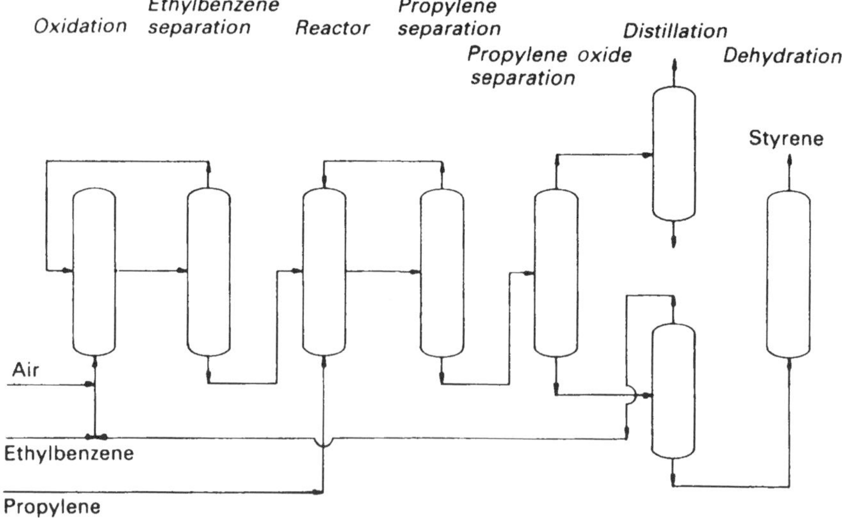

FIGURE 110 Styrene from ethylbenzene as a co-product of propylene oxide

The reactor effluent is distilled and any unreacted ethylbenzene is recycled. The ethylbenzene hydroperoxide is concentrated and fed together with excess propylene into an epoxidation reactor. In the presence of a metallic catalyst, propylene oxide and α-methylbenzyl alcohol are formed. The reaction takes place in the liquid phase at 100–130°C and a pressure of 17–55 bar. A range of different catalyst systems have been proposed for the epoxidation but the most popular are molybdenum naphthenate, coordination compounds of molybdenum hexacarbonyl and oxyacetyllactonate dissolved in the ethylbenzene–α-methylbenzyl alcohol mixture, or titanium on silica gel.

The reaction mixture (containing propylene oxide, ethylbenzene, propylene, catalyst, α-methylbenzyl alcohol and acetophenone) is separated by multi-distillation. Any unreacted propylene is removed overhead and recycled, before the propylene oxide is separated from the remaining product stream. After washing with a caustic solution to remove the catalyst, any unreacted ethylbenzene is isolated and recycled. To improve yields, acetophenone is hydrogenated to α-methylbenzyl alcohol over a catalyst consisting of a mixture of copper and zinc oxides in a slurry or fixed-bed reactor at 80–130°C and 84 bar pressure. This stream is combined with the α-methylbenzyl alcohol from the mixture and dehydrated to styrene at 200–250°C and 35 bar pressure over a titanium dioxide or silica gel catalyst in the vapour phase. The resultant styrene is further purified before use. The styrene to propylene oxide ratio is between 2.2 and 2.7 in commercial plants compared to a theoretical value of 1.8.

Although this process produces styrene at a higher cost than the ethylbenzene route, the propylene oxide which is co-produced can make the operation profitable, depending on the value of both products.

High-purity styrene is more easily generated by the propylene oxide route because of its easier boiling-point separation from other co-products. Treatment and disposal costs of waste streams, such as catalyst–caustic solutions and oxidation off-gases, are higher than in the ethylbenzene process.

Reaction

$$C_6H_5CH_2CH_3 + O_2 \rightarrow C_6H_5C(OOH)HCH_3$$

$$C_6H_5C(OOH)HCH_3 + CH_3CH{=}CH_2 \rightarrow C_6H_5CH(OH)CH_3 + CH_3CHOCH_2$$

$$C_6H_5CH(OH)CH_3 \rightarrow C_6H_5CH{=}CH_2 + H_2O$$

Raw material requirements and yield

Raw materials required per tonne of styrene:

Ethylbenzene	263kg

Yield 90%

PROPERTIES

Colourless, mobile, oily liquid with an odour of hyacinths. Polymerizes slowly in light, but rapidly on heating. Miscible with most organic solvents in any ratio but only sparingly soluble in water. Forms an azeotropic mixture at standard pressure of 66wt% styrene and 34wt% water with a minimum boiling point of 94.8°C. Flammable, burning with a sooty flame.

Molecular Weight	104.15
Density at 20°C	0.905
Freezing Point	–30.6°C
Boiling Point	145.2°C
Autoignition Temperature	490°C
Explosive Limits in air	
lower	1.1vol%
upper	6.1vol%
Flash Point Closed Cup	31°C
Vapour Density (air = 1)	3.6

Exposure Limit COSHH (Schedule 1)	250ppm 15 minutes (Maximum Exposure Limit)
	100ppm 8 hour TWA (Maximum Exposure Limit)

Being reviewed under the Existing Substances Regulations.

Exposure Limit ACGIH	100ppm 15 minutes TLV-STEL (skin)
	50ppm 8 hour TWA-TLV

Exposure Limit of TLV-TWA 20ppm and TLV-STEL 40ppm proposed.

GRADES

Technical 99%, inhibited with 10–15ppm *p-tert*-butyl catechol (TBC) with trace amounts of oxygen to make it effective. Polymer and rubber grades are available.

INTERNATIONAL CLASSIFICATIONS

UN No.	2055
CAS Reg No.	100–42–5
EINECS No.	202–851–5
EC No.	601–026–00–0
Description	Flammable liquid
Emergency Action Code	3Y
Packing Group	III
HI (Kemler Code)	39

APPLICATIONS

The importance of styrene has increased rapidly since World War II until today it is a major monomer. The most important outlet for styrene, which accounts for 64% of total demand, can be broken down into 49% for the manufacture of general-purpose and high-impact polystyrene and 15% going to expanded polystyrene.

Two other important derivatives of styrene, utilizing approximately 12% of styrene demand each, are ABS/SAN resins and SBR rubbers and latexes. Unsaturated polyester resins used in reinforced plastics consume a further 5% of styrene, while other copolymers account for the remainder.

The world market for styrene grew at around 7–8% in 1997, due to increased demand for polystyrene in such outlets as video cassettes, disks and food packaging. In 1998, growth is forecast to be around 3% due to the slow-down in Asia. In the 1990s, the most rapid increase in capacity has taken place in the Far East.

HEALTH AND HANDLING

Styrene vapour is slightly toxic and has an irritating effect on the eyes and respiratory tract. Prolonged skin contact from contaminated clothing can cause irritation and blistering. When handling the product, goggles, rubber gloves, boots and protective clothing must be worn. Smoking, eating and drinking must be forbidden in all work areas.

Stainless steel containers are normally used for storage. Rubber- and copper-containing materials must be avoided because they cause discoloration. Copper is also a polymerization inhibitor.

Styrene polymerizes readily and must not be stored unless it is inhibited and then only for short periods. The storage period decreases rapidly as the surrounding temperature rises, so containers must be kept as cool as possible, with an alarm system which triggers above 30°C. Regular checks on inhibitor levels must be carried out to ensure that minimum levels of 10–15ppm are maintained by the addition of extra inhibitor if necessary. The rate of polymerization is accelerated by sunlight, peroxides and strong acids. At temperatures above 52°C the inhibitor is no longer effective.

Styrene is highly flammable and forms explosive mixtures in air. As styrene can hold high static electric charges, grounding measures are required during transfer to prevent an electrical discharge. Storage areas must be well ventilated and monitored regularly for vapour leaks to reduce the fire hazard.

In the event of spills, extinguish all forms of ignition and absorb with dry sand or earth. Residues can be washed away with large quantities of water. Carbon dioxide, foam or dry chemical are effective to extinguish fires. Firefighting and clean-up personnel must wear protective clothing and self-contained breathing apparatus.

The storage and transportation of styrene are controlled in most European countries and containers must be labelled in accordance with local regulations.

MAJOR PLANTS

Plants with capacities greater than 370 000 tonnes per year:

BASF	Antwerp	Belgium
Elf Atochem	Gonfreville	France
BASF	Ludwigshafen	Germany
BP Chemicals	Marl	Germany
Dow Chemical	Terneuzen	Netherlands
Shell Nederland Chemie	Moerdijk	Netherlands
EniChem	Mantua	Italy
Complex	Shevchenko	Russia
Amoco	Texas City	US
Chevron Chemicals	St. James	US
Cos-Mar	Carville	US
Dow Chemical	Freeport	US
Huntsman Chemical	Bayport	US
Lyondell	Channelview	US
Sterling Chemicals	Texas City	US
Shell Canada	Scotford	Canada
SADAF	Al Jubail	Saudi Arabia
Asahi Chemical	Mizushima	Japan
Idemitsu Petrochemical	Tokayama	Japan
Yukong Arco	Ulsan	South Korea

MAJOR LICENSORS

ABB Lummus Global
Arco Technology/Koppers
Badger/Fina Technology
BASF
CDTECH
Cosden Technology/Badger
Dow Chemical
Glitsch
Halcon-Scientific Design
Lurgi
Monsanto
Shell Development
Uhde
UOP

Terephthalic Acid (TPA) & Dimethyl Terephthalate (DMT)

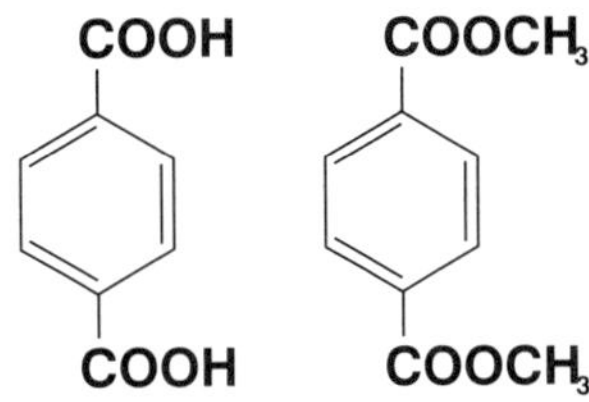

SYNONYMS

TEREPHTHALIC ACID 1,4-benzenedicarboxylic acid

DIMETHYL TEREPHTHALATE dimethyl *p*-phthalate, 1,4-benzenedicarboxylic acid dimethyl ester

Terephthalic acid (TPA) was not produced in any quantity until the 1950s, when ICI and Du Pont commercialized polyester fibres.

Initially, the acid was produced by the oxidation of *p*-xylene using dilute nitric acid, but later, in order to reduce the volume of acid required, air was used. Both these routes have been replaced by the catalytic, liquid-phase air oxidation of *p*-xylene, Amoco's process being the dominant one in use.

Although many variants of this process exist, all use a cobalt salt in acetic acid solvent as catalyst, promoted by sodium bromide. Acetaldehyde, paraldehyde or methyl ethyl ketone have been suggested as alternative promoters and these are co-converted to acetic acid.

Alternative routes developed in Japan and based on earlier technology by Henkel, use phthalic anhydride as the starting material. In the presence of carbon dioxide and a catalyst consisting of cadmium oxide or zinc oxide, the potassium salt of phthalic anhydride is disproportionated to dimethyl terephthalate (DMT). TPA is recovered from the ester by treatment with sulphuric acid.

Mitsubishi Gas Chemical has researched a process based on phthalic anhydride derived from toluene. Boron trifluoride, a complex between toluene and hydrogen fluoride, can be carbonylated with carbon monoxide to form a *p*-tolualdehyde complex. *p*-Tolualdehyde obtained after decomposition of the complex can be oxidized in an aqueous solution using manganese–bromine catalysts to give terephthalic acid. Although toluene is a cheaper feedstock than *p*-xylene and acetic acid is not required, these advantages are counterbalanced by the complex and expensive handling required by carbon monoxide and the catalyst. The process has not been commercialized.

Considerable effort has been made by numerous producers to upgrade the purity of TPA made by these routes so that it will reach the standard demanded by fibre manufacturers. Because of difficulties with purification of TPA, esterification procedures have been developed in an attempt to obtain fibre-grade DMT.

To minimize by-product formation, esterification is carried out at high temperatures and pressure and the ester is separated by two-stage vacuum distillation. In spite of the use of specialized techniques, most TPA is used as polymer-grade TPA or converted to DMT.

The majority of fibre-grade DMT is manufactured by the Dynamit Nobel technology based on *p*-xylene using oxidation and esterification steps.

Although DMT and TPA can be used for the production of polyester, TPA offers cost advantages over the ester due to:

- higher yield of polyester;
- lower ethylene glycol requirement;
- no recovery of by-product methyl alcohol.

Increasingly, polyester resins are being made directly from TPA to avoid the recovery and handling of by-product methyl alcohol and to increase productivity due to the lower molecular weight of the acid over the ester.

Processes now obsolete have involved the oxidation of naphthalene to phthalic anhydride, its conversion to dipotassium *o*-phthalate and subsequent isomerism to dipotassium terephthalate in the presence of carbon dioxide, as well as the rearrangement of benzene carboxylic acids, such as phthalic acid and benzoic acid.

p-Xylene is the dominant raw material used for the production of TPA and DMT, all TPA and over 90% of DMT coming from this aromatic. Amoco, including its joint venture partners, is the world's largest producer with over 70% of installed world capacity.

TPA plant capacities range from 50 000 to 810 000 tonnes per year, while those of DMT lie between 24 000 and 460 000 tonnes per year.

PROCESSES

1. Liquid-phase oxidation of *p*-xylene to TPA

A solution of *p*-xylene in acetic acid, catalyst and compressed air are continuously fed into a reactor. The catalyst is usually a cobalt salt, or cobalt–manganese salts promoted by bromine, hydrogen bromide, sodium bromate or tetrabromoethane. Acetaldehyde, paraldehyde or methyl ethyl ketone have been suggested as alternative promoters and these are co-converted to acetic acid. (*See Figure 111*)

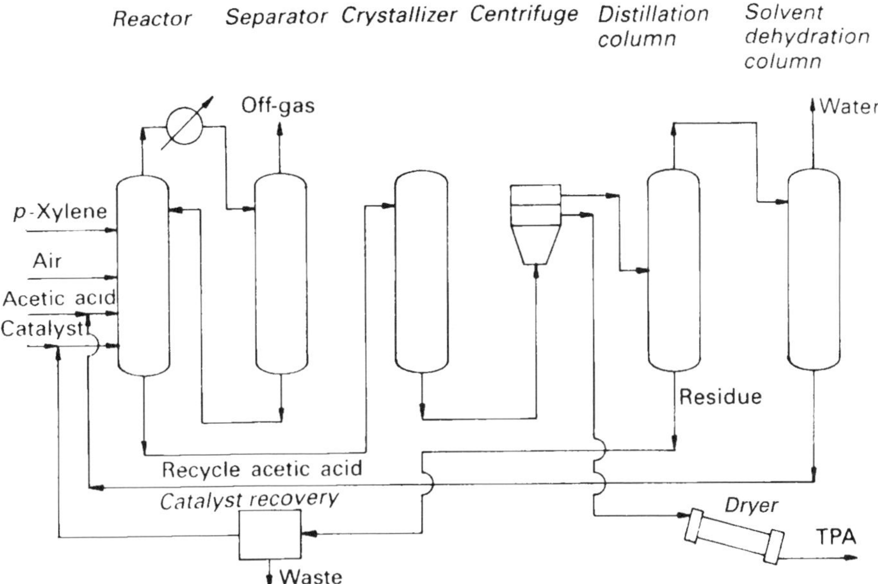

FIGURE 111 Terephthalic acid from *p*-xylene by oxidation

The reaction takes place in the liquid phase at a temperature of 175–230°C and pressure of 15–30 bar; the heat generated being removed by refluxing and condensing acetic acid. The *p*-xylene is fully oxidized but air is present in excess to ensure high *p*-xylene conversion and to reduce by-product formation.

Because of its limited solubility in acetic acid, the TPA produced forms a slurry which is removed from the base of the reactor. Pressure is reduced and unreacted xylene, acetic acid and water are flashed off using heat generated by the exothermic reaction, before centrifuging to recover the crystals from the mother liquor and catalyst.

After repeated washing with pure acetic acid, the TPA crystals are dried. Acetic acid is recovered from the mother liquor by distillation and any water is removed prior to recycle. Residues from the distillation column can be processed to recover the catalyst.

Many modifications to the process have been proposed, such as the use of pure oxygen and methyl ethyl ketone as activators to improve the purity of the TPA formed. Most of the remaining impurities are removed by washing with hot acetic acid. In spite of such treatment, the TPA obtained does not usually meet the standard required for fibre-grade material.

Reaction

$$p\text{-}(CH_3)_2C_6H_4 + 3O_2 \rightarrow p\text{-}(HOOC)_2C_6H_4 + 2H_2O$$

Raw material requirements and yield

Raw materials required per tonne of TPA (technical):

p-Xylene	650kg
Acetic acid	57kg

Yield 95%

2. From terephthalic acid by esterification to DMT

To produce DMT, TPA is esterified with methyl alcohol. The process can be carried out in one or two stages. Crude TPA and excess methyl alcohol are preheated and pressurized prior to being fed into an esterification reactor. The liquid-phase reaction is carried out at 250–300°C and although a catalyst is not required, one is frequently used to improve yields obtained. The crude DMT formed is recovered and impurities are catalytically oxidized with air and removed prior to distillation.

Reaction

$$C_6H_4(COOH)_2 + 2CH_3OH \rightarrow C_6H_4(COOCH_3)_2 + 2H_2O$$

Raw material requirements and yield

Raw materials required per tonne of DMT (fibre-grade):

p-Xylene	580kg
Methyl alcohol	370kg
Acetic acid	50kg

Yield 95%

3. Dimethyl terephthalate from *p*-xylene and methyl alcohol

p-Xylene, air and catalyst are fed into an oxidizer vessel where they are mixed with recycle *p*-methyl toluate. The catalyst consists of a cobalt salt, but combinations of cobalt and manganese salts are also used. Oxidation takes place at 140–170°C at a pressure of 4–8 bar, heat produced by the reaction being removed by vaporization of the *p*-xylene and water. The aqueous mixture is condensed in a separator and the *p*-xylene is recycled while water is run to waste. (*See Figure 112*)

The oxidate, containing *p*-toluic acid and monomethyl terephthalate, flows into a tower containing methyl alcohol. The esterification is carried out in the liquid phase at 250–280°C and a pressure of 20–25 bar. Excess methyl alcohol from the esterifier is condensed from the top of the tower, distilled, separated from water and recycled.

The esterification mixture is passed to a distillation column where *p*-methyl toluate is collected overhead and returned to the oxidizer. The bottoms are vacuum

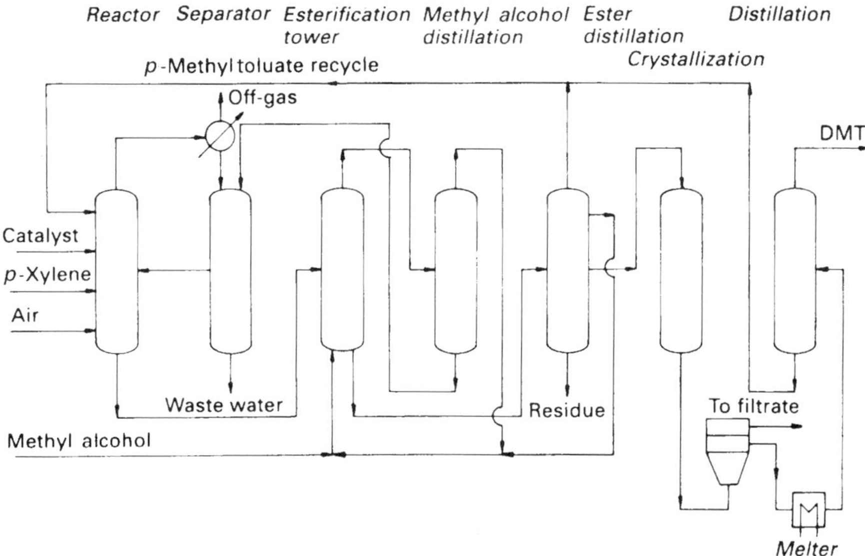

FIGURE 112 Dimethyl terephthalate from *p*-xylene and methyl alcohol

distilled and crude DMT is obtained overhead. Purification is carried out by two-stage crystallization from methyl alcohol followed by distillation.

Reaction

Oxidation.

$p\text{-}(CH_3)_2C_6H_4 + 3O_2 \rightarrow C_6H_4(COOH)_2 + 2H_2O$

Esterification

$C_6H_4(COOH)_2 + 2CH_3OH \rightarrow C_6H_4(COOCH_3)_2 + 2H_2O$

Raw materials required and yields

Raw materials required per tonne of DMT (fibre-grade):

p-Xylene	610kg
Methyl alcohol	400kg

Yield 87%

4. Purification of TPA

Several different methods of purifying TPA to fibre-grade material have been utilized or based on hydrogenation over a noble metal catalyst at 250°C under slight pressure. Various impurities present in the crude slurry are hydrogenated in the liquid phase to colourless products which remain in solution when TPA is crystallized out.

Pure DMT can be hydrolysed at 150–200°C in the presence of aqueous potassium chloride to yield polymer-grade TPA.

PROPERTIES

TPA White crystalline powder with slightly acidic odour. Insoluble in water but slightly soluble in ethyl alcohol. Soluble in ether and acetic acid. Reacts with alkalis and pyridene.

DMT Colourless crystals. Insoluble in water, but soluble in ether and hot ethyl alcohol.

	TPA	*DMT*
Molecular Weight	166.14	194.19
Specific Gravity at 20°C	1.51	1.04
Melting Point	402°C sublimes	140.6°C
Boiling Point		284°C
Autoignition Temperature	496°C	570°C
Explosive Limits in air		
dust	0.05g/litre	0.05g/litre
Flash Point Open Cup	271°C	154°C
Vapour Density (air = 1)	5.74	
Exposure Limit COSHH		10mg/m^3 (15 mins) 5mg/m^3 8 hour TWA
Exposure Limit ACGIH (powdered material–total dust)	10mg/m^3	No limit set

GRADES

Technical 95%, fibre 99.6%.

INTERNATIONAL CLASSIFICATIONS

TPA

UN No.	Not allocated
CAS Reg No.	584–84–9

DMT

UN No.	Not allocated
CAS Reg No.	120–61–6
EINECS No.	204–411–8

APPLICATIONS

The major outlet for TPA and DMT is in the production of polyester fibres. Around 75% of world consumption is utilized for this purpose. As the fibre market is mature, this percentage is expected to decline within the next few years. Approximately 10% is used for the manufacture of food and beverage containers, and 7% for polyester film which finds outlets in photography, computers, audio-visual equipment and packaging. The remainder is consumed by technical plastics such as polyester resins for electrical fittings, polybutylene terephthalate and poly-acrylate resins.

Future demand is expected to grow at up to 2% per year but this will be dependent on the prospects for the fibre market.

HEALTH AND HANDLING

Both TPA and DMT are low-risk toxic hazards to health. Dust is an irritant to the eyes, skin and lungs and prolonged exposure should be avoided. Dust masks, protective gloves and eye protection to exclude dust should be worn. Any contaminated clothing must be laundered before reuse.

Both products can be stored in closed containers in a cool, well-ventilated area with local exhaust to remove fine particles which may exist in the air. Dust build-up must be avoided as severe explosions can occur especially at levels of 0.05g/litre.

Electrical equipment must be earthed where powder is handled to prevent static build-up. Spilled material can be swept up and disposed of according to local regulations. Clean-up staff should wear protective clothing.

Carbon dioxide or dry chemicals are recommended for fighting fires, and protective clothing and breathing apparatus must be worn on these occasions.

MAJOR PLANTS

TPA

Plants with capacities greater than 250 000 tonnes per year:

ICI	Wilton	UK
Amoco	Charleston	US
	Decatur	US
Du Pont	Wilmington	US
Arabian Industrial Fibre	Yanbu	Saudi Arabia
Mitsubishi Chemical	Kurosaki	Japan
Mitsui Petrochemical Industries	Iwakuni	Japan
Mizushima Petrochemical	Mizushima	Japan
Samsung Petrochemical	Ulsan	South Korea
China Petrochemical	Kaohsiung	Taiwan
	Linyuan	Taiwan
Fibre Intermediates	Tao Liao	Taiwan
Formosa Chemical & Fibre	Changhua	Taiwan
Amoco	Kuantan	Malaysia

DMT

Plants with capacities greater than 100 000 tonnes per year:

Hoechst	Ausberg	Germany
	Offenbach	Germany
Huels	Niederkassel	Germany
	Steyerberg	Germany
Hoechst	Vlissingen	Netherlands
Montefibre	Acerra	Italy
Complex	Mogilev	Russia
Cape Industries	Wilmington	US
Du Pont	Old Hickory	US
	Wilmington	US
Eastman Kodak	Columbia	US
	Kingsport	US
Petrocel	Altamira	Mexico
Temex	Cosoleacaque	Mexico
Teijin	Ehime	Japan
Toray Industries	Mizushima	Japan

MAJOR LICENSORS

Amoco
Du Pont
Dynamit Nobel
Eastman Chemical
Glitsch
Hercules
Hoechst
Huels
Inventa
Mitsubishi Chemical
Mitsui Petrochemical
Teijin
Toray Industries

Toluene

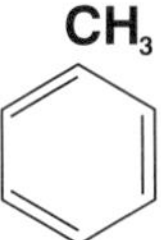

SYNONYMS

TOLUENE toluol, methylbenzene, phenylmethane

Toluene, like benzene, is available from two different sources: petroleum, which is the prime raw material, and light oil, formed by the carbonization of coal. It is obtained from reformates or, since the mid-1960s, produced along with benzene, xylenes and C_9 aromatics by catalytic reforming of naphtha streams.

Because petroleum contains a mixture of aromatics, if toluene is the desired end product, hydroforming streams which are rich in dimethylcyclopentane, methylcyclohexane, and ethylcyclopentane are chosen. The reformate is extracted with a solvent to give a mixture of aromatics which are then separated by fractionation. Typically the composition of the reformate is 13% toluene, 18% xylenes and 3% benzene. Toluene is also present together with other aromatics in pyrolysis gasoline.

The production of toluene, which takes place at refineries and petrochemical complexes, is either used directly for gasoline blending to raise the octane content or processed into pure toluene feedstock. Unfortunately, toluene has few uses as a petrochemical raw material apart from the production of tolylene diisocyanate, benzoic acid, benzyl chloride, nitrobenzene and phenol. Because of the volumes of toluene present in benzene–toluene–xylene (BTX) streams, large quantities are hydrodealkylated to the more valuable benzene. The proportions used in gasoline blending or for conversion to benzene are flexible depending on their economic values.

The supply pattern for toluene used as a petrochemical is as follows: 71% comes from reformate, 25% from pyrolysis gasolines and only 2% from coal and light oil. Small quantities are recovered as a by-product of styrene manufacture. A significant amount of toluene is not isolated but added (mixed with other aromatics) directly to motor fuels.

Capacities range from 10 000 to 800 000 tonnes per year.

PROCESSES

From petroleum by catalytic reforming

A hydrocarbon mixture rich in aromatics (obtained by distillation from crude petroleum and natural gasoline) is passed over a dehydrogenation catalyst. Operating conditions vary but temperatures of around 530°C and pressures of 17–21 bar are commonly employed. After cooling by passage through heat exchangers, the reaction gases are compressed and hydrogen is recycled to the reactor. High hydrogen concentrations help to maintain catalyst activity by reducing coking. (*See Figure 113*)

Light hydrocarbon gases are removed by fractionation before the reformate is sent to a tower where an aromatic-rich fraction is obtained overhead. Toluene can be recovered from the middle cut by azeotropic distillation, extractive distillation or solvent extraction. Solvent extraction is the usual route employed to recover toluene. Solvents with a higher boiling point than toluene are used and the toluene–solvent mixture is separated into its components in a recovery column. The solvent is returned to the extraction unit (see Benzene).

From the mid-1960s, pyrolysis gasoline produced by steam cracking became available as another source of toluene. Because of its instability, the pyrolysis stream must be hydrogenated before toluene can be extracted (see Benzene).

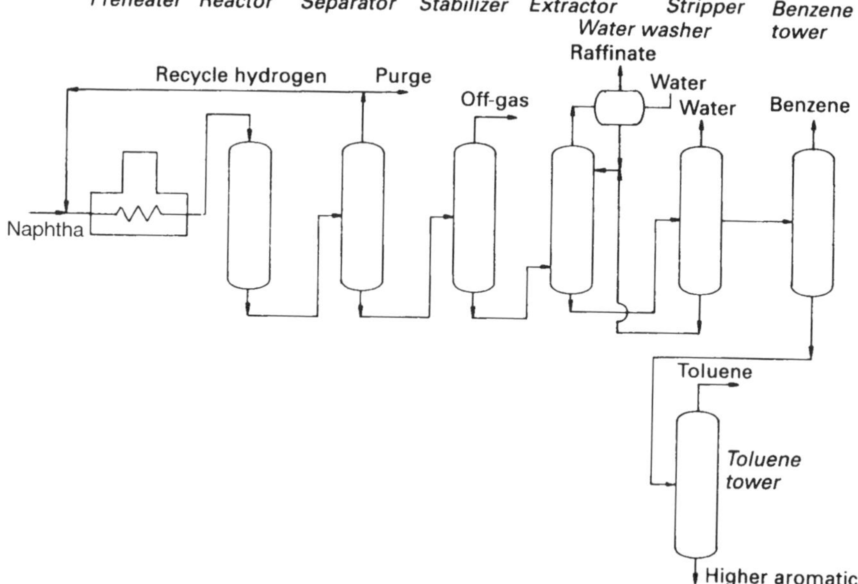

FIGURE 113 Toluene from petroleum by catalytic reforming

Reaction and yield

$C_6H_{11}CH_3 \rightarrow C_6H_5CH_3 + 3H_2$

Yield 85–90%

OTHER PROCESSES

From coal

Light oil, obtained as a by-product from the high-temperature carbonization of coal, contains between 12 and 20vol% of toluene. Aromatics are removed by absorption with a high-boiling petroleum fraction and recovered by steam distillation (see Benzene).

Sulphuric acid is used to take out unsaturated compounds and sulphur. Any remaining acid is neutralized and the crude toluene is washed with water before being fractionated to yield the pure product.

PROPERTIES

Colourless, mobile liquid with an odour similar to that of benzene. Highly flammable burning with a sooty flame. Soluble in benzene, ethyl alcohol, ketones and most organic solvents, but only slightly soluble in water.

Molecular Weight	92.13
Density at 20°C	0.867
Melting Point	–94.99°C
Boiling Point	110.6°C
Autoignition Temperature	552°C
Explosive Limits in air	
lower	1.3vol%
upper	6.8vol%
Flash Point Closed Cup	4.4°C
Vapour Density (air = 1)	3.14
Exposure Limit COSHH	150ppm 15 minutes 50ppm 8 hour TWA
Exposure Limit ACGIH	50ppm TLV-TWA (skin)

GRADES

Nitration 99%, technical >90%.

INTERNATIONAL CLASSIFICATIONS

UN No.	1294
CAS Reg No.	108–88–3
EINECS No.	203–625–9
EC No.	601–021–00–3
Description	Flammable liquid
Packing Group	II
Emergency Action Code	3YE
HI (Kemler Code)	33

APPLICATIONS

The major outlet for toluene is in gasoline, because it increases the octane-rating. Catalytic reformate, which is rich in aromatics, is used to upgrade gasoline and toluene is the preferred blending component. In the US, in order to meet Clean Air Act requirements, toluene has been replaced by ethyl alcohol.

Of the tonnage worldwide used as a chemical, 54% is converted to benzene; the percentage varying with the price of benzene because it is a more expensive process than catalytic reforming. Approximately 16% of toluene is used as a solvent, a declining outlet as its use in paints has diminished due to legislation to reduce vapour emissions.

The fourth major outlet, accounting for 8% of total demand worldwide, goes into the manufacture of tolylene diisocyanate, which is used in polyurethane foams.

Other minor uses include the manufacture of caprolactam, phenol and gasoline additives. Explosives, which once formed the major outlet for toluene, have been replaced by cheaper modern alternatives. World demand for toluene is expected to grow very slowly.

HEALTH AND HAZARDS

Toluene is an irritant to the eyes, skin, nose and lungs. It has a degreasing effect on the skin and contact should be avoided. The greatest danger is caused by inhalation as it is a more powerful narcotic and has a greater acute toxicity than benzene. Exposure leads to drowsiness, confusion and headaches. Because toluene does not lead to the harmful blood changes caused by benzene, it is preferred industrially where it can be used. It is not considered to be a carcinogen.

Toluene should be stored in closed containers vented to the atmosphere away from working areas. It is not corrosive to metals, and iron, mild steel, copper or aluminium are suitable materials for their fabrication. Contact with strong oxidizing agents must be avoided.

Great care must be exercised when handling toluene to prevent skin contact, and protective clothing and good ventilation are essential. Goggles, neoprene aprons and gloves and protective clothing must be worn at all times. Any contaminated clothing must be removed and dried out in a safe area away from personnel. Equipment must be non-sparking and earthed to prevent static build-up.

Leakage of toluene into soil or water is considered to be a serious pollution problem. Spills must be absorbed with earth and sand and disposed of in accordance with local regulations. Flushing into sewers or waterways is not permitted because of its flammability. Toluene is a dangerous fire hazard as its heavy vapours can lead to a flashback. Fires can be extinguished with carbon dioxide, dry chemical or alcohol foam. Water streams tend to scatter the flames and should not be used. Clean-up and firefighting staff must wear protective clothing and self-contained breathing apparatus.

The movement and labelling of toluene is closely controlled because of its toxicity and flammability.

MAJOR PLANTS

Plants with capacities greater than 200 000 tonnes per year:

DEA	Wesseling	Germany
Exxon Chemical	Botlek	Netherlands
Dow Chemical	Terneuzen	Netherlands
Agip Petroli	Priolo	Italy
CEPSA	Algeciras	Spain
ICI	Middlesbrough	UK
Amoco	Texas City	US
BP Chemicals	Alliance	US
	Lima	US
Chevron Chemicals	Port Arthur	US
Exxon Chemical	Baytown	US
Koch Refining	Corpus Christi	US
Mobil Chemical	Beaumont	US
Phillips Petroleum	Guyama	Puerto Rico
Sun Chemicals	Sarnia	Canada
Hess Oil	St Croix	Virgin Islands
Pemex	La Cangrejera	Mexico
Copene	Camaraci	Brazil
Idemitsu Petrochemical	Hokkaido	Japan
Mitsubishi Oil	Kurashi	Japan
Chinese Petroleum	Kaoshsiung	Taiwan
	Linyuan	Taiwan
Honam Oil	Yeochon	South Korea
Yukong	Ulsan	South Korea

Most toluene capacities are approximate as operations offer considerable flexibility.

MAJOR LICENSORS

BP/UOP
Glitsch
HRI Inc.
IFP
Krupp-Koppers
Leunawerke
Lurgi
Shell Development
Snamprogetti

2,4-Tolylene Diisocyanate (TDI) & Diphenylmethane Diisocyanate (MDI)

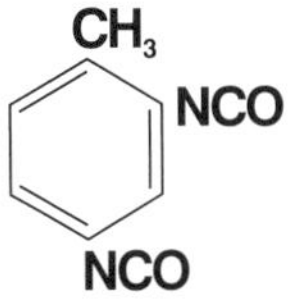

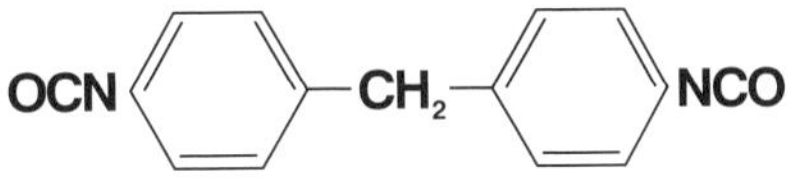

SYNONYMS

2,4-TOLYLENE DIISOCYANATE 2,4-tolyl diisocyanate, 2,4-toluene diisocyanate, 2,4-diisocyanate, 2,4-diisocyanotoluene

DIPHENYL METHANE DIISOCYANATE diphenylmethane-4,4-diisocyanate, 4,4-methylenebis(phenyl isocyanate), methylene di-*p*-phenyl isocyanate

Isocyanates have become important only since the 1950s, when polyurethanes resins became commercial products. The two isocyanates which form the basis of the industry are tolylene diisocyanate (TDI) and diphenylmethane diisocyanate (MDI). Some aliphatic amines are used in the paint and lacquer industry, or as UV retardants against yellowing in polymers.

Industrial preparation of isocyanates is by the reaction of the corresponding amine with phosgene in an aromatic solvent. The choice of catalyst and reaction conditions in these processes are particularly important because of the reactivity of isocyanates and hence the possibility of side reactions.

TDI can be produced directly from dinitrotoluene by liquid-phase carbonylation with *o*-dichlorobenzene. The major advantage of this route is that it avoids the use of phosgene (undesirable because of its volatility and toxicity), as well as the waste recovery problems associated with hydrochloric acid.

One-step processes based on the reaction between dinitrotoluene and carbon monoxide have been developed, but severe reaction conditions have led to their abandonment. They have been replaced by two-step routes using selenium catalysts. MDI is produced by the condensation of aniline with formaldehyde.

Currently, the most important route for the manufacture of TDI is from toluene via the phosgene route, followed by dinitrotoluene carbonylation and a smaller amount from toluene diamine. The world's major producers are Union Carbide in the US and Bayer in Western Europe.

Capacities range from 12 000 to 140 000 tonnes per year for TDI, and 20 000 to 245 000 tonnes per year for MDI.

PROCESSES

From 2,4-tolylenediamine and phosgene

Toluene is nitrated directly with a mixture of nitric and sulphuric acids to nitrotoluene which is catalytically hydrogenated in the liquid phase to the corresponding diamine. (*See Figure 114*)

A solution of the diamine, either the 2,4 or 2,6 isomer or a mixture of both in an aromatic solvent (usually xylene, monochlorobenzene or *o*-dichlorobenzene), is mixed with a solution of phosgene in the same solvent in a reaction vessel. The temperature of the reactor is kept around 60°C. A 50–200% excess of phosgene is used to reduce side reactions.

The resultant slurry is digested in stages for several hours in three phosgenerators and the temperature is raised progressively to 180°C as more phosgene is added until the phosgenation is complete. Dry hydrogen chloride gas is blown into the mixture to reduce the activity of the free amine. Hydrogen chloride is removed by blowing an inert gas through the solution at a temperature of 110–115°C. Alternatively, phosgene can be dissolved in the same solvent and the reaction carried out at 0–5°C to minimize side reactions.

The solution of crude 2,4-tolylene diisocyanate is fractionated to remove any unreacted phosgene and solvent for recycling and hydrogen chloride before being distilled under slight vacuum. The product is collected overhead.

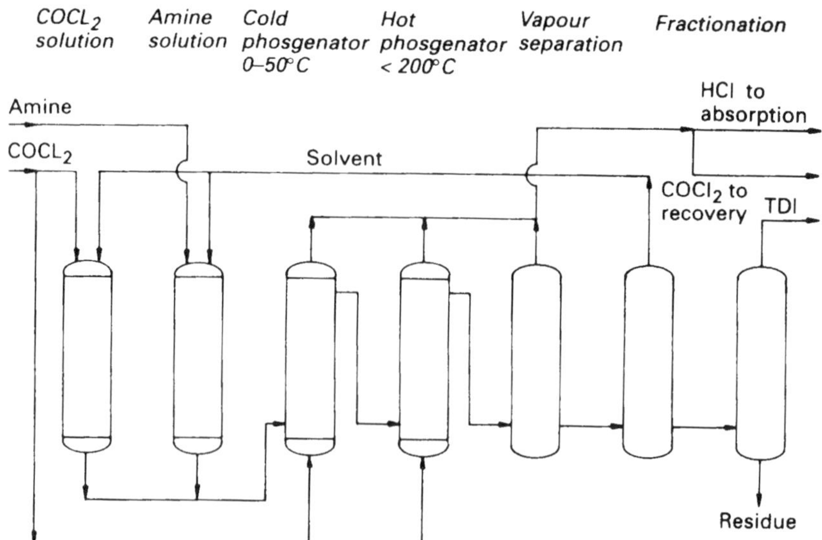

FIGURE 114 Tolylene diisocyanate from 2,4-tolylenediamine and phosgene

MDI can be made from the diamine, formed by the reaction of aniline and formaldehyde, by the same route as TDI.

Reaction

$C_6H_3CH_3(NH_2)_2 + 2COCl_2 \rightarrow C_6H_3CH_3(NHCOCl)_2 + 2HCl$

$C_6H_3CH_3(NHCOCl)_2 \rightarrow C_6H_3CH_3(NCO)_2 + 2HCl$

Raw material requirements and yield

Raw materials required per tonne of tolylene diisocyanate:

2,4-Tolylenediamine	875kg
Phosgene	1420kg

Yield 85%

PROPERTIES

TDI

White to pale yellow liquid with a pungent odour. Soluble in aromatic and chlorinated aromatic hydrocarbons, ether, acetone, and esters. Reacts with water, ammonia and alcohols with the evolution of carbon dioxide and oxidizing agents.

MDI

White to pale yellow flakes with a slightly musty odour.

	TDI	*MDI*
Molecular Weight	174.16	250.3
Specific Gravity at 20°C	1.22	1.19
Melting Point	14°C	37.2°C
Boiling Point	251°C	230°C
Autoignition Temperature	620°C	232°C
Flammability Limits in air		
lower	0.9vol%	Not listed
upper	9.5vol%	Not listed
Flash Point Open Cup	132°C	199°C
Vapour Density (air = 1)	6.0	8.5
Exposure Limit COSHH (Schedule 1)		
15 minutes STEL (Maximum Exposure Limit)	0.02mg/m^3	0.02mg/m^3
8 hour TWA (Maximum Exposure Limit)	0.07mg/m^3	0.07mg/m^3
Exposure Limit ACGIH		
TLV-STEL	0.02ppm	None established
TLV-TWA	0.005ppm	0.005ppm

GRADES

Pure 2,4 isomer 99%, technical mixtures of 2,4 and 2,6 isomers 65:35 and 80:20.

INTERNATIONAL CLASSIFICATIONS

TDI

UN No.	2078
CAS Reg No. (2–4D)	584–84–9
EINECS No. (2–4D)	209–544–5
EC No.	615–006–00–4
Description	Toxic substance
Packing Group	II
Emergency Action Code	2XE

MDI

UN No.	2489
CAS Reg No.	101–68–8
EINECS No.	Not allocated
EC No.	615–005–01–9
Description	Harmful substance Poison B (DOT)
Packing Group	III
Emergency Action Code	2X

APPLICATIONS

The major outlet for TDI, accounting for approximately 90% of total world demand, is in the manufacture of flexible polyurethane foams which are used in furniture, buildings and transportation. Other small outlets include rigid foams and adhesives, and sealants.

The most important outlet for MDI is rigid foams where their good insulating properties are utilized in construction and insulation applications. Automotive, appliances and other durable goods, together with adhesives, are growing sectors for MDI-based materials.

MDI and 80–20 TDI are used for the production of polyurethane coatings. Polyurethane elastomers, produced by the reaction of TDI or MDI with short-chain polyols, can be used in the manufacture of polyurethane fibres. Globally, growth rates of around 4% per year for TDI, and 6% for MDI are forecast.

HEALTH AND HANDLING

All isocyanates are hazardous and must be handled with care. The vapour causes irritation to the eyes, which could lead to permanent damage, and to the lungs, leading to symptoms of asthma and bronchitis even at low concentrations. Skin contact must be avoided, as redness, blistering and possible sensitization can result.

Personnel working with isocyanates should receive annual medical checks of pulmonary functions. Sensitized workers must keep away from all contact with TDI and MDI.

Store in sealed, stainless steel containers at room temperature in a well-ventilated area and away from compounds containing active hydrogen such as water. Exposure to moisture leads to the formation of carbon dioxide and the pressure of the gas formed could lead to the rupture of containers. To prevent moisture contamination storage under an inert gas is advised, preferably outside.

MDI is stable if kept in closed containers at a temperature below 5°C, but it has an average shelf-life of two months. In the presence of moisture, toxic compounds such as hydrogen cyanide may be formed.

Full protective clothing and self-contained breathing apparatus must be worn when dealing with spills. They should be absorbed on vermiculite or special absorbents prior to being disposed of according to local regulations. Any contaminated clothing must be removed as quickly as possible and laundered before reuse.

In the event of fire, carbon dioxide, dry chemical or water spray can be used, but as water will generate carbon dioxide, firefighters must wear self-contained breathing apparatus.

The handling and transportation of isocyanates is regulated in all developed countries.

MAJOR PLANTS

TDI

Plants with capacities greater than 30 000 tonnes per year:

Rhone-Poulenc	La Madeleine	France
	Pont de Claix	France
Bayer	Brunsbuttel	Germany
	Dormagen	Germany
EniChem	Porto Marghera	Italy
BASF	Geismar	US
Bayer	Baytown	US
	New Martinsville	US
Dow Chemical	Freeport	US
ICI	Geismar	US
Lyondell	Lake Charles	US
Petronor	Camaraci	Brazil
Mitsui Toatsu Chemical	Omuta	Japan
Takeda Chemical	Kashima	Japan
Tinguang Chemical	Lanzhou	China

MDI

Plants with capacities greater than 60 000 tonnes per year:

BASF	Antwerp	Belgium
Bayer	Krefeld	Germany
Dow Chemical	Stade	Germany
ICI	Rozenburg	Netherlands
ICI	Fleetwood	UK
BASF	Geismar	US
Dow Chemical	La Porte	US
ICI	Geismar	US
Nippon Polyurethane	Shin-Nanyo	Japan
Sumitomo/Bayer Polyurethane	Niihama	Japan

MAJOR LICENSORS

BASF
Bayer
Dow Chemical
ICI
Mitsui Petrochemical
Rhone-Poulenc

Trichloroethylene (TCE)

$Cl(H)C=CCl_2$

SYNONYMS

TRICHLOROETHYLENE ethinyl trichloride, trichloroethene, TCE, ethylene trichloride

Trichloroethylene can be produced commercially from either ethylene or acetylene. In practice, as the chlorination of acetylene leads to the production of 1,1,2,2,-tetrachloroethane, this product is dehydrochlorinated in either the liquid or vapour phase to trichloroethylene.

In most Western countries, ethylene or ethylene dichloride is generally employed as the starting material. There are two routes, either chlorination followed by dehydrochlorination or oxychlorination. The latter route has the advantage that minimal amounts of by-product hydrogen chloride are produced.

Ethylene dichloride is preferred as a feedstock to ethylene because of its availability, lower cost, higher selectivity and yield. Aliphatic chlorinated residues obtained from chlorination processes (which present considerable disposal problems) are alternative feedstocks. This option is increasingly attractive as the problem of residue destruction becomes more difficult and costly.

Other routes to trichloroethylene which have been investigated include the chlorination of ethane, pyrolysis of chloroform or carbon tetrachloride and the hydrohalogenation of perchloroethylene.

Capacities range from 10 000 to 170 000 tonnes per year.

PROCESSES

1. From ethylene dichloride

Perchloroethylene and trichloroethylene are co-produced by the chlorination–oxychlorination of ethylene dichloride. The ratio of perchloroethylene to trichloroethylene can be varied by altering the chlorine–ethylene dichloride balance (see Perchloroethylene). (*See Figure 115*)

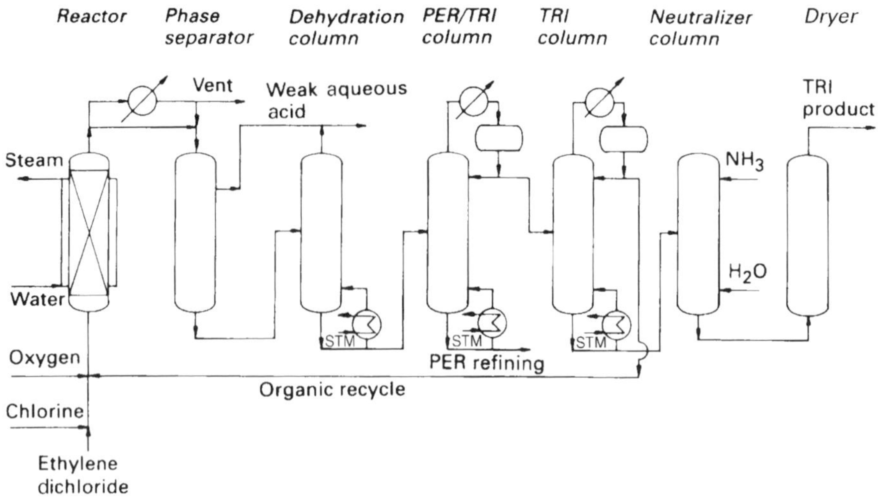

FIGURE 115 Trichloroethylene from ethylene dichloride

Reaction

$$C_2H_4Cl_2 + 2Cl_2 \rightarrow C_2HCl_3 + 3HCl$$

$$2C_2H_4Cl_2 + 1\tfrac{1}{2}Cl_2 + 1\tfrac{3}{4}O_2 \rightarrow C_2HCl_3 + 2CCl_4 + 3\tfrac{1}{2}H_2O$$

Raw material requirements and yield

Raw materials required per tonne of products: see Perchloroethylene.

2. From acetylene by chlorination

Acetylene and chlorine, preheated to around 80°C, are fed separately into a reactor where tetrachloroethane is formed. The reaction mixture is distilled to recover the tetrachloroethane.

The tetrachloroethane is dehydrochlorinated to trichloroethylene and hydrogen chloride at 250–300°C in a catalytic reactor. The catalyst usually used consists of 30% barium chloride deposited on active carbon. Heat from the exothermic reaction distills the trichloroethylene formed overhead and it is collected as an azeotrope. After removal of the hydrogen chloride by degassing, trichloroethylene is separated from the heavy ends by distillation.

Reaction

$$C_2H_2 + 2Cl_2 \rightarrow C_2H_2Cl_4$$

$$C_2H_2Cl_4 \rightarrow C_2HCl_3 + HCl$$

Raw material requirements and yield

Raw materials required per tonne of trichloroethylene:

Acetylene	210kg

Yield 90%

3. From hydrocarbons or chlorinated wastes by chlorolysis

Chloroethylenes can be produced from hydrocarbons or C_2 chlorohydrocarbon wastes by chlorination–oxychlorination.

Chlorine and hydrogen are recovered from waste chlorocarbons by pyrolysis in the presence of air at 1050°C and 6 bar pressure in a reactor. The mixed gases are fed midway into a chlorination–oxychlorination–dehydrochlorination reactor. Molten salt containing copper oxychloride and potassium chloride enters at the top. Air and the hydrocarbons are introduced towards the base. The reaction takes place at 430°C and 6 bar. The hydrocarbon feed is oxidized to chloroethylenes and the copper oxychloride is converted to cuprous and cupric chloride. Off-gases from the oxidation reactor are used to raise the molten salts to an oxidation reactor where they are oxidized with air back to copper oxychloride. Chlorine is released and recycled.

The reaction stream from the base of the reactor is washed with dilute alkali to remove carbon dioxide before the perchloroethylene and trichloroethylene formed are separated by distillation.

PROPERTIES

Colourless, non-flammable liquid with a sweet odour similar to chloroform. Soluble in ethyl alcohol and ether.

Molecular Weight	131.4
Density at 20°C	1.464
Melting Point	–87.1°C
Boiling Point	86.7°C
Autoignition Temperature	410°C
Flammability Limits in air	
lower	7.9vol%
upper	10.5vol%
Vapour Density (air = 1)	4.53

Exposure Limit COSHH (Schedule 1) 150ppm 15 minutes (Maximum Exposure Limit)
100ppm 8 hour TWA (Maximum Exposure Limit)

Being reviewed under the Existing Substances Regulations.

Exposure Limit ACGIH 100ppm TLV-STEL
50ppm TLV-TWA

Classified A5, not suspected as a human carcinogen.

GRADES

Technical 99%.

INTERNATIONAL CLASSIFICATIONS

UN No.	1710
CAS Reg No.	79–01–6
EINECS No.	201–167–4
EC No.	602–027–00–9
Description	Toxic substance
Packing Group	III
Emergency Action Code	2Z
HI (Kemler Code)	60

APPLICATIONS

Being an excellent solvent, trichloroethylene's major outlet is in the dry cleaning of fabrics where it is popular because of its low toxicity and non-flammability. It has a number of uses in the textile industry as a solvent for sizing and desizing, in textile finishes and for scouring.

Other outlets which utilize its solvent properties include metal degreasing, electroplating cleaning, extraction of oils and fats, paint removal, dissolving rubber and sulphur recovery. As a chemical intermediate, it is used in the manufacture of chlorinated fluorocarbons.

HEALTH AND HANDLING

Trichloroethylene vapour is irritating to the nose, throat and eyes; moderate exposure can cause headaches, irritability, fatigue and nausea. Over-exposure affects the central nervous system causing damage to the liver, kidneys, lungs and heart. The liquid defats the skin and dermatitis can result. The effects of trichloroethylene are enhanced by caffeine, alcohol and some drugs. In certain individuals, the synergistic effects of trichloroethylene and alcohol result in a red, blotchy skin rash. Staff handling the product must wear full protective clothing (to prevent skin contact), boots and chemical safety goggles. Contact lenses can concentrate the vapour and must be avoided. Only trained personnel should handle trichloroethylene.

Store in a cool, dry, well-ventilated area away from sunlight, strong alkalis, barium, lithium, sodium, aluminium and magnesium. If trichloroethylene is used as a degreaser, care must be taken to prevent the build-up of small particles of aluminium. Trichloroethylene is stable but at elevated temperatures, especially when moist, it can decompose to give off toxic fumes. The product is normally stabilized. Levels should be monitored regularly and good stock control practised.

If a spill occurs, personnel should be evacuated before clean-up staff contain the liquid to prevent pollution of waterways and sewers. Trichloroethylene should be placed in containers for disposal. It is non-flammable.

MAJOR PLANTS

Plants with capacities greater than 30 000 tonnes per year:

Elf Atochem	St. Auban	France
Solvay	Tavaux	France
EniChem	Assemini	Italy
ICI	Runcorn	UK
Dow Chemical	Freeport	US
PPG Industries	Lake Charles	US
Asahi Chemical	Chiba	Japan
Kanto Denka	Mizushima	Japan
Toagosei Chemical	Tokushima	Japan

MAJOR LICENSORS

ABB Lummus Global
Diamond Shamrock
Elf Atochem
Hoechst
John Brown
PPG Industries
Stauffer Chemical
Toagosei Chemical
Vulcan Materials

Urea

$$H_2N-\underset{\displaystyle \overset{\|}{O}}{C}-NH_2$$

SYNONYMS

UREA carbamide, carbamimidic acid, carbonyl diamide

Initially urea was produced industrially by the hydration of calcium cyanamide, but the ready availability of ammonia and improvements in high-pressure technology led to the development of ammonia–carbon dioxide based processes. In the two-step process, ammonium carbamate is formed by the reaction of ammonia with carbon dioxide, and this intermediate is then dehydrated to urea. Heat produced by the exothermic carbamate step is used in the endothermic urea step. Depending on the operating conditions used, conversion rates between 55 and 70% based on carbon dioxide are obtained.

Several commercial processes exist based on this reaction and, since the 1970s, improvements have concentrated on reducing energy consumption.

Urea processes can be categorized under two main groupings:

- external solution recycle systems;
- internal solution stripping systems.

The major difference between the two is that in the former energy is saved by high carbon dioxide conversion rates, while the latter reduces net energy requirements by optimizing heat recovery. Conventional processes have generally been replaced by stripping processes, only two of the former having some importance. Recent research has concentrated on combining these two elements to reduce process costs.

Capacities range from 100 000 to 2 000 000 tonnes per year.

PROCESSES

1. From ammonia and carbon dioxide

Ammonia and carbon dioxide are fed separately through a feed pump and compressor respectively into the urea synthesis reactor. Ammonia is present in a 3–3.5:1 molar ratio excess to promote urea formation. The reaction takes place at a temperature within the range of 180–210°C and a pressure of 150 bar; around 70% of the ammonium carbamate formed is decomposed to urea. (*See Figure 116*)

The reaction mixture (containing urea, ammonia, ammonium carbamate and water) is cooled to about 150°C and fed to the top of a stripper tower. Excess ammonia is separated from the feed at the top of the tower. The solution passes through a decomposer where fresh carbon dioxide decomposes the carbamate to its constituents. This process is frequently carried out in two or three decomposers operating at reducing pressures. The aqueous urea solution is concentrated either by direct evaporation or by crystallization followed by centrifuging. The crystals can be melted to yield pure urea. Granular or prilled urea is produced in the granulator or prill tower directly from the urea melt. The mother liquor remaining after centrifuging is used in the manufacture of fertilizers.

Several different techniques have been developed to decompose the unconverted carbamate and recover ammonia and carbon dioxide. All are aimed at efficient recovery of heat and a reduction in energy costs.

By increasing the conversion rate of carbon dioxide to urea in the urea reactor, the amount of heat subsequently required to transform the unconverted ammonium carbamate is reduced. In the stripping process, undecomposed ammonium carbamate from the urea reactor is broken down using ammonia present in the solution. The vapours pass to a condenser under high pressure where, at a temperature of 200°C, 85% conversion is achieved. Heat from the reaction is used to generate high-pressure steam which can be used for processing, thus effecting considerable energy savings.

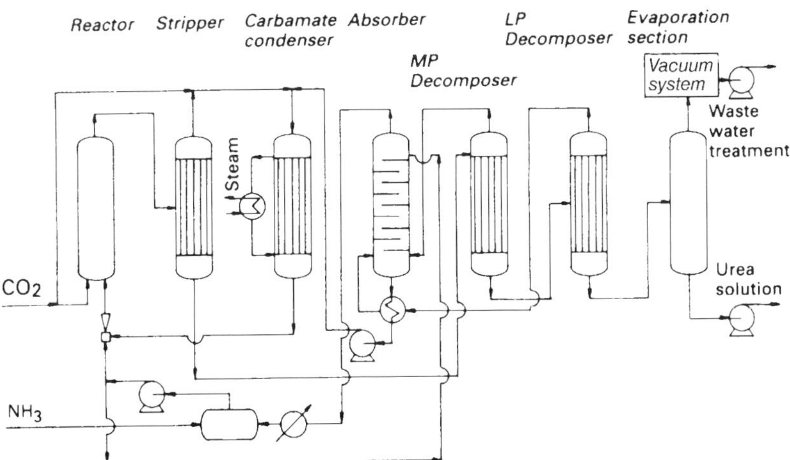

FIGURE 116 Urea from ammonia and carbon dioxide

Reaction

$2NH_3 + CO_2 \rightarrow NH_4COONH_2$

$NH_4COONH_2 \rightarrow NH_2CONH_2 + H_2O$

Raw material requirements and yield

Raw materials required per tonne of urea:

Ammonia (100%)	575kg
Carbon dioxide (100%)	760kg

Yield 80%

PROPERTIES

White hygroscopic crystals, which are either odourless or have a slight ammonia smell. Soluble in water, ethyl alcohol and benzene. Decomposes to ammonia and cyanic acid on heating.

Molecular Weight	60.06
Density at 20°C	1.323
Melting Point	132.7°C
Boiling Point	Decomposes
Exposure Limit COSHH	Not listed
Exposure Limit ACGIH	Not listed

GRADES

Technical, fertilizer (granular, prilled and in solution).

INTERNATIONAL CLASSIFICATIONS

UN No.	Not allocated
CAS Reg No.	57–13–6
EINECS No.	200–315–5
EC No.	Not allocated

APPLICATIONS

The major outlet for urea, accounting for around 80% of total consumption, is as a fertilizer. It is used either prilled or as granules, or for direct soil treatment as a 70% aqueous solution. Demand for granules has increased with the growth in mechanical and aerial spreading of fertilizers, for bulk blending and deep dressing which is used to distribute fertilizer directly into the soil by means of an applicator.

Urea is used in the manufacture of urea–formaldehyde resins produced by the condensation reaction between urea and formaldehyde. These resins find outlets in adhesives for paper, board, plywood, surface-coatings moulding resins and textile processing. They are also used to coat textiles, paper and leather. Another outlet is for the synthesis of melamine employed for the production of melamine–formaldehyde resins. These are used in adhesives and paints, and for laminates, moulding compounds, impregnating paper and textiles. They find uses as a fire retardant for polyurethane foams.

Urea is a constituent of cattle feeds. It is a useful viscosity modifier for casein or starch-based paper coatings and explosive stabilizer. Small quantities are used as an intermediate in the manufacture of polyurethanes, pharmaceuticals, toothpaste, cosmetics, flame-proofing agents, sulfamic acid and fabric softeners.

HEALTH AND HANDLING

Urea is not considered to be harmful at normal temperature, but the dust may irritate the skin, eyes and nose leading to coughing, headaches and nausea. When handling the material, goggles and a dust mask should be worn. Contact lenses must not be worn. Prolonged exposure can lead to inflammation of the kidneys and bladder.

Store in a cool, dry, explosion-proof area. Urea is stable at room temperature but will react violently with sodium hypochlorite. Contact with oxidizing agents and alkalis must be avoided. Good ventilation is needed to reduce dust build-up.

Spills should be scooped up and placed in a container for disposal, care being taken not to generate dust, and wetting the material before clean-up will help reduce dust levels. Staff should wear gloves, gauntlets, aprons, boots and goggles to prevent skin contact. Any contaminated clothing must be laundered before reuse.

Urea is not flammable but will decompose on heating to give off toxic fumes. Firefighters should use carbon dioxide to blanket flames and wear self-contained breathing apparatus and protection against skin contact.

MAJOR PLANTS

Plants with capacities greater than 800 000 tonnes per year:

Complex	Vrataa	Bulgaria
Zakłady Azotowe	Pufawy	Poland
Complex	Berezniki	Russia
Complex	Kemerovo	Russia

Complex	Tolyatti	Russia
Complex	Cherkassy	Ukraine
Complex	Gorlovka	Ukraine
CF Industries	Donaldsonville	US
Uniol	Kenai	US
Petrobras-Fafen	Araucaria	Brazil
Nitroven	El Tablazo	Venezuela
Petrochemical Industries Corp.	Shuaiba	Kuwait
Qatar Fertiliser	Umm Said	Qatar
Razi Chemical	Bandar	Iran
State	Khor al Zubair	Iraq
Krishak Bharati	Hazira	India
Rashtriya Chemicals	Thai Vaishet	India
Pakistan Fertilizers	Dharki	Pakistan
Pupuk Kalimatan	Bontang	Indonesia
Pupuk Sriwidjaja	Palembang	Indonesia
	Sumatra	Indonesia

MAJOR LICENSORS

Chemico
Mitsui Petrochemical
Snamprogetti
Stamicarbon
TEC-MTC ACSA
Toyo Engineering

Vinyl Acetate

$CH_3COOCH = CH_2$

SYNONYMS

VINYL ACETATE ethenyl ethanoate, VAM, acetic acid ethenyl ester, acetic acid vinyl ester

Ethylene has become the preferred feedstock for the manufacture of vinyl acetate, largely replacing the earlier acetylene-based processes. Although two acetylene routes were developed, only the gas-phase process is in use owing to problems with purification of the vinyl acetate produced by the liquid-phase reaction, which utilized catalysts based on mercury salts.

Ethylene-based processes can be carried out in either the liquid or gaseous phase, but because of corrosion problems and by-product formation, the former route is no longer used. National Distillers in the US and Bayer and Hoechst in Germany have developed gas-phase ethylene-based processes, the German version is the more widely used.

An earlier process developed by Celanese, but now obsolete, is the addition of acetic anhydride to acetaldehyde to form ethylidene diacetate. This is cleaved to give vinyl acetate and acetic acid. The higher cost of acetic anhydride over acetic acid made it uneconomic.

A process involving the reductive carbonylation of methyl acetate to ethylidene diacetate (in the presence of rhodium or palladium catalysts) followed by pyrolysis to vinyl acetate has been developed by Halcon. Although the raw materials can easily be produced, this technology is not currently competitive with ethylene-based routes.

Capacity increases in the 1990s have been based on the ethylene gas-phase process. Acetylene has mostly been replaced by the cheaper and readily available ethylene until today most vinyl acetate production is based on the reaction of ethylene with acetic acid and oxygen in the presence of heterogeneous palladium-containing catalysts.

Capacities range from 25 000 to 340 000 tonnes per year.

PROCESSES

1. From ethylene, acetic acid and oxygen

In the gas-phase process, ethylene gas saturated with acetic acid is heated to 120°C, and the resultant vapours are mixed with oxygen before entering a multi-tubular reactor. The concentration of oxygen is adjusted, depending on operating temperature and pressure, to avoid the gas mixture from igniting. The tubes contain a catalyst (consisting of finely divided palladium promoted with potassium acetate) deposited on aluminium trioxide, aluminium–silicon oxides or activated charcoal, which act as the carrier. Other activators which can be used include gold, platinum, rhodium or cadmium acetate. (*See Figure 117*)

The reactor temperature is maintained at 175–200°C with a pressure of 5–9 bar. Heat produced by the exothermic reaction is removed and used to generate steam. Gases leaving the reactor are immediately cooled in a heat exchanger, the heat being used to warm up the incoming recycle gases. The gaseous–liquid mixture is separated under pressure to give an aqueous vinyl acetate phase, from which the aqueous phase is taken off. After washing separately with glycol and sodium carbonate, the gases are passed through a desorber where carbon dioxide is removed, and any unreacted ethylene is returned to the reactor.

The liquid mixture enters a second separator where the pressure is reduced and dissolved gases released. These gases are fed into a distillation column and vinyl acetate is separated from light ends, mainly acetaldehyde. The liquid is distilled in a second column; acetic acid is condensed and separated from other by-products

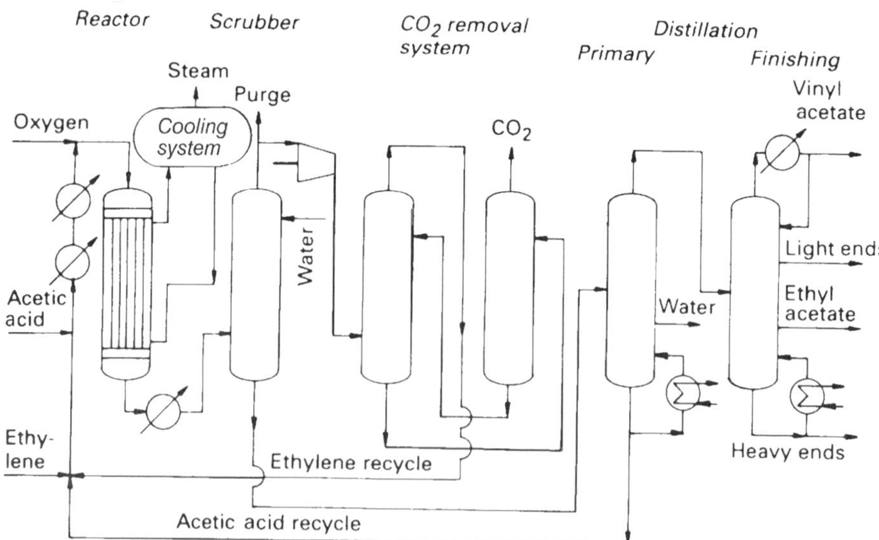

FIGURE 117 Vinyl acetate from ethylene, acetic acid and oxygen

and water before being recycled. Usually, polymerization inhibitors (such as hydroquinone or diphenylamine) are added prior to distillation to prevent polymer formation. Waste products are incinerated.

Reaction

$$C_2H_4 + CH_3COOH + \tfrac{1}{2}O_2 \rightarrow CH_3COOCH{=}CH_2 + H_2O$$

Raw material requirements and yield

Raw materials required per tonne of vinyl acetate:

Ethylene	330kg
Acetic acid	730kg

Yield 95%

2. From acetylene and acetic acid

A mixture of acetic acid vapour and preheated acetylene are passed over a fluid-bed catalyst consisting of zinc acetate deposited on activated carbon in a tubular reactor. A tubular reactor allows better temperature control as well as facilitating loading and unloading of the catalyst. It has largely replaced the fluidized-bed reactor which was originally used. Acetylene, which is present in slight excess, must be free from sulphur and phosphorus compounds. The exothermic reaction takes place at 160–200°C and 1 bar pressure, and excess heat is removed by oil circulation around the tubes. (*See Figure 118*)

After multi-stage cooling, the effluent is fed to a distillation column where unreacted acetylene and low-boiling by-products are removed overhead. After purification the acetylene is recycled to the reactor. Vinyl acetate is removed from the

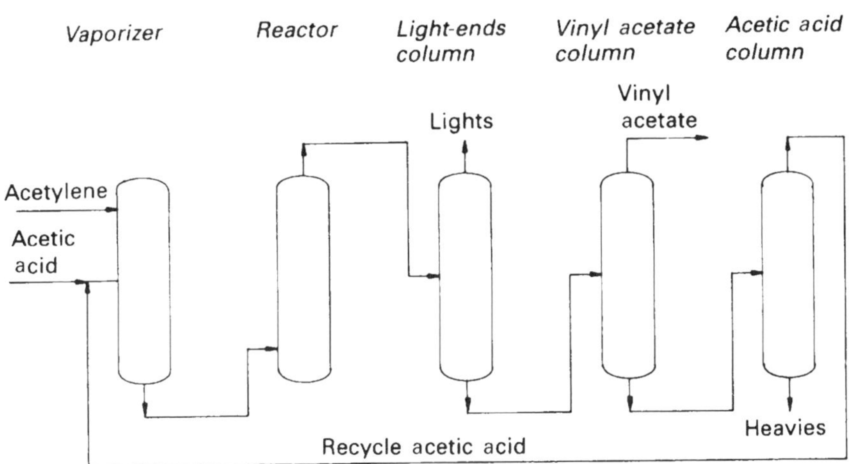

FIGURE 118 Vinyl acetate from acetylene and acetic acid

reaction mixture by distillation. Acetic acid is recovered from the base of the column and returned to the input acid stream. Waste gases are burnt.

Reaction

$$HC{\equiv}CH + CH_3COOH \rightarrow CH_3COOCH{=}CH_2$$

Raw material requirements and yield

Raw materials required per tonne of vinyl acetate:

Acetylene	325kg
Acetic acid	705kg

Yield

on acetylene	90–95%
on acetic acid	95–98%

OTHER PROCESSES

Considerable research has been carried out on the reaction of methyl acetate with carbon monoxide to form ethylidene diacetate, and its cleavage to vinyl acetate and acetic acid. Alternative feedstocks, such as acetic anhydride, dimethyl ether and methyl alcohol have been employed, and the acetic acid formed can be esterified to methyl acetate. The initial reaction takes place under pressure at a temperature of 150°C in the presence of rhodium salts with methylamine and methyl iodide.

BP Chemicals claims that its latest ethylene-based technology utilizing palladium catalysts in a redesigned fluid-bed reactor doubles capacity.

PROPERTIES

Colourless, volatile, flammable liquid with a sweet ethereal odour. Slightly soluble in water but soluble in many organic liquids, such as alcohols, ketones and esters. It polymerizes readily to polyvinyl acetate.

Molecular Weight	86.09
Density at 20°C	0.931
Melting Point	–92.8°C
Freezing Point	–100°C
Autoignition Temperature	385°C
Explosive Limits in air	
lower	2.3vol%
upper	13.4vol%
Flash Point Open Cup	–8°C
Vapour Density (air = 1)	2.98
Exposure Limits COSHH	20ppm 15 minutes 10ppm 8 hour TWA

Exposure Limits ACGIH (Proposed TLV-STEL 10ppm) Classified A3, animal carcinogen.	15ppm 15 minutes TLV-STEL 10ppm 8 hour TLV-TWA

GRADES

Technical 99.9% stabilized with 8–12ppm hydroquinone to prevent spontaneous polymerization.

INTERNATIONAL CLASSIFICATIONS

UN No.	1301
CAS Reg No.	108–05–4
EINECS No.	203–545–4
EC No.	607–023–00–0
Description	Flammable liquid
Packing Group	II
Emergency Action Code	3YE
HI (Kemler Code)	339

APPLICATIONS

In the US, vinyl acetate's main outlet is for the manufacture of polyvinylacetate thermoplastic polymers and copolymers (with acrylate esters, vinyl chloride, acrylonitrile, and maleate and fumaric esters).

Plasticized polyvinylacetate and acetate–acrylate copolymers are increasingly being used as aqueous dispersions for emulsion paints and 23% of vinyl acetate demand goes into this outlet. Adhesives consume another 23%, the polymers and copolymers being employed in paper, wood, leather, textile and concrete applications, and in the solid form as hot-melt adhesives. Vinyl acetate polymers and copolymers utilize 16% of the monomer. Outlets include the manufacture of transparent film, flooring and tubing.

Around 10% of vinyl acetate is used in the paper and paperboard industry. Polyvinyl alcohol, which is used in the manufacture of fibres, accounts for 18% of total demand.

Vinyl acetate is also used as a chewing gum base, and ethylene–vinyl acetate copolymers have replaced natural starches and glues for book binding and packaging. It is also a valuable chemical intermediate especially for the production of halogenated derivatives and thiazoles.

Vinyl acetate is expected to increase at 3% per year in the US, with adhesives and paints being the fastest growing sectors due to reformulation to meet new environmental levels for solvent emissions.

HEALTH AND HANDLING

Vinyl acetate vapour is irritating to the eyes and high concentrations cause dizziness and drowsiness. Contact with the liquid will defat the skin leading to blisters.

Storage containers of stainless steel, aluminium or mild steel with a baked phenolic plastic lining are recommended, blanketed with gas containing between 5 and 21% of oxygen. Inert gases should not be used as hydroquinone cannot function as an inhibitor in the absence of oxygen. At low temperatures, the addition of stabilizers is not necessary and vinyl acetate can be stored under nitrogen. When the material is inhibited, storage for more than six months is not advised and good stock rotation must be employed. Inhibitor levels must be checked at regular intervals and additions made when necessary.

Contact with copper, zinc, lead or their alloys can lead to contamination of the vinyl acetate. The containers should be stored in a cool, well-ventilated, explosion-proof area away from sunlight, acids, alkalis or amines.

Because of vinyl acetate's flammability and high electrical resistivity, all containers must be earthed to prevent static build-up, and care must be taken when handling. PVC gloves and goggles should be worn. Spills should be contained with sand or earth and prevented from reaching drains because of the risk of explosion.

Vinyl acetate is a dangerous fire and explosion hazard because it will polymerize violently when exposed to heat. In the event of fire, evacuate personnel and extinguish all forms of ignition. Carbon dioxide, dry chemical or foam can be used to blanket the flames. The heavy vapour can roll long distances along the ground and flashback is a hazard. Firefighters should wear protective clothing and self-contained breathing apparatus.

MAJOR PLANTS

Plants with capacities greater than 100 000 tonnes per year:

Paradies Acetiques	Pardies	France
Hoechst	Frankfurt	Germany
BP Chemicals	Baglan Bay	UK
Du Pont	La Porte	US
Hoechst Celanese	Bay City	US
	Clear Lake	US

Millenium Petrochemicals	Deer Park	US
	La Porte	US
Union Carbide	Texas City	US
Kuraray	Okayama	Japan
Nippon Synthetic Chemical	Mizushima	Japan
Showa Denko	Oita	Japan
Asian Acetyls	Ulsan	South Korea
Dairen Chemical	Linyuan	Taiwan
Sichuan Chemical	Changshou	China
Hoechst-Acetyl	Palau-Ayer	Singapore

MAJOR LICENSORS

Bayer
BP Chemicals
Denki Kagaku
Hoechst/Uhde
Kuraray Chemical
National Distillers
Nippon Synthetic Chemical
Millenium

Vinyl Chloride (VCM)

$CH_2{=}CHCl$

SYNONYMS

VINYL CHLORIDE chloroethylene, VCM

Vinyl chloride monomer (VCM), one of the major intermediates used in the petrochemical industry, has a single important end use – as monomer in the production of polyvinyl chloride (PVC) polymers and vinyl copolymers.

First produced industrially by the catalytic hydrochlorination of acetylene, high energy costs have caused this route to be largely replaced by the chlorination and subsequently the oxychlorination of ethylene.

In the chlorination process, ethylene dichloride (EDC) produced as an intermediate, is cracked to vinyl chloride. With the development of the balanced process, the hydrogen chloride by-product from the decomposition of ethylene dichloride is reacted with acetylene to make more vinyl chloride and avoid the waste of the expensive chlorine molecule.

The development of a commercial oxychlorination process to ethylene dichloride utilizing only cheaper ethylene revolutionized the manufacture of vinyl chloride. In this integrated balanced process, vinyl chloride is made by the thermal cracking of ethylene dichloride which is produced by the chlorination or oxychlorination of ethylene. Hydrogen chloride produced by the cracking process, in the presence of oxygen, is utilized by the oxychlorination reaction to produce more ethylene dichloride. Most of present day capacity is based on ethylene and most plants use the ethylene–EDC–oxy–EDC–vinyl chloride integrated balanced process because of its economic superiority over other routes. Nearly all new oxychlorination plants are oxygen based and many of the existing air-based units have been converted. The great advantage is the reduction in vent gas volumes discharged by the process.

New paths to vinyl chloride directly from ethylene have been researched but suffer from problems of control and operation. One, involving the reaction of ethylene with an aqueous solution of cuprous chloride and iodine, has been developed but not commercialized.

The ready availability of ethane has resulted in much research to find alternative routes to vinyl chloride which bypass ethylene. ABB Lummus Global has introduced a chlorination–oxychlorination–dehydrochlorination process – the Transcat process – in which ethane is converted to vinyl chloride in a molten salt reactor with yields of 80%. With the development of new catalysts the ethane route could become the process of the future.

Vinyl chloride can be recovered from chloroethanes by catalytic dehydrochlorination or direct catalytic dechlorination with ethylene.

Although historically vinyl chloride plants were to be found in the US and Europe, new capacity is being built outside these traditional areas. Most vinyl chloride plants are adjacent to an ethylene dichloride facility and to sources of ethylene and chlorine so that the processes can be fully integrated, thus avoiding the difficult and expensive shipping of ethylene dichloride.

Worldwide around 95% of vinyl chloride is produced from ethylene and the remainder is obtained from acetylene.

Capacities range from 30 000 to 590 000 tonnes per year.

PROCESSES

1. From ethylene by chlorination–oxychlorination

The integrated balanced chlorination–oxychlorination process proceeds via three principal stages: the chlorination of ethylene in the liquid or vapour phase; the cracking of the ethylene dichloride formed to vinyl chloride and hydrogen chloride; and finally the oxychlorination of ethylene where the hydrogen chloride produced is combined with ethylene and oxygen to yield ethylene dichloride and water. (*See Figure 119*)

In the first stage of the process, ethylene and chlorine are sparged into liquid ethylene dichloride containing ferric chloride as the catalyst. The temperature of the reactor is maintained at 40–70°C by internal or external heat exchangers. The effluent is cooled and any unreacted gases are separated before ethylene dichloride is obtained by fractionation.

In the second stage, dry ethylene dichloride gas is fed into a furnace containing one or more tubes constructed from chromium–nickel alloys. The gases are cracked at a temperature of 450–550°C and pressures ranging between 1 and 40 bar, with a residence time of 10–20 seconds. Pumice or charcoal is used as the catalyst and chlorine is fed into the reaction zone to act as a promoter.

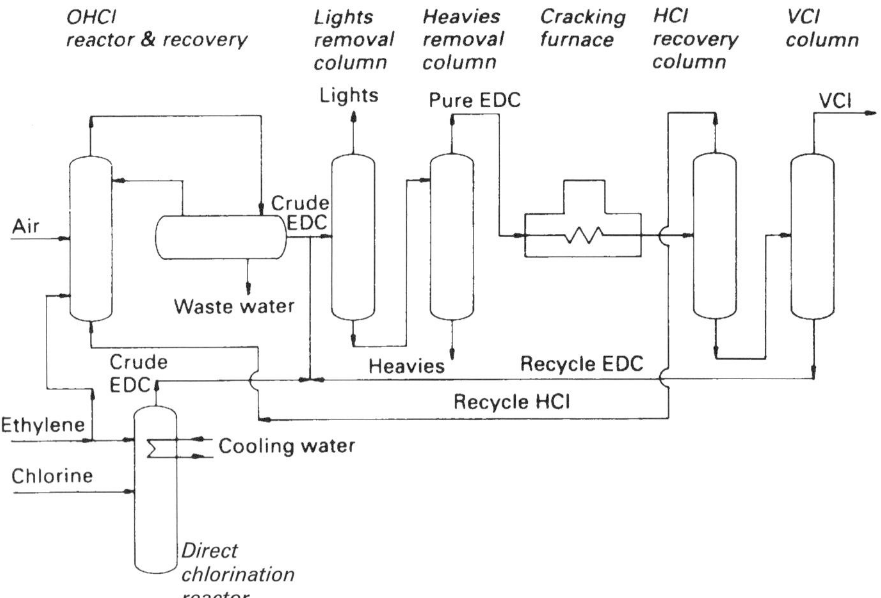

FIGURE 119 Vinyl chloride from ethylene by chlorination–oxychlorination

The exit gases leaving the reactor are cooled rapidly in a quench tower to limit further formation of by-products, particularly coke. Heat from the quench tower is used to generate medium-pressure steam.

Hydrogen chloride is removed by distillation in the first column. Vinyl chloride obtained from the top of the second distillation column is washed with dilute sodium hydroxide before being purified by extractive distillation with acetonitrile. Light-boiling compounds are separated from the bottoms of the vinyl chloride column by distillation prior to the recovery of ethylene dichloride from the remaining heavy ends. The ethylene dichloride is recycled to the cracking furnace.

In the third stage, hydrogen chloride produced by the cracking of ethylene dichloride, ethylene and air are fed into a fluidized-bed oxychlorination reactor. The highly exothermic reaction takes place at moderate temperature and pressure. The reaction can be carried out in the liquid or gaseous phase, but industrially the latter is the more important route. It proceeds with or without a catalyst. Although the catalytic route has a higher conversion rate, non-catalytic processes are favoured because of the time-consuming shutdown and costs of catalyst replacement.

The reaction gases are quenched, and water and ethylene dichloride are condensed. Any unreacted gases are recycled to the reactor. Ethylene dichloride is washed with sodium hydroxide to remove any acid. After separation from water and dehydration, it joins the feed into the cracking furnace for conversion to vinyl chloride.

Reaction

$$CH_2{=}CH_2 + Cl_2 \rightarrow C_2H_4Cl_2$$

$$C_2H_4Cl_2 \rightarrow C_2H_3Cl + HCl$$

$$2CH_2{=}CH_2 + 4HCl + O_2 \rightarrow 2C_2H_4Cl_2 + 2H_2O$$

Raw material requirements and yield

Raw materials required per tonne of vinyl chloride:

Ethylene	470kg
Chlorine	607kg
Oxygen	138kg

Yield 90%

2. From acetylene and hydrogen chloride

The catalysed reaction between acetylene and hydrogen chloride can be carried out in the liquid or vapour phase, but the latter is more commonly employed. Acetylene must not contain catalyst poisons such as sulphur, phosphorus or arsenic; the amounts of unsaturated hydrocarbons are kept to a minimum to reduce catalyst inactivation. Hydrogen chloride must be free from chlorine which could cause explosions and dry to prevent corrosion to the plant. (*See Figure 120*)

Purified acetylene and dry hydrogen chloride, in slight excess, are mixed prior to being fed into a fixed-bed multi-tube reactor containing 2–10wt% of mercuric

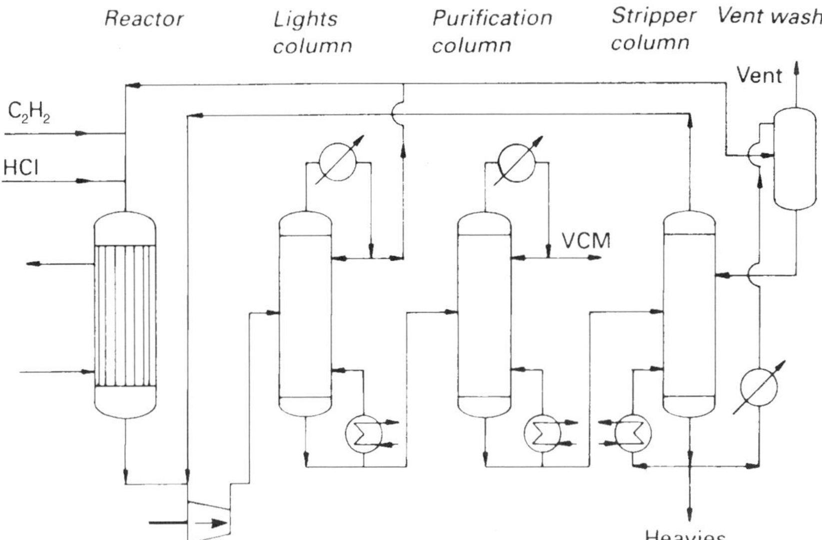

FIGURE 120 Vinyl chloride from acetylene and hydrogen chloride

chloride absorbed onto a carrier such as activated carbon. In order to reduce the volatility of mercury, cerium oxide, copper and thorium chlorides can be added to the catalyst system. Hot spots, which cause catalyst inactivity, are minimized by occasional reversal of the gas flow through the reactor.

The reaction is highly exothermic and the heat generated is removed by an external coolant circulating between the tubes to maintain the temperature at around 200°C. Contact time is up to 1 second and pressure is maintained at 1–3 bar.

The exit gases from the reactor are cooled by passing them through a heat exchanger before being compressed. Unreacted acetylene and hydrogen chloride are removed in the gaseous phase by fractionation. The remaining liquid is distilled in a second column and vinyl chloride is collected overhead. Heavy products from the base of the column are stripped to remove any residual vinyl chloride and acetylene prior to incineration.

Heat treatment or steam desorption are used to recover mercury from the spent catalyst prior to the carbon being reactivated.

In the balanced process, unpurified acetylene, obtained by the cracking of naphtha or methane, is fed into a reactor with hydrogen chloride where it is converted to vinyl chloride. Any ethylene in the gaseous mixture then passes into the chlorinator where ethylene dichloride is produced. After cracking to vinyl chloride, the hydrogen chloride by-product is used in the hydrochlorination of acetylene.

By balancing the gases so that equimolecular amounts of ethylene and acetylene are present, the process can utilize all of the expensive chlorine molecule. Although this process avoids the expensive separation of ethylene and acetylene from cracked gases, the feedstock is still more expensive than in ethylene-based routes.

Reaction

$$C_2H_2 + HCl \rightarrow CH_2{=}CHCl$$

Raw material requirements and yield

Raw materials required per tonne of vinyl chloride:

Acetylene	435kg
Hydrogen chloride	595kg
Mercuric chloride	1kg

Yield 95%

OTHER PROCESSES

A number of routes to the production of vinyl chloride from ethane have been developed. The main attraction is the ready availability of the feedstock and avoidance of the intermediate step to ethylene with consequent savings in costs.

Conversion of ethane can be carried out by direct chlorination or oxychlorination but yields per pass are low in order to avoid excess by-product formation. Catalysts consisting of copper and potassium chlorides have been proposed. The Transcat process utilizes ethane, ethylene or chlorinated wastes to produce chloroethylenes and chloromethanes (see Trichloroethylene).

Two processes for the recovery of vinyl chloride from trichloroethanes are in use: either direct catalytic dehydrochlorination or catalytic dechlorination with ethylene. They could be of value where quantities of unwanted 1,1,2-trichloroethane are available.

Research has been carried out on photochemically induced dehydroxychlorination of ethylene dichloride. Light from mercury, thallium or tungsten lamps is used to excite the ethylene dichloride molecule leading to the liberation of chlorine which can be used to start a free chain radical reaction. Monochromatic light from lasers is an effective alternative. When this process is used after thermal cracking, higher conversion rates can be obtained with lower energy consumption. The process is not yet commercial.

PROPERTIES

Colourless, flammable gas with a sweetish odour. Partially soluble in water. Soluble in most organic liquids.

Molecular Weight	62.5
Density at 13°C (liquid)	0.911
Boiling Point	–13.9°C
Freezing Point	–158.8°C
Autoignition Temperature	472°C
Explosive Limits in air	
lower	3.6vol%
upper	33vol%
Flash Point Open Cup	–78°C
Vapour Density (air = 1)	2.2

Exposure Limits COSHH (Schedule 1) 7ppm 8 hour TWA (Maximum Exposure Limit)
Defined carcinogen.
Overriding annual maximum exposure limit 3ppm. Exposure should be recorded as the TWA of vinyl chloride in the atmosphere of a working area over a period of one year.

Exposure Limit ACGIH 5ppm TLV-TWA
Classified A1, confirmed human carcinogen.
Contact with the liquid is prohibited and monitoring is required.

OSHA have set PEL levels at 0.5ppm over an 8 hour period and exposure above this must be regulated by extensive record keeping.

GRADES

Inhibited 99%.

INTERNATIONAL CLASSIFICATIONS

UN No.	1086
CAS Reg No.	75–01–4
EINECS No.	200–831–0
EC No.	602–023–00–7
Description	Flammable gas
Packing Group	Not allocated
Emergency Action Code	2WE
HI (Kemler Code)	239

APPLICATIONS

The main outlet for vinyl chloride, amounting to almost 99% of demand, is in the manufacture of polyvinyl chloride, one of the major industrial polymers in the world. These resins are used in the automotive, building, electrical, packaging, and toy industries.

The other use, accounting for 1% of total demand, is for the production of chlorinated solvents such as trichloromethane which is declining due to environmental legislation and vinylidene chloride.

Future growth of vinyl chloride will depend on the demand for PVC resins; worldwide, it is forecast to increase by about 5% per year to 2000, with higher rates in Asia and the Middle East.

HEALTH AND HANDLING

Vinyl chloride gas is irritating to the eyes and skin. Contact lenses are a particular hazard as they absorb and concentrate the vapours. Prolonged exposure depresses the central nervous system and affects the lungs, liver and kidneys. It is a carcinogen. Flame-resistant protective clothing must always be worn to prevent skin contact together with goggles and a self-contained respirator.

Vinyl chloride is stable if stored in closed, air-tight, pressurized containers at room temperature in a well-ventilated area. Sunlight, exposure to air, sources of ignition and sparks must be avoided. Vinyl chloride is incompatible with hydroquinone, copper, aluminium or other polymerization catalysts. It will polymerize rapidly if heated with a polymerization catalyst or if the concentration of inhibitor is too low. Good stock control is essential.

Spills should be treated as an emergency and any staff in the vicinity must be evacuated. After containment, dispose of it in accordance with local legislation. It is important that clean-up personnel are fully protected against liquid contact and inhalation of the vapour.

Vinyl chloride is a severe fire and explosion hazard. Being heavier than air, the vapour can roll along the ground, particularly in underground piping systems, and lead to explosions. Fires should be blanketed with carbon dioxide or foam. As the product decomposes to hydrogen chloride, phosgene and carbon monoxide, self-contained breathing apparatus and protective clothing must be worn.

MAJOR PLANTS

Plants with capacities greater than 400 000 tonnes per year:

EVC	Tessenderlo	Belgium
Elf Atochem	Lavera	France
Rovin	Botlek	Netherlands
Norsk Hydro	Rafnes	Norway
CONDEA Vista	Lake Charles	US
Dow Chemical	Oyster Creek	US
	Plaquemine	US
Formosa Plastics	Baton Rouge	US
Geon	La Porte	US
Georgia Gulf	Plaquemine	US
Oxymar	Corpus Christi	US
	Deer Park	US
PPG Industries	Lake Charles	US
Westlake	Calvert City	US
Dow Chemical	Fort Saskatchewan	Canada
Kaneka Corp.	Takasago	Japan
Tosoh Corp.	Shin-Nanyo	Japan
LG Chemical	Yeochon	South Korea
Formosa Plastics	Jenwu	Taiwan

MAJOR LICENSORS

ABB Lummus Global
Dow Chemical
EVC
Geon
Halcon-Scientific Design
Hoechst
John Brown
Mitsui Petrochemical
Monsanto
PPG Industries
Solvay
Stauffer Chemical

Xylene

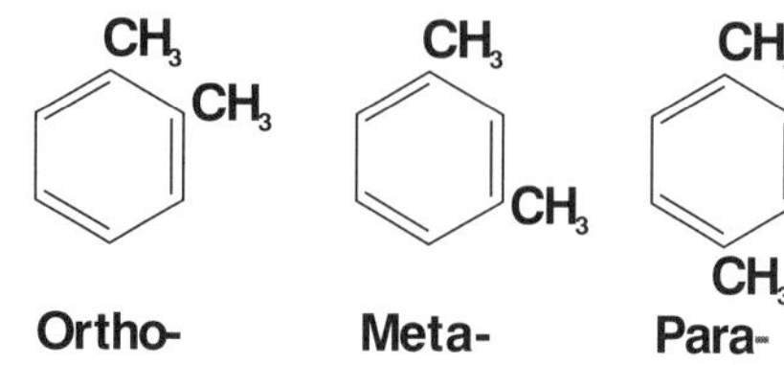

SYNONYMS

XYLENE dimethylbenzene, dimethyl benzol, methyl toluene, xylol, xylole

Xylenes are obtained from refinery reformate streams or from C_8 fractions, a by-product from pyrolysis gasoline. There are three isomers – ortho, meta and para – which differ in the position of the two methyl substituents on the benzene ring. Although similar concentration and separation techniques are used as for benzene and toluene, difficulties arise due to the close boiling points of the three xylene isomers.

Ethylbenzene, which is always co-present except in xylene obtained by the disproportionation of toluene, adds further complications owing to the closeness of its boiling point to that of *m*- and *p*-xylene. Some companies which have a high demand for xylenes have found the disproportionation of surplus toluene to benzene and xylene economic, because xylene obtained by this route does not contain ethylbenzene. The product cost of isomer separation from disproportionated xylenes is less expensive than reformate xylene, which offsets the higher cost of this process over conventional routes.

The relative percentages of C_8 isomers varies widely depending on the source, with the ratios of ethylbenzene to xylene showing the greatest range. Typical percentage compositions of aromatics from various sources are shown below.

Component	*Catalytic Reforming*	*Steam Cracking*	*Disproportionation*
Ethylbenzene	21%	52%	0%
p-Xylene	18%	10%	26%
m-Xylene	39%	25%	50%
o-Xylene	22%	13%	24%

In the US, the C_8 cut from catalytic reforming is the most widely used as an isomer feedstock because of its availability and competitive price. In Europe, catalytic reforming is of less importance due to the quantity of pyrolysis gasoline, a by-product from naphtha plants (see Benzene). The increasing volume of ethylene being produced from naphtha cracking has led to significant amounts of aromatics as by-products being generated.

Xylene can be recovered from C_8 and C_9 streams (obtained from catalytic cracking or pyrolysis gasoline) by a variety of different methods, depending on which of the three isomers is required. Extractive distillation, azeotropic distillation, crystallization, and solid absorption have all been employed.

Solvents can be used to extract the high concentration of aromatics followed by distillation to separate the individual components. Crystallization techniques, which take advantage of the spread of melting points to separate the xylene isomers, are usually directed towards the isolation of *p*-xylene. Extraction processes are associated with isomerization.

p-Xylene is the most important of the xylene isomers, whereas *m*-xylene, which accounts for 40% of the C_8 reformate cut, does not have a major market. After recovering *p*- and *o*-xylene, the residual stream (containing a high proportion of *m*-xylene as well as ethylbenzene and some *o* and *p*-xylene) can be used as a solvent, in gasoline or isomerized. This process increases the proportion of *p*-xylene in the residual stream which can then be recycled and recovered by crystallization. The demand breakdown for chemical production of mixed xylene streams is 68% *p*-xylene, 24% *o*-xylene and 6% for ethyl benzene.

Xylenes have increased their importance as chemical raw materials, with recent effort directed to improve process yields and catalyst life, and to the use of larger reactors. Globally, about 83% of *p*-xylene is extracted from reformate, 10% is produced by disproportionation and the remainder is obtained from pyrolysis gas. Only about 1% is obtained from coal.

Capacities of *o*-xylene plants range from 8000 to 200 000 tonnes per year, while those of *p*-xylene are between 30 000 and 1 000 000 tonnes per year.

PROCESSES

1. From petroleum

Xylene is obtained from aromatic-rich streams from refinery streams or as a by-product of naphtha cracking. The following techniques are used to separate the isomers (see Benzene and Toluene). (*See Figure 121*)

Crystallization

As *p*-xylene is the most important of the isomers, many commercial processes for the production of *p*-xylene use crystallization to separate this isomer from the xylene stream.

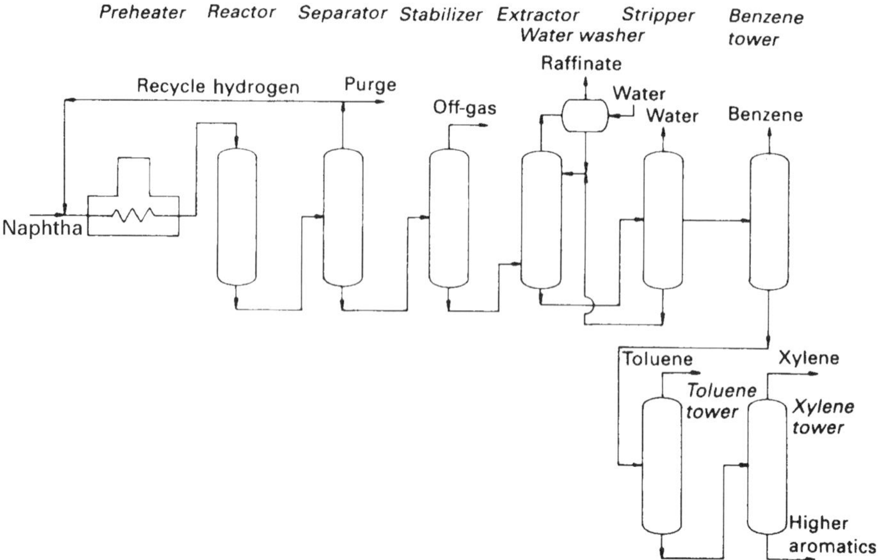

FIGURE 121 Xylene from petroleum by catalytic reforming

The mixed xylene feed must be dried by passing it through an alumina or silver gel bed so that the water content does not exceed 10ppm. Usually, two beds are used so that one is in operation while the other is being regenerated by electrical heating. Otherwise water would clog the centrifuges and filters during the crystallization process.

The feed is precooled, using propane and ethylene as refrigerants, before entering the first crystallizator tank which is maintained at –20°C. Crystals formed on the surface walls are scrapped off by rotating blades before being centrifuged. *p*-Xylene of up to 85% purity can be obtained.

Crystals from the first stage are melted and recrystallized. As the liquor in the second stage has a much lower viscosity that of the first, the crystals drain more easily when centrifuged. *p*-Xylene of 99% purity is obtained. The centrifugate is used to cool fresh feed before being recycled for further recovery.

Several companies have introduced a variant into their crystallization process. In the Phillips process, the second-stage slurry is fed into a purification column where countercurrent contact between the high-purity reflux liquid and the cold crystals results in a partial freezing of the reflux liquid and the removal of lower-melting-point impurities. Flow through the column is achieved by a pulse pump, and pure molten *p*-xylene is obtained from the bottom of the column.

Solid absorption

A significant improvement in *p*-xylene recovery was made by the UOP Parex continuous process. *p*-Xylene is extracted from a C_8 reformate or from a C_8 mixed aromatic streams by absorption in a fixed bed of solid absorbent of the molecular-sieve type. To obtain a moving bed of absorbent, the inlet and outlet points for both the aromatics stream and the deabsorbant liquid are continuously moved by a multi-outlet valve. The absorbent and deabsorbant beds are reusable. Absorption takes place in the liquid phase within a temperature range of 150–175°C and at 10 bar. By washing with a deabsorbant having a boiling point different from C_8 aromatics, *p*-xylene can be recovered by distillation. A purity in excess of 99.5% is claimed.

Because no refrigeration is involved and high recovery on a single-pass operation is obtained, *p*-xylene producers have been able to increase their production capacity at minimal cost. Where *o*-xylene is required, it is separated from *p*-xylene by fractionation.

Isomerization

There are four different types of isomerization process, each of which employs a different catalyst system.

The catalysts are:

- noble metal (platinum) on a silica–alumina support;
- non-noble metal on a silica–alumina support;
- silica–alumina;
- hydrofluoric acid and boron trifluoride ($HF–BF_3$).

The isomerization feed consists of a C_8 aromatic mixture deficient in *p*-xylene, obtained from *p*-xylene recycle catalytic reformate or pyrolysis gasoline. Even feedstocks containing up to 40% ethylbenzene are not detrimental, because the platinum catalyst is effective in converting them to xylene. The platinum-based catalytic process takes place at 450°C and 1 bar in the presence of hydrogen to maintain catalyst activity.

If a non-noble metal catalyst is used, the reaction temperature ranges from 400–500°C with a pressure of 1–10 bar, the processes operating at the higher temperature requiring hydrogen. Coking occurs on the catalyst which has to be regenerated at frequent intervals.

Maruzen has developed a silica–alumina catalyst which operates in the presence of hydrogen and steam at a temperature of 460–560°C and atmospheric pressure. The major difference between the noble and non-noble catalysts in these processes is their ability to isomerize ethylbenzene. Noble metals isomerize ethylbenzene in the presence of hydrogen, while non-noble metals disproportionate and transalkylate it.

The isomerized effluent is cooled in a heat exchanger and the liquid and gas mixture is separated in a flash drum. The vapour phase, rich in hydrogen, is recycled and the liquid phase passes to a fractionation column where light ends are removed and *p*-xylene is crystallized from the heart cut.

The HF–BF_3 process, which unlike the other isomerization routes operates in the liquid phase, uses *m*-xylene (separated as described below) as the feedstock. This comes into contact with HF–BF_3 in a diluent, the reaction taking place at a temperature of around 110°C. The *m*-xylene–HF–BF_3 acts as an isomerization catalyst.

The isomerized mixture is sent to the extractor where fractionation removes the diluent which is recycled. After the removal of light ends, *o*-xylene is obtained by distillation and *p*-xylene by crystallization. Because of lower operating conditions and the lack of *p*-xylene in the isomerization feed, a higher *p*-xylene concentration is produced with lower xylene disproportionation and transalkyation than with vapour-phase isomerization processes. The process is very flexible and conditions can be adjusted to optimize the desired isomer.

m-xylene

m-Xylene can be recovered from the filtrate from *p*-xylene crystallization processes at up to 85% purity. Japan Gas Chemical has introduced a process which extracts m-xylene from C_8 aromatics using a combination of hydrogen fluoride and boron trifluoride in a diluent. The catalyst and C_8 aromatic feedstock are fed into a multistage reactor where the liquid-phase reaction takes place at around 0–10°C. The *m*-xylene–HF–BF_3 complex formed is decomposed by heating at a temperature below 65°C into *m*-xylene and gaseous HF–BF_3 which is recycled. Pure *m*-xylene is obtained by fractionation.

2. From toluene by disproportionation and transalkylation

Several processes have been developed for the disproportionation and transalkylation of toluene and C_9 aromatics to benzene and xylene. (*See Figure 122*)

Toluene, either separately or with a mixture of C_9 aromatics having a low saturates content, is fed into a reactor containing a fixed-bed catalyst. In the Tatoray process (Toray Industries/UOP), the catalyst is based on a noble metal or rare earth in conjunction with alumina. The reaction takes place in the presence of hydrogen which helps to minimize deposition on the catalyst. Operating pressure ranges are 10–50 bar and a reaction temperature of 350–530°C. High conversion rates per pass are obtained, and the catalyst has a life of approximately 3 years.

After cooling, hydrogen is recovered from the reactor effluent and recycled, and the aromatic mixture is separated by distillation. With a toluene feed, yields of high-purity benzene and xylene in a molar ratio ranging from 1:1 to 10:1 (containing less than 1ppm of saturated hydrocarbons) can be obtained.

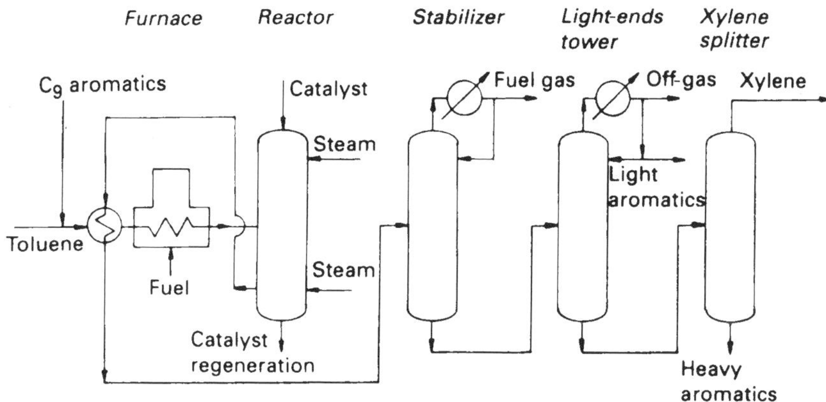

FIGURE 122 Xylene from toluene and C_9 aromatics

An alternative process developed by Arco Technology uses an inexpensive non-noble catalyst based on silica–alumina. The vaporized feed enters the reactor where it mixes with the hot downward-circulating catalyst. Reaction conditions are 430°C and atmospheric pressure. Towards the base of the reactor, the exit gases are separated from the catalyst and are cooled in a heat exchanger. Hydrogen is not required and the catalyst is regenerated by burning off any coke deposits. Using a toluene feedstock, benzene and xylene are produced in a molar ratio of 1.5:1.

The major advantage of these routes is that the xylene produced contains negligible quantities of ethylbenzene, which facilitates xylene isomer separation.

Comparison of percentage product yields from toluene for various processes:

Product	*Tatoray*	*ARCO*	*Mobil*
Benzene	41%	40–10%	44%
Xylenes	56%	55–84%	50%
C_{10+} aromatics	3%	5–6%	6%

Reaction

$$2C_6H_5CH_3 \rightarrow C_6H_6 + C_6H_4(CH_3)_2$$

$$C_6H_5CH_3 + C_6H_3(CH_3)_3 \rightarrow 2C_6H_4(CH_3)_2$$

Raw material requirements and yield

Average yields of products per tonne of toluene:

Benzene	416kg
Xylenes	557kg
C_{10+}	13kg
Light ends	16kg

Yield 95–97%

PROPERTIES

o-Xylene: colourless liquid.
m-Xylene: colourless liquid.
p-Xylene: colourless liquid or crystals.

All are flammable and toxic. Soluble in ethyl alcohol but insoluble in water.

	Mixed xylenes	o-*Xylene*	m-*Xylene*	p-*Xylene*
Molecular Weight	106.16	106.16	106.16	106.16
Density at 20°C	0.86	0.876	0.86	0.857
Melting Point	–25°C	–25.2°C	–48°C	–13.3°C
Boiling Point	138.5°C	144.4°C	139°C	138.4°C
Autoignition Temperature	464°C	465°C	530°C	530°C
Flammability Limits in air (vol%)				
lower	1	1.1	1.1	1.1
upper	7	6.4	6.4	6.6
Flash Point Closed Cup	27–32°C	17.2°C	25°C	25°C
Vapour Density (air = 1)	3.7	3.7	3.7	3.7
Exposure Limit COSHH				
ppm 15 minutes	150	150	150	150
ppm 8 hour TWA	100	100	100	100
Exposure Limit ACGIH				
ppm 15 minutes TLV-STEL	150	150	150	150
ppm 8 hour TLV-TWA	100	100	100	100

GRADES

Technical 95%.

INTERNATIONAL CLASSIFICATIONS

UN Number	1307
CAS Reg No.	
m-xylene	108–38–3
o-xylene	95–47–6
p-xylene	106–42–3
EINECS No.	
m-xylene	203–576–3
o-xylene	202–422–2
p-xylene	203–396–5
EC No.	601–222–00–9
Description	Flammable substance
Packing Group	
flash point <23°C; boiling point >35°C	II
flash point >23°C, <64°C; boiling point >35°C	III
Emergency Action Code	
flash point <23°C; boiling point >35°C	3YE
flash point >23°C, <64°C; boiling point >35°C	3Y
HI (Kemeler Code)	
flash point <23°C; boiling point >35°C	33
flash point >23°C, <64°C; boiling point >35°C	30

Mixed xylenes

UN No.	1307
CAS Reg No.	1330–20–7
EINECS No.	215–535–7
EC No.	601–222–00–9
Description	Flammable substance

APPLICATIONS

The largest outlet for xylenes is as a gasoline octane improver. Chemical uses for the xylene isomers are in the plastics and fibres markets and each isomer has one major outlet.

p-Xylene is the most important of the isomers and it is used to make terephthalic acid and dimethyl terephthalate, the raw materials for the production of polyester fibre and film.

Almost all *o*-xylene is used in the manufacture of phthalic anhydride which is converted to plasticizers, alkyd and polyester resins. *m*-Xylene is the least important of the isomers and only small quantities are consumed for the production of isophthalic acid which finds an outlet in plasticizers.

World demand for xylenes is expected to grow at 7% per year between 1996 and 2000. Future growth for the individual isomers will depend on the demand for their respective end products.

HEALTH AND HANDLING

The absorption of xylene takes place chiefly by breathing the vapours which cause irritation to the eyes, nose and throat. Skin irritation is more serious than that caused by benzene or toluene. The acute toxicity of xylenes is slightly greater than those of benzene and toluene but the long-term effects are less than those of benzene. Excessive exposure should be avoided and good ventilation is essential. Protective clothing, including rubber gloves and goggles, must be worn.

Xylenes are not corrosive to metals and storage containers of iron, mild steel, copper or aluminium may be used, but it does have a deleterious action on rubber seals and valves. Storage containers should be kept in a cool, ventilated area away from strong oxidizing agents. Equipment must be earthed to prevent static build-up. The lower explosion limit of xylene vapour in air is 11 000ppm.

Spills can be contained and absorbed using dry earth, sand or vermiculite, care being taken to ensure that xylene does not get into waterways or sewers. Waste must be disposed of in accordance with local regulations.

Dry foam or carbon dioxide should be used to fight fires as water can scatter the flames. There is a potentially dangerous fire hazard from flashback as xylene vapours are heavier than air.

Because of its flammability and toxicity, the labelling and transportation of xylene is closely controlled.

MAJOR PLANTS

o-*Xylene*

Plants with capacities greater than 100 000 tonnes per year:

Exxon Chemical	Botlek	Netherlands
Deutsche Shell	Godorf	Germany
Complex	Omsk	Russia
Complex	Ufa	Russia
Exxon Chemical	Baytown	US
Koch Refining	Corpus Christi	US
Lyondell-Citgo	Houston	US
Kohap Chemical	Ulsan	South Korea
Chinese Petroleum	Kaohsiung	Taiwan

p-*Xylene*

Plants with capacities greater than 300 000 tonnes per year:

ICI	Wilton	UK
Koch Refining	Corpus Christi	US
Amoco	Decatur	US
	Texas City	US
Chevron Chemicals	Pascagoula	US
Exxon Chemical	Baytown	US
Phillips Puerto Rico	Guayama	Puerto Rico
Japan Energy	Chita	Japan
Mitsubishi Oil	Mizushima	Japan
Oita Paraxylene	Oita	Japan
Yukong	Ulsan	South Korea
LG-Caltex	Yeochon	South Korea
Kohap Chemical	Ulsan	South Korea
Yangze Petrochemical	Nanjiung	China
Mobil Oil	Jurong	Singapore
Singapore Aromatics	Pulau Ayer	Singapore

MAJOR LICENSORS

Process	Licensor
Extraction/separation/ isomerization	*Chevron*
	Englehard
	Exxon
	Glitsch Technology
	HRI Inc.
	ICI
	IFP
	Krupp-Koppers
	Leunawerke
	Lurgi
	Maruzen
	Mitsubishi Gas Chemical
	Phillips Petroleum
	Shell Development
	Snamprogetti
	UOP
Disproportionation	*Arco Technology*
	Chevron
	Mobil Oil
	Toray Industries/UOP (Tatoray)
	UOP
Transalkylation	*Arco Technology*
	BASF
	Mobil
Petroleum gas conversion	*BP/UOP*

Transportation of Dangerous Goods

The increased movement of chemicals has led to possible dangers to man and the environment. Although laws were passed in individual countries to control their transportation, it was not until the 1950s that the United Nations (UN) set up the International Committee of Experts on the Transportation of Dangerous Goods to investigate the situation. Under its auspices, a number of recommendations have been published which formed the basis of national and international regulations concerning the movement of chemicals by land, sea, rail and inland waterways. UN code numbers have been assigned to individual substances for ease of identification. This overcomes the problems of language and similarity of names.

In addition to the UN number, a class must also be attributed signifying the hazard involved. These classes are:

1. Explosives;
2. Gases: compressed, liquefied, dissolved under pressure or refrigerated;
3. Flammable liquids;
4. Flammable solids or substances liable to spontaneous combustion or which in contact with water emit flammable gases;
5. Oxidizing substances;
6. Poisonous substances;
7. Radioactive;
8. Corrosive;
9. Miscellaneous dangerous substances.

Specific guidance for different types of transport has been devolved to the international organizations involved. Although sea and air transportation are governed by international groups, road and rail movements tend to be subject to national laws. The International Maritime Organisation has issued its IMDG (International Maritime Dangerous Goods) code for the movement of dangerous substances by sea and the International Air Transport Association sets out a similar code (IATA-DGR) for air transportation.

The Carriage of Goods by Rail (RID) regulations are used in most of Europe while road movement is subject to the European Agreement Concerning the International Carriage of Dangerous Goods by Road (ADR). The movement of dangerous substances along the Rhine is controlled by the Central Commission for the Navigation of the Rhine.

The European Commission has harmonized the legislation to avoid the commercial and administrative chaos if every country had its own regulations. This would also have impeded the free movement of goods. The first legislation was enacted in 1967 under 67/432/EEC and since that date there have been a large number of amendments.

The objective of the Directive is to harmonize the classification, labelling and packaging of chemical substances and to provide a mechanism for the notification of new substances. Dangerous substances in the European Inventory of Existing Commercial Chemical Substances (EINECS) have a seven digit code number allotted starting at 200–001–8. The European List of Notified Chemical Substances (ELINCS) also has a seven digit code system starting at 400–010–9.

Manufacturers, importers and distributors, obliged to label all dangerous substances since January 1986, may experience difficulty in identifying the substance for which uniform labelling has been agreed by the Union. However, the CAS (Chemical Abstracts Service) number has been included under the *International classifications* section which should enable the reader to identify rapidly those substances which have been classified. Most official publications contain cross-reference tables which are useful where either language or synonyms cause problems. As the Directive is frequently updated, substances not currently listed may be included at a later date. Reference must always be made to the latest legislation.

Some of the various national and international organizations involved in transportation, health and safety matters are listed (pages 474–478) to whom reference should be made for more information. The Health and Safety Executive in London run a very useful help and advice telephone line, details of which are given under *Health and Safety*.

Transportation

UN Committee of Experts on Transportation of Dangerous Goods
United Nations
New York NY 10017
USA
Tel: (1) 212 963 1234

US Department of Transportation
Special Programs Administration
Office of Hazardous Materials Transportation
400 Seventh Street SW DHM-1
Washington DC 20590
USA
Tel: (1) 202 366 4900

Department of the Environment, Transport and Regions
Eland House
Bressenden Place
London SW1E 5DU
England
Tel: 0171 890 3333

Dangerous Goods Advisory Service (DAGAS)
Room 6/15
Laboratory of the Government Chemist
Queens Road
Teddington
Middlesex TW11 0LY
England
Tel: 0181 943 7413

AIR

International Air Transport Association (IATA)
2000 Peal Street
Montreal
Canada H3A 2RU
Tel: (1) 514 844 6311

Civil Aviation Authority (CAA)
Dangerous Goods Office
1W Aviation House
Gatwick Airport South
West Sussex RH6 0YR
England
Tel: 01293 573800

SEA

International Maritime Organisation (IMO)
4 Albert Embankment
London SE1 7SR
England
Tel: 0171 735 7611

Central Commission for the Navigation of the Rhine (CCR)
Palais du Rhin
2 Place de la Republique
F 6700 Strasbourg
France
Tel: (33) 388 32 35 84

Maritime and Coastguard Agency
Dangerous Goods Division
Spring Place
105 Commercial Road
Southampton SO15 1EG
England
Tel: 01703 329181

ROAD

Department of Trade and Industry (DTI)
Transport Department
Dangerous Goods Section
Great Minster House
76 Marsham Street
London SW1P 4DR
England
Tel: 0171 271 4537

British International Freight Association
Redfern House
Browells Lane
Feltham
Middlesex TW13 7EP
England
Tel: 0181 844 2266

RAIL

International Union of Railways (UIC)
14 Rue Jean Rey
F 75015 Paris
France
Tel: (33) 142 73 01 20

Railtrack PLC
Standards Manager, Safety and Standards
Railtrack House
Euston Square
London NW1 2EE
England
Tel: 0171 557 8779

Eurotunnel
Freight Division
PO Box 2000
Folkestone CT18 8XY
England
Tel: 01303 273300

ESR–IRS (ESRIN)
Via Galileo Gaililei
100044 Fracati
Italy
Tel: (39) 6 941801

Health and Safety

Health & Safety Executive (HSE)
Hazardous Substances Division
Rose Court
2 Southwark Bridge
London SE1 9HS
England
Tel: 0171 717 6597
HSE Infoline: 0541 545500
(for assistance and advice Monday–Friday, 8.30–17.30)

American Conference of Governmental Industrial Hygienists (ACGIH)
Technical Affairs Office
1330 Kemper Meadow Drive
Cincinnati OH 45226
USA
Tel: (1) 513 742 2020

National Institute for Occupational Safety and Health (NIOSH)
Robert A Taft Laboratories
4676 Columbia Parkway
Cincinnati OH 45226
USA
Tel: (1) 513 533 8326

Occupational Safety and Health Administration (OSHA)
1825K Street N W
Washington DC 20006
USA
Tel: (1) 202 634 7943

Other Organizations

Chemical Industries Association
Kings Buildings
Smith Square
London SW1P 3JJ
England
Tel: 0171 834 3399

AEA Technology PLC
AEA Environment
National Chemical Emergency Centre F6
Culham
Abingdon
Oxfordshire OX14 3BD
England
Tel: 01235 463060

International Register of Potentially Toxic Chemicals (IRPTC)
Palais des Nations
CH 1221 Geneva 10
Switzerland
Tel: (41) 22 98 58 50

European Commission
DG XI
Environment, Consumer Protection and Nuclear Safety
200 Rue de la Loi
B 1049 Brussels
Belgium
Tel: (32) 2 322 235 1111

Licensors' Index

Subject Index